Isaac Newton

THE LAST SORCERER

Also by Michael White

Stephen Hawking: A Life in Science (with John Gribbin)
Einstein: A Life in Science (with John Gribbin)
Darwin: A Life in Science (with John Gribbin)
Asimov: The Unauthorised Life
Breakthrough: The Hunt for the Breast Cancer Gene (with Kevin Davies)
Mozart
Lennon
Newton
Galileo
The Science of the X-Files
Life Out There

Isaac Newton
THE LAST SORCERER

Michael White

FOURTH ESTATE • *London*

This paperback edition first published in 1998

First published in Great Britain in 1997 by
Fourth Estate Limited
6 Salem Road
London, W2 4BU

3 5 7 9 10 8 6 4 2

A catalogue record for this book is
available from the British Library

ISBN 1–85702–706–X

Number XXV from *LAO TZU: TAO TE CHING*
translated by D. C. Lau (Penguin Classics 1963) © D. C. Lau 1963.
Reproduced by permission of Penguin Books.

Typeset by Rowland Phototypesetting Ltd,
Bury St Edmunds, Suffolk
Printed in Great Britain by
Clays Ltd, St Ives plc

To Bill Hamilton: for getting the ball rolling

Contents

Isaac Newton

THE LAST SORCERER

Truth Revealed

This strange spirit, who was tempted by the Devil to believe he could reach all the secrets of God and Nature by the pure power of mind – Copernicus and Faustus in one.

MAYNARD KEYNES[1]

According to a list of the most influential people in history, *The 100*, Isaac Newton ranks number 2 – after Muhammad and ahead of Jesus Christ.[2] This position is justified by his unparalleled contributions to science – principles that have moulded the modern world. Yet Newton was not the man that history has claimed him to be. More than any other scientist in history, Newton's image has been protected by his disciples and by generations of biographers who have produced inaccurate and sometimes totally false accounts of his life. Not until the 1930s did the real Isaac Newton begin to emerge from the mists of history into the light of critical analysis. Amazingly, it has taken since then to shrug off the final deceits of those who wished to perpetuate the myth that Newton was in some way omnipotent, beyond the baser mundanities of human existence; that he was the pure, distilled essence of scientific inquiry – genius unsullied.

The hagiographic accounts began within a year of Newton's death. William Stukeley, who is today better remembered as a scholar of Druidism and ancient mythology, was Newton's first biographer. His *Memoirs of Sir Isaac Newton's Life*, written during the 1720s, is a devotional account of his hero's life, based uniquely upon first-hand experience.[3] Stukeley knew Newton well during the final decade of his life, and because of this the *Memoirs* is an important book. But, like many of Newton's later biographers, Stukeley was blinkered by adoration: he saw Newton as a demigod, almost immortal and utterly without fault.

Sir David Brewster's *Memoirs of the Life, Writings, and Discoveries of Sir Isaac Newton*, published in 1855, is a worthy successor to Stukeley and contributes much to our understanding of Newton, but again it is tarnished by the author's lack of objectivity. Brewster, like others, ignored evidence that did not fit his image of Newton; he decided to paint a one-sided view that merely reinforced the image that Newton himself tried to establish for posterity, without questioning the many contradictions in the scientist's long and complex life.

There is no question of the greatness of Newton's work, nor of his intellect. But, just as his most famous work, the *Principia Mathematica*, is a highly sophisticated and complex description of the mechanistic workings of the universe based upon simple, easily understood truths, so too was his personality far more twisted and convoluted than orthodox historians of science would have us believe.

Newton was above all a secretive man, a man coiled in upon himself, detached from the world, and for long periods of his life he was secluded from the everyday current of affairs. For much of his working life he studied and experimented alone in his college rooms and in his laboratory nearby. In many respects, he was nonconformist from an early age, shunning the simple rural life of his family, living in self-imposed isolation at university, refusing to take holy orders. He subscribed to Arianism – the doctrine of an heretical sect which denied the principle of the Holy Trinity – when public awareness of such beliefs would have wrecked his career. And, most importantly of all, he was an alchemist.

By the time biographers came to consider his life, Newton was dead and the need to hide his religious leanings had gone. But what stuck in the craw of those early biographers was a body of material found in Newton's vast library and within his huge collection of papers and notebooks that made it very clear that the most respected scientist in history, the model for the scientific method, had spent more of his life intensely involved with alchemy than he had delving into the clear blue waters of pure science. It also confirmed what a few of Newton's close friends had known during his lifetime: that he had expended a vast amount of his time studying the chronology of the Bible, examining prophecy, investigating natural magic, and, most of all, attempting to unravel the hermetic secrets – the *prisca sapientia*.

Newton's early biographers found it impossible to reconcile these opposites and were forced to gloss over any ignominious or disturbing findings they unearthed, putting them down to aberrations, eccentricities: 'the obvious production of a fool and a knave' was how David Brewster described Newton's vast collection of alchemical writings.[4]

The real life story of Isaac Newton, the neurotic, the obsessive, driven mystic, began to emerge only in 1936, when a collection of Newton's papers, considered to be of 'no scientific value' when offered to Cambridge University some fifty years earlier, was purchased at Sotheby's by the distinguished economist and Newton scholar John Maynard Keynes. (He bequeathed it to King's College, Cambridge, when he died ten years later.)

After studying the contents of Newton's secret papers – those documents, manuscripts and notebooks ignored by the hagiographers – in 1942 Keynes delivered a lecture to the Royal Society Club in which he portrayed an altogether different and highly controversial image of history's most renowned and exalted scientist:

> In the eighteenth century and since, Newton came to be thought of as the first and greatest of the modern age of scientists, a rationalist, one who taught us to think on the lines of cold and untinctured reason. I do not see him in this light. I do not think that any one who has pored over the contents of that box which he packed up when he left Cambridge in 1696 and which, though partly dispersed, have come down to us, can see him like that. Newton was not the first of the age of reason. He was the last of the magicians, the last of the Babylonians and Sumerians, the last great mind which looked out on the visible and intellectual world with the same eyes as those who began to build our intellectual inheritance rather less than 10,000 years ago. Isaac Newton, a posthumous child born with no father on Christmas Day, 1642, was the last wonder-child to whom the Magi could do sincere and appropriate homage.[5]

Keynes was obviously enthralled by what he had discovered, and, fortunately for us, he lived in an age that could accept such findings. What he found raised two linked questions about Newton. First, if the creator of modern mechanical theory had spent the majority of his time involved with alchemical experiments, what else might be hidden about him? Second, did Newton's work in alchemy influence his purely scientific work?

The first of these problems was relatively easy to answer. Newton was known to have been a difficult man, a man who had been damaged emotionally by childhood trauma, a supreme egotist who had been involved in well-publicised battles with a number of contemporaries. But, before Keynes's revelation, biographers had barely alluded to these facts. Until 1936, most Newton biographers were content to rely upon the opinions of William Stukeley. Only gradually did others begin to question the old authorities and to dig a little deeper.

What has been unearthed does not always paint a pretty picture. The reality of human character rarely does. However, the newly revealed Newton, the broader-canvas Newton, is a human Newton – a man whom we should be proud to accept for his peculiarities and failings as we are for his unique skills and talents. As Sir Christopher Wren, his contemporary, put it, 'Neither need we fear to diminish a miracle by explaining it.'[6]

What has been gradually revealed is the image of a genius who sought knowledge in everything he came across, a man who was driven to investigate all facets of life he encountered, everything that puzzled him. Such voraciousness drove him to self-inflicted injury, nervous breakdown, to a state in which he almost lost his mind, and possibly even to occult practices and the black arts. But the work that emerged from these explorations changed the world.

The other major question provoked by the Keynes papers – whether or not there was cross-fertilisation between Newton's alchemical studies and his scientific researches – was a much more difficult problem to address and remains a question that is far from being resolved completely.

Not least of the problems facing any serious research into what Newton was doing is the fact that he left behind over a million words on the subject of alchemy. Beyond that has lain the problem of deciphering such a mass of material written largely in code, in Latin and in Newton's tiny handwriting. The task has occupied scholars for sixty years and is ongoing. The late American scholar Betty Jo Dobbs produced a vast body of work providing a detailed analysis of Newton's alchemical experiments gathered together in two academic works, *The Foundations of Newton's Alchemy* (1975) and *The*

Janus Faces of Genius: The Role of Alchemy in Newton's Thought (1991). Others have begun to analyse Newton's vast collection of writings on biblical prophecy and his ideas on a range of subjects from astrology to numerology.[7] But for the lay reader there remains the added difficulty of understanding the mental processes behind seventeenth-century alchemy. It is not easy to empathise with a mentality that is, on so many levels, quite alien to the late-twentieth-century mind.

In the following pages I will discuss both sides of the argument, for and against alchemical influence upon Newton's scientific work. But, based upon the evidence available, my conclusion is unequivocal: the influence of Newton's researches in alchemy was the key to his world-changing discoveries in science. His alchemical work and his science were inextricably linked.

Newton himself said, 'A man may imagine things that are false, but he can only understand things that are true.'[8] The no man's land between imagining and understanding is, at times, the natural home of the biographer; but by demythologising truths that have long been veiled in secrecy this no man's land becomes narrower. Newton the towering intellect, the pioneer and father of modern science, can now stand alongside Newton the mystic, the emotionally desiccated obsessive and the self-proclaimed, but deluded, discoverer of the philosophers' stone – divested but undiminished.

CHAPTER 1

Desertion

Nature, and Nature's Laws lay hide in Night.
God said, Let Newton be! *and All was* Light.
ALEXANDER POPE[1]

In the days before the English Civil War, Woolsthorpe was a peaceful Lincolnshire village, and even when, for a time, the world seemed turned upside down by internecine struggle the village survived the traumas almost unscathed. A few hundred yards beyond the village, up on the Great North Road (today the A1), the soldiers of the King and those of Parliament clanked their way towards cannon blast and bloody death during the bleak winter of 1642–3; but few men from the village became embroiled in the fighting, and the nearest battles were several miles away.

Woolsthorpe (or Wulsthorpe as it was once known) is an ancient settlement, nestled in a hollow on the west side of the river Witham, about seven miles from the nearest sizeable town, Grantham. Newton's first biographer, William Stukeley, described the village as having a good prospect eastwards, with a view of the Roman road and the Hermen-Street going over the fields to the east of Colsterworth: 'There can be no finer country than this,' he declared.[2]

During the seventeenth century, Woolsthorpe was little more than a collection of small farms and humble country dwellings clustered around the manor house. The area offered poor opportunities as arable land and would sustain only a two-field rotation, which meant that fields were left fallow half the time, so the locals eked out a frugal existence largely from sheep farming.

The Newtons, of which there were many scattered around the Grantham region, had for several generations before Isaac's birth been viewed as being one cut above the local populace, existing on the social cusp between yeomen and lower gentry.* This was all thanks to Isaac's great-great-grandfather, one John Newton, of the nearby village of Westby, who, according to community records and evidence pieced together from wills and tax demands, managed miraculously to ascend the social order from peasant to yeoman during his lifetime.[4] In fact John Newton of Westby did so well that he was able to leave substantial inheritances and dowries for his children – including, for his son Richard, sixty acres of some of the best land of the area, situated in the village of Woolsthorpe, bought shortly before the old man's death in 1562.

John Newton's descendants were neither so aspiring nor so successful. Although the impetus he had provided placed them in good stead, none of them until Isaac made much of an impression in any area of life or improved their social standing to the same degree. The Newton men married relatively well between John and Isaac Newton senior (Isaac's father) – a period of perhaps a century. Although this nudged them slowly upward through the grades of yeoman, none of them was educated formally and it is a startling fact that Isaac Newton senior (like many of his class) could not sign his own name. Yet his son became President of the Royal Society and Lucasian Professor at Cambridge University. Perhaps because of this confused social position of his family and ancestors, class and standing always meant a great deal to Newton.

Isaac's great-grandfather, Richard Newton, bequeathed the sixty acres in Woolsthorpe to his son Robert, who was born around 1570, and it was he who purchased the manor house standing nearby. According to local records, the property had changed hands by sale four times in the preceding century and was in a dilapidated state when it was acquired by the Newtons. Basic repairs were carried out

---

* At the time of the Civil War, the historian Thomas Fuller wrote, 'The yeomanry is an estate of people almost peculiar to England, France and Italy and like a die which hath no points between sink and ace, nobility and peasantry . . . The yeoman wears russet clothes, but makes golden payment, having tin in his buttons and silver in his pockets . . . In his own country he is a main man in juries. He seldom goes abroad and his credit stretches further than his travel.'[3]

within a few years, and it became the family home for this, perhaps the most prosperous, branch of the local Newton family.*

It was with the next generation that the Newtons gained a further modicum of social elevation and a smattering of academic credibility, when Isaac senior, Robert's son, married into the Ayscough family – respected local lower gentry who sent their sons to Oxford and Cambridge universities and whose family members found their way into parsonages and lectureships. When the illiterate but propertied Isaac Newton senior married Hannah Ayscough, whose family had fallen upon hard times and were in danger of sliding down the social scale, it was a match of convenience as much as an auspicious melting of genes and environment: a cocktail to change the world. Thanks to his aspiring forebear John Newton, Isaac had money – in December 1639 Robert Newton had settled the entire Woolsthorpe estate on him. Hannah had breeding. Both families were therefore satisfied, and in April 1642 the couple were married. Hannah took the name Ayscough-Newton.

The winter of that year was bleak both for the Newtons and for the country as a whole – England had slid into a savage civil war. By the time Hannah and Isaac were married, King Charles I had left London, never to return as England's acknowledged sovereign, and had headed north. His queen, Henrietta Maria, adored by her doting husband but loathed by many of his subjects, had been sent to Europe for her own safety. During the summer and autumn of 1642, what had begun as petty skirmishes and political and religious wrangling developed into full-scale civil war, with the royalists camped first at York and then at Oxford. The battle of Edgehill, one of the most famous of the war, had been fought at the end of October and had gone the royalists' way; the country was gripped by battle fever. Within the space of a few years, England had been transformed from a nation at peace, existing beyond the turmoil of the Thirty Years War which had ravaged mainland Europe since 1618, into a nation

---

* The house is T-shaped, two-bedroomed and built of local grey stone. The off-centre door in the cross-stroke of the 'T' leads into a narrow hall with doors to a single room on either side and a central staircase leading to the bedroom in which Newton was born. The west-facing garden contains what is purported to be a descendant of the famous apple-tree of lore.

in which brother had taken up arms against brother and lifelong friendships had been shattered by the taking of sides in the dispute – for the King or for Parliament.

Most biographers of Newton, from Stukeley to recent times, have assumed that the Newtons had royalist leanings. They have based this opinion upon the family's class and social aspirations, reasoning that, as upwardly mobile lower gentry, they would favour the status quo and disapprove of attacks upon the traditional monarchical system. This may have been so, but the sides in the Civil War were not defined clearly along class lines. There were many noblemen who fought on the side of Parliament, and many of the lower orders supported the King. Furthermore, the many complex reasons for the dispute included not only political preoccupations but religious issues, which for some would have been more important. For many historians, the Civil War had its foundations in the decisions of Henry VIII and his immediate descendants and was as much to do with the ideological clash between Rome and the Church of England as with the position and powers of Parliament. The Newton family would not have shared Charles's sympathy towards Catholicism, and indeed in later life Isaac was positively anti-Catholic.

The political views of the Newtons during the Civil War were not recorded. The fact that Hannah's brother, William Ayscough, maintained his position as rector in the nearby village of Burton Coggles during and after the war neither confirms nor refutes the Ayscoughs' Cavalier sympathies: it merely shows that he, like many others, including perhaps the Newtons, bent with the wind.

For all the turmoil the Civil War wrought on the people of England, at the time Hannah Newton was far more concerned with immediate problems caused by a domestic tragedy a few days before the battle of Edgehill: her husband, Isaac, had died leaving her heavily pregnant. What caused his death is unclear. He had just turned thirty-six and appears to have been ill for some time beforehand. We know this because of the introduction to his will, which reads, 'In the name of God amen the first day of October (anno Dom 1642) I Isaac Newton of Woolsthorpe in the parish of Colsterworth in the county of Lincoln yeoman sick of body but of good and perfect memory . . .'[5]

Very little is known of Isaac senior. Misinterpretations have been made of his character based largely on the research of the eighteenth-century writer Thomas Maude, who claimed that Hannah's husband was a wastrel.[6] Maude had actually confused him with a relative, another John Newton, and it seems from the contents of the will that Isaac had actually managed well his newly acquired estate and had taken seriously his responsibilities before marrying Hannah. He was illiterate, but seventeenth-century farmers had little real need for learning, and he left the estate pretty much as he inherited it; Hannah and her child were well provided for.

During those miserable days between the death of her husband and the birth of their child, we can only assume that Hannah did her best to maintain the farm and to prepare herself for the coming event. She went into labour late on Christmas Eve and almost certainly gave birth in the room in which the child was conceived – the bedroom to the left of the top of the stairs. Hannah's mother, Margery, travelled from the nearby village of Market Overton to supervise the birth, and two women from Woolsthorpe were paid a few pennies to help. Sometime soon after 2 a.m. on Christmas morning a son – Isaac – was born.

By Newton's own account, offered late in life, he was born premature. This may have been true, but he was fond of mythologising his childhood and, for complex reasons, he encouraged the idea that there had been something miraculous about his birth. Also, Newton quite naturally did everything he could to pre-empt any rumours that he may have been born illegitimate. The records do not give an exact date for the marriage of Isaac and Hannah, but an unkind analysis, ignoring the declared prematurity, would conclude that Newton was conceived out of wedlock. Most revealing is the fact that when, at the time of his knighthood in 1705, Newton was asked to draft a genealogy for the College of Heralds, he pushed back the date of his parents' marriage to 1639 (the year in which his grandfather acquired the manor house).[7] This could have been a genuine mistake, but Newton's deep-rooted need for secrecy, impeccable social credentials and high-caste moral attitudes (along with the convenience of the changed date) mean that the real reason may not be so accidental.

According to John Conduitt, husband of Newton's half-niece Catherine Barton and a collector of personal anecdotes about his famous relative, Isaac was a tiny baby who, so the legend goes, was small enough to fit into a quart pot. 'Sir I. N. told me', Conduitt recalled, 'that he had been told that when he was born he was so little they could put him into a quart pot & so weakly that he was forced to have a bolster all around his neck to keep it on his shoulders.' It is an appealing story, and one supported by others that Newton passed on to Conduitt. In an elaboration of the story, the two women attending Hannah at the time of Newton's birth were sent to the home of one Lady Pakenham to obtain medicines. Apparently, Newton was so frail that on their way back to the manor the two women 'sat down on a stile sure the child would be dead before they could get back'.[8]

Such tales have become part of the legend of Newton's life, supported by the testimony of many of the villagers whom Stukeley interviewed during the late 1720s for his biography of Newton. Unlike many interpretations of Newton's later exploits, they are at least plausible. But it is typical of the adulatory style of Newton's early biographers that so much is made of the *miraculous* nature of his early survival. A description of Newton's infancy from Sir David Brewster, writing in 1855, serves to illustrate:

> Providence, however, disappointed their fears, and that frail tenement which seemed scarcely able to imprison its immortal mind, was destined to enjoy a vigorous maturity, and to survive even the average term of human existence.[9]

Newton would have strongly approved of such a description, which adds still more weight to the self-image he so much treasured.

Of the first three years of Newton's life almost nothing is recorded. We know from a tiny scrap of parchment unearthed by Stukeley that the baby Isaac was baptised on 1 January 1643. It is easy to conjure up a romantic image of Hannah on New Year's Day trudging through the snow with her feeble baby wrapped in swaddling-clothes on their way to the local church for the christening ceremony, but it is almost certain that the village vicar would have visited the manor to conduct the service.

After this there is a three-year period of blankness. As the Civil War raged the length and breadth of the country, Hannah and her son continued to live at the manor house. Their employees tilled the land and carried out the annual lambing, the shearing, the milking and the feeding, while Hannah dealt with the many bureaucratic aspects of the business and supervised sales of animals and the maintenance of farm stocks. It would also have been natural for Hannah's parents to play a significant role in helping their widowed daughter. Hannah and her son were not rich, but they were comfortably off. At the time of Isaac senior's death, the deeds of the manor house and the surrounding lands, along with goods and chattels valued at £459.12s. 4d., 234 sheep, 46 head of cattle and several barns full of oats, barley and malt were all bequeathed to Hannah. To put this into perspective, the average farmer of the region owned a flock of between 35 and 40 sheep, and the will of a typical yeoman contained goods worth little more than £100. During the 1640s, a workman could expect to earn in the region of one shilling and sixpence a week.

By the winter of 1645 King Charles was holed up in Oxford, effectively under siege by Cromwell's army. In June of that disastrous year for the royalists, his troops suffered their worst military defeat at the battle of Naseby. England was still far from regicide, but the forces that, four years later, would lead to this singular event were already coalescing. Lincolnshire continued to pass through the upheaval relatively unscathed, making Woolsthorpe a haven of solitude and anonymity for advancing or retreating armies. Throughout the Civil War, troops were away from their garrisons for months at a time and relied upon the hospitality of town and country folk alike; stories of villages and towns refusing to accommodate troops of either side are rare. Isaac would have seen soldiers of both sides passing through the village, and there may have been occasions when troops stayed in the houses beyond the fields of his little sanctuary, or even at the manor itself. If Hannah accommodated royalists or Roundheads, no record has been passed on to us, and Newton never mentioned such a thing, but it would not have been surprising.

The worst of the fighting was over by the summer of 1646, but for Isaac a far more significant event had transformed his life. At the beginning of the year, soon after his third birthday, his mother had decided to remarry.

Barnabas Smith was the rector of North Witham, a hamlet just over a mile from Woolsthorpe. Little is known about him, but what is known does not paint a pretty picture. He was successful academically but seems to have displayed only a passing interest in learning. The son of a wealthy landowner, he attended Lincoln College, Oxford, where he graduated in 1601. He collected books, but by all accounts did not often read them; he made a half-hearted effort to start a notebook in which he intended collecting his thoughts on a variety of theological subjects, but gave up after a few pages. Both the books and the notebook did eventually find serious use, however, as they were passed on to Isaac as part of his inheritance upon Smith's death. The books – some 250 of them – may have led Newton into serious collecting himself and could well have introduced him to a number of the theological subjects which later preoccupied him to and beyond the point of obsession. Nor was the notebook wasted: the mostly blank pages ended up covered in Isaac's earliest scribblings on the subject of gravitation and the formulation of the calculus. With a barely disguised dig at his stepfather, Newton never referred to Smith's hand-me-down as a notebook, but, doubtless knowing its history, called it the 'Waste Book'.

Smith was sixty-three years old when the widowed Hannah Newton first caught his eye. Hannah was around thirty. (There are no surviving official records giving her exact date of birth.) By then he had been rector at North Witham for over thirty-five years, the rectorship having been bought for him by his father in 1610 as the source of a convenient annuity. According to a visiting bishop who had come to check up on the new rector twelve months after his arrival in North Witham, Smith was a non-resident and, presumably on good behaviour, but 'inhospitable'.

For Smith the rectorship was little more than a dalliance. By the time of his proposal to Hannah in 1645 he commanded an independent income of over £500 per annum – a considerable sum in the seventeenth century, to which his clergyman's stipend would have added little. Perhaps he had no need to be 'hospitable'. During his rectorship, he certainly appears to have sailed calmly through the upheavals in Church doctrine created by the Civil War. Between the start of the first Civil War and the end of the second, many Anglican clergymen chose banishment from their living over conformity to the constantly changing tide of theological fashion. Smith, however, went with the flow.

Smith's first wife had died only six months earlier, in June 1645, and it may have been for this reason that his initial approach to Hannah was businesslike even for the time. Instead of attempting to woo or even talk to her face to face, he paid a servant a day's wages to deliver a letter of proposal.

Whatever Smith's reasons for making such a decidedly unromantic proposal, Hannah did not at first reply. Instead, she consulted with her brother, William – who, as incumbent in a nearby village, must have known Smith – and a family conclave was convened to weigh up the pros and cons of the match.

The terms and conditions of the proposed arrangement were negotiated as a business transaction, and the eventual agreement seems equitable, but little thought appears to have been spared for the pawn in this game – Hannah's three-year-old son. The deal was that Hannah and Smith would marry and she would naturally move to North Witham, but Isaac would stay at the Newton home. In return, some land to the value of £50 was to be signed over for him to inherit at the age of twenty-one, and the house in Woolsthorpe was to be refurbished completely.

What is so surprising is not so much that Smith did not want Hannah's son to live with them but that Hannah should go along with these terms. Even accepting that nothing was known of psychology in the seventeenth century, that a mother would willingly trade her son for a new life strikes the twentieth-century observer as totally heartless.

Contemporary accounts of Hannah's character provide us with very little that is believable about her or helpful in reading her character: John Conduitt tells us, 'She was a woman of so extraordinary an understanding and a virtue that those who ... think that a soul like Sir Isaac Newton's could be formed by anything less than the immediate operation of a Divine Creator might be apt to ascribe to her many of those extraordinary qualities with which it was endowed.'[10] Giving her the benefit of the doubt, however, we can only speculate that she agreed to the arrangement reluctantly and primarily for her son's future. Smith was old and wealthy, and Hannah doubtless thought he would not live long. Because his first marriage had been childless, upon his death she would, she might assume, inherit everything, and after the short period of separation Isaac would benefit greatly from the union. But it was hardly as if the Newtons were destitute. By the standards of the day they were doing very nicely indeed. Did she really need to agree?

We only have two pieces of evidence to demonstrate Hannah's love for her son. First is the fact that she made him sole heir to the estate and in her will she left her body to be buried as Isaac 'shall think fit'.[11] The other is a scrap of a letter written in her barely literate hand during Isaac's undergraduate days which reads:

> Isack
> received your letter and I perceive you letter from mee with your cloth but none to you your sisters present thai love to you with my motherly lov you and prayers to god for you I your loving mother
>
> Hannah[12]

It would appear from this that Hannah loved her son as any mother would, and, rather than their separation being something she agreed to readily for financial gain or convenience, it is far more likely that Isaac's fate was decided by others within the Newton family. Smith may not have been enthusiastic about the idea of having another man's son in his house; he may have seen Isaac as a threat to his nascent relationship with Hannah; the boy might even have been perceived as a disruptive influence: but this does not mean that Smith insisted that Isaac remain at the manor. It is quite possible that the suggestion came from the Newtons: by keeping the three-year-old at the family home, they could maintain control of the estate, keeping it out of Smith's hands.

Perhaps Newton never learned the real reason why he was left at the manor; he certainly never mentioned it later in life. But, after the various arguments had been mulled over, his mother accepted Smith's offer under the negotiated conditions and on 27 January 1646 her list of surnames increased by one, to Ayscough-Newton-Smith.

Unfortunately for Newton the boy, but conceivably of the utmost importance to the advancement of human knowledge, any hopes that Hannah might have had for a short marriage followed by great wealth were disappointed. Barnabas Smith lived for almost another eight years, dying at the age of seventy-one in 1653 after fathering three children: Mary (Marie), born in 1647; Benjamin, born in 1651; and Hannah, born in 1652. Although circumstances would later result in these half-siblings doing little to damage the value of Isaac's inheritance, the enforced separation from his mother at such an impressionable age has long been recognised as one of the key factors in shaping Newton's character.

Isaac had been raised solely by his mother, and there is little doubt that until her departure they were almost inseparable; his dependency upon her would have been far greater than if both his parents had survived. Furthermore, it was not death which effectively deprived Newton of his mother: she was taken by another man and continued to live close by. To rub salt into Isaac's emotional wounds, he never knew when or if his mother would turn up. Sometimes she would reappear for an hour or for an afternoon, but always she would go again – and always to the other man, the hated Barnabas Smith. We know from clues left in his notebooks and personal papers that Newton loathed his stepfather and, to a lesser extent, grew to resent his mother.

Of Newton's earliest writings only a few fragments remain, but these provide some revealing insights into the psychology of the boy. Four notebooks survive from his days in Lincolnshire and his first year at Trinity College, Cambridge. Of these, the most interesting are a notebook kept at the Fitzwilliam Museum, Cambridge, and a schoolboy exercise book in the possession of the Pierpont Morgan Library in New York, which has become known as the Morgan Notebook.

Newton began writing in the Fitzwilliam Notebook sometime during the early summer of 1662, when he was a nineteen-year-old undergraduate. At first he treated the book as a form of confessional, purging himself by writing out his felonies against the Lord. In shorthand, he drew up two lists: of sins committed 'Before Whitsunday 1662' and of those 'Since Whitsunday 1662'. The first list, stretching back to his childhood, contains forty-five transgressions; the other, more recent, set contains nine 'sins'. Most startling of this earlier batch are numbers 13 and 14: 'Threatening my father and mother Smith to burne them and the house over them' and 'Wishing death and hoping it to some'.[13] What is so surprising about these entries is not so much the violence of Isaac's feelings as a child – that could be expected – but the fact that he remembered his anger so clearly and felt compelled to confess it so many years later, long after the principal object of his hatred was dead and buried.

Perhaps even more fascinating is the window into Newton's psychology offered by the schoolboy exercise book, the Morgan Notebook. This contains collections of scribblings, ideas jotted down from books Isaac read, lists and other notes probably dating from about the age of ten to his early teenage years. Most pertinent to his mental state is an alphabetical list of word associations based upon a contemporary book, Father Francis Gregory's *Nomenclatura*. Newton placed these associations under different headings. Under the heading 'Of Kindred and Titles', we have F: 'Father' (out of Gregory) followed by 'Fornicator' and 'Flatterer' (both from Newton). Under B we have 'Brother' (Gregory) followed by 'Bastard', 'Blasphemer', 'Brawler', 'Bedlam', 'Beggar' and finally 'Benjamite' (all from Newton – whose half-brother, it should be remembered, was named Benjamin).* W begins innocently with Gregory's 'Wife' and 'Wedlock', but is completed by Newton's 'Whore'.[14]

Although he was later attentive and caring towards his mother, and nursed her during her final illness in 1679, the psychological scars of her remarriage clearly dug deep and almost certainly affected Newton's future relationships with women. Although unquantifiable, the effect of such trauma moulds different individuals in different ways according to their particular circumstances, and in Newton's case we may match the damage to the personality which emerged from the wreckage.

During the eight years in which Hannah lived in North Witham, Isaac remained in the care of his grandparents, James and Margery Ayscough, who had taken up residence at the manor. As Newton never mentioned his grandparents later in life, it would appear there was little love lost between them. The Ayscoughs probably did their best, but they could never have replaced Hannah. From a state where he could do no wrong, bonded intimately with his gentle mother, he was inexplicably deserted and thrust into the care of two elderly people. The three-year-old Isaac's instinctive reaction to being left in this situation would be to feel overwhelming guilt, to imagine he had somehow done something terribly wrong. Interestingly, the Morgan Notebook includes O: 'Orphan' (Gregory) followed by 'Offender' (Newton). When, eight years later, Hannah returned to stay, the eleven-year-old pre-pubescent was again disturbed emotionally. Was his mother's sudden return a reward? Had he then been right all along to feel guilty over some unknown and unknowable deed? Had he now served his punishment?

---

* Benjamin means 'favourite son', so here Newton is also referring to one who supports or loves a favourite son, Benjamin – a Benjamite.

Upon Hannah's return, his parents moved back to their home village of Market Overton, a few miles distant, and Hannah perhaps assumed that her and Isaac's lives could pick up where they had left off in 1646. But it was not to be. They were certainly wealthier and more secure, but the rift between mother and son was now too wide ever to heal properly. Not only was the fact of her desertion irreversible, but she was returning to the manor with three young children in tow.

Apart from the word associations of the Morgan Notebook, little evidence remains of Isaac's feelings towards his half-siblings, but we can understand the resentment he must have harboured for this threesome whose father he despised and with whom he could only associate betrayal and desertion. In his school Latin exercise book (one of the four documents surviving from his youth) we have his comment, 'I have my brother to entreat',[15] in which a number of biographers have detected a hint of sarcasm – at least, now his mother had returned, he had a brother to talk to – but nothing else on the subject survives.

Although by all accounts Newton was a quiet child, he also possessed a malicious streak, as is evidenced by his private outbursts against Smith recorded in the Fitzwilliam Notebook. Hannah may well have been relieved when a year after her return he was old enough to attend King's School in Grantham, seven miles away – a distance much too great to travel each day.

Established sometime during the 1520s, by the time Newton arrived King's School had for some 130 years provided a solid grounding in what were then considered to be the basics of education: Latin, Greek and Bible studies. In 1654 the headmaster was Henry Stokes, a graduate of Pembroke Hall, Cambridge, and a man who later exerted a profound influence upon the course of Newton's early academic career.

If the years immediately following Hannah Newton's departure for a new life with Smith and the confusion of her return moulded Isaac's emotional make-up, then those between his arrival in Grantham and his leaving for Cambridge in 1661 laid the foundations of his intellectual outlook. He had two major influences in this new life. The first was the routine of a formal education; the other was his new home environment as a lodger with the family of the local apothecary, the Clarks, who had close links with Grantham School and provided accommodation for a succession of pupils in their apartment above the apothecary shop next to the George Inn on the High Street.

At first, school was of little interest to Isaac, as is shown by his lacklustre and completely unexceptional academic status. He was quick to learn but was also a natural autodidact, ignored by most of his teachers and disliked by the other boys. Pupils were expected to learn the core curriculum of classical languages and scriptural studies parrot-fashion. It all required little imagination and offered no inspiration for inquisitiveness. It is, to the modern mind, astonishing that Newton had no formal mathematical training until he entered Cambridge (and even then mathematics was not part of the standard curriculum during his first years as an undergraduate). To compensate for this dull fare, Isaac first read the books handed down to him by his stepfather and later those he found in the library of St Wulfram's church in Grantham – a long, narrow room above the church porch. Most of these texts were dry fodder indeed: theological tracts and Puritan propaganda that Newton was encouraged to read by a Puritan divine and lecturer at the school named John Angell.

These theology books and the encouragement of Angell led Newton into a religious doctrine he maintained for the rest of his life, but they did not provide the intellectual meat he needed. Fortunately, other books came his way. The most important in leading him to scientific inquiry was *The Mysteries of Nature and Art* by John Bate, which Isaac discovered when he was about thirteen. He was totally captivated by it and spent 2½d. on an exercise book into which he copied out long passages.

Bate's book, first published in 1634, was full of detailed instructions for making wonderful machines and devices, and it was from following these that the teenage Newton was able to design and build working mechanical models for which he gained something of a reputation as a schoolboy. Some seventy years later, Stukeley was able to find a few old folk who still remembered Newton's miraculous models – windmills that actually worked, into which the boy sometimes placed a mouse to turn the sails; kites; perfectly functioning sundials; and paper lanterns with which he found his way to school on dark winter mornings. The ancient villagers to whom Stukeley referred knew nothing of Bate's book, which might go some way to account for such hyperbole as 'Newton's innate fire was soon excited. He penetrated beyond the superficial view of the thing ... He obtained so exact a notion of the mechanism of it, that he made a true and perfect model of it in wood; and it was said to be as clean a piece of workmanship as the original.'[16]

Model-building provided a suitably insular pastime for a boy who appears to have had no friends at school. According to Stukeley's interviewees, Newton tried to interest his schoolmates in his cerebral activities; rather than being content to watch his contemporaries indulge in what Stukeley calls 'trifling sports', Isaac apparently tried to 'teach them ... to play philosophically'.[17]

It is easy to detect here the personality of a boy crying out for attention and companionship but simply unable to communicate with others of his age. He had been an only child, and under the best of circumstances such an upbringing can cause children to have difficulty in adjusting when they first encounter others of similar age at school. Stukeley and his followers have tried to imply that the matter went deeper than this: that there was something totally other-worldly about Newton as a boy. And no doubt he was exceptional; he was certainly a talented youth, even if at this age he showed little interest in the official curriculum. Thanks to Stukeley's first-hand accounts, we know Isaac could draw and write well. 'Sir Isaac furnished his whole room with pictures of his own making, which probably he copied from prints, as well as from life,' claimed one interviewee.[18] Another recalled in her dotage that Newton had written a poem for her, which she could still recall from memory.[19] And, although his mathematical talent had not yet emerged, his interest in mechanical devices illustrates that the skill and curiosity of the scientist, the talent for constructing experiments and testing ideas, was already awakened.

What eventually transformed his unhappy relationship with official learning was a seemingly trivial event. On the way to school one morning, one of the boys in his class (according to some historians, Clark's stepson, Arthur) kicked Isaac hard in the stomach. What provoked the attack is open to conjecture, but it is significant that the bully was one place above Newton in their class ratings. Enraged, Isaac challenged the other, much larger, boy to a fight after school. According to John Conduitt, who popularised the tale:

[A]s soon as the school was over he challenged the boy to fight, & they went out together into the church yard, the schoolmaster's son came to them whilst they were fighting & clapped one on the back & winked at the other to encourage them both. Though Sir Isaac was not so lusty as his antagonist he had so much more spirit & resolution that he beat him 'til he declared he would fight no more, upon which the schoolmaster's son bad him use him like a coward, & rub his nose against the wall & accordingly Sir Isaac pulled him along by the ears and thrust his face against the side of the church.[20]

Still not content, before leaving the bully to nurse his wounds, Newton declared he would not rest until he had overtaken his combatant's academic position. According to Conduitt, Isaac not only overtook the bully but, within a short time, rose to first place in the school.

The rooms Newton shared with the Clarks above the apothecary shop must have been crowded at this time; Mr Clark and his wife had three children from his wife's previous marriage – Catherine, Eduard and Arthur Storer. Newton probably shared a room with one or both of the brothers, and in any event, given the size of apartments above town-centre shops of the seventeenth century, living conditions must have seemed very cramped to a boy brought up in a manor house set in acres of open space. Yet, by all accounts, Isaac was content living with the Clarks. If Arthur Storer was indeed his antagonist until the fight in the churchyard, then we can imagine the schoolboy arguments and rivalries within the Clark household when the adults were out or busy in the shop. It is easier to imagine the rows and recriminations after the fight, when Arthur and Isaac returned home, one with cuts and bruises all over his face, the other still rigid with anger.

Fortunately for Isaac, the Clarks appear to have been a very placid couple who raised their children with a distinctly far-sighted liberalism quite atypical of the time. Furthermore, Clark – proud of his position as an apothecary and, according to contemporaries, a cheerful, open man – encouraged the inquisitive Newton to watch him at work and to ask questions.

To Newton, bored with school and searching for something to stimulate his intellect, the apothecary – a repository for chemicals from which remedies and medicines of all descriptions were con-cocted and sold to the public – was a place full of wonders. On the shelves around the walls of the shop stood jar upon jar of strange-coloured powders and liquids – yellow sulphur, silver mercury, red lead oxide. The shop provided him with his earliest experience of the possibilities of chemistry. It also offered an opportunity to con-duct his own experiments.

We know from his surviving notebooks that Newton did not simply watch Mr Clark go about his business but transcribed remedies and cures from books he discovered alongside the chemical jars. He may have even devised his own recipes. In these journals we find descrip-tions of how to produce paints and pigments, methods by which glass may be cut with chemicals, and 'a bait to catch fish'. We also encounter cures for various illnesses – such as that for fistulas (here meaning surgically produced openings into the body), which involved 'drinking twice or thrice a day a . . . small portion of mint and wormwood and 300 millipedes well beaten (when their heads are pulled off) in a mortar . . . & suspended in 4 gallons of ale in its fermentation'.[21] Newton was evidently a hypochondriac from an early age and was fond of concocting remedies which he both used upon himself and offered to others. He listed over 200 different human ailments in the Morgan Notebook under the heading 'Of Diseases'.

As well as receiving his earliest knowledge of primitive chemistry from Clark the apothecary, Newton acquired from him an introduc-tion to the concept of brotherhood.* Along with all other members of his profession (which at the time existed in a anachronistic limbo: part shopkeeping, part quack medicine), Clark was a member of the Society of Apothecaries. Perhaps, upon his return from regular meetings of the society at its then premises in Water Lane in London,

---

* What I here refer to as 'primitive chemistry' was quite distinct from alchemy. Clark's employment of chemistry would have been driven by commerce – the concoction of remedies and mixtures for everyday use. Alchemy was, and remains, a mystical art, shrouded in ritual and used to produce exotica such as the mythical philosophers' stone and magic elixirs. Clark was a tradesman and probably feared notions of magic or alchemical systems, or else looked upon them as crazy delusions. For a detailed explanation of alchemy and how it differs from traditional chemistry, see Chapter 6: 'The Search for the Philosophers' Stone'.

the affable Clark would hint at the proceedings and glamorise the rules and regulations of the organisation to the ever-inquisitive Newton. In this way he not only inspired the boy to delve into the arcane world of cures, remedies and recipes but provided him with another valued piece of knowledge: the concept that there existed brotherhoods through which individuals could communicate and circulate information.

Primitive chemistry and the charms of the apothecary's world were not the only distractions in the Clarks' home. Living under the same roof was Mr Clark's stepdaughter, Catherine Storer, the only female other than his mother and later his half-niece Catherine Barton to whom Newton is known to have been emotionally attached.*

It is difficult to assess accurately how important Catherine Storer was to Newton, because we have only her account of their relationship – conveyed to Stukeley shortly before Newton's death. By this time Newton was a world-renowned scientist and, aside from the fact that she was in her early eighties and doubtless romanticising her own past, for Catherine to exaggerate her place in the great man's boyhood affections would have been quite natural.

They were certainly close friends. This much is demonstrated by their writing to each other during Newton's early days in Cambridge. Further evidence comes from a conversation Stukeley recalled having with Newton shortly before the scientist's death. Newton, he claimed, expressed a desire to return to live out his days in Woolsthorpe and showed a particular interest in acquiring a property near to where Catherine had once lived.[22] However, Catherine Storer's suggestions to Stukeley that she and Isaac were sweethearts, and that Newton had at one time seriously considered passing up his academic career in order to marry her are most probably pure fantasy. In his memoirs, Stukeley recounted Catherine Storer's tale, saying:

Sir Isaac and she being thus brought up together, it is said that he entertained a love for her, nor does she deny it. But her portion being not considerable, and he being [a] fellow of a college, it was incompatible with his fortunes to marry, perhaps

---

* Catherine Storer became Catherine Bakon and, later, Catherine Vincent by successive marriages.

his studies too. It is certain he always had a kindness for her, visited her whenever in the country, in both her husbands' days, and gave her forty shillings upon a time, when it was of service to her. She is a little woman, but we may with ease discern that she has been very handsome.[23]

Catherine may have harboured hopes, but any spark of romantic interest that Isaac might have shown her was soon extinguished. As his academic performance improved, he was drawn to the attention of his headmaster, Henry Stokes, who saw in him a talent he could not allow to go to waste.

No record of Newton's academic progress survives, but it is safe to assume that by the time the boy was sixteen Stokes was already viewing him as a likely candidate for university entrance. What Hannah's initial reaction to her son's progress might have been is unknown, but late in 1658, as Stokes was about to suggest that her son should consider a university education, Hannah decided to remove him from King's School.

Hannah had shown little consideration for education, and it had been at the insistence of her brother, the Cambridge-educated William Ayscough, that Isaac had attended an elementary school while living with his grandparents. Ironically, it could have been Stokes's enthusiasm that prompted Hannah to remove Isaac from the school. She saw little need for her son to be educated; her husband had demonstrated how the farm could be managed even without the benefit of literacy.

At first Hannah had her way. For most of 1659 Isaac lived at the manor with his mother and Barnabas Smith's children. But, in the notebook started in 1662, the list of his 'sins' during the period in which he lived there indicates that it was a time fraught with bitterness and family arguments.

He was, for the most part, an obedient and respectful son, but the stress of being taken away from an environment in which he was blossoming and the threat of having his life ruined again by the wishes of his mother were evidently too much. The Fitzwilliam Notebook lists his crimes as 'Refusing to go to the close at my mother's command', 'Striking many', 'Peevishness with my mother', 'With my sister', 'Punching my sister' and 'Falling out with the servants'. The signs of strain are clear.

Whether it was to show deliberately how bad he was at farm duties or through genuine inability and absent-mindedness, he did not perform his duties at all well. Stukeley tells us that:

When at home if his mother ordered him into the fields to look after the sheep, the corn, or upon any rural employment, it went on very heavily through his manage [i.e. he did not conduct the task well]. His chief delight was to sit under a tree, with a book in his hands, or to busy himself with his knife in cutting wood for models of somewhat or other that struck his fancy, or he would go to the running stream, and make little millwheels to put into the water.[24]

His lack of interest even brought an admonition from the authorities. The records of the manor court of the nearby village of Colsterworth show that on 28 October 1659 an Isaac Newton was fined 3s. 4d. 'for suffering his sheep to break the stubs of 23 of loes [loose?, meaning unenclosed] furlongs'. On top of this, he was obliged to pay 1s. on each of two further counts, 'for suffering his swine to trespass in the corn fields' and 'for suffering his fence belonging to his yards to be out of repair'.[25]

Following this, Hannah decided that her son should be supervised by a servant from the household who would look after him and give the boy proper instruction. Predictably, the idea failed because Newton quickly turned the situation to his advantage, allowing the servant to do all the work while he sloped off to read or to pursue other interests.

Each Saturday, Isaac set off dutifully with the servant to Grantham to sell the farm produce and to purchase supplies for the following week. Arriving at the Saracen's Head, the inn in Westgate, he would instruct the servant to continue with the business of the day while he went off to visit Mr Clark at his shop in the High Street.

What drew Newton there was a collection of books that Clark had acquired from his recently deceased brother, Dr Joseph Clark, the usher (assistant teacher) of King's School. The apothecary himself was interested in the collection, but had little time to read. Perhaps Newton had offered to catalogue the books in exchange for the chance to read them; be that as it may, somehow he managed to persuade Clark to allow him to spend almost all of each Saturday in the back room behind the shop in solitary bliss studying texts on physics, anatomy, botany, philosophy and mathematics – his first real exposure to these things. From Bate's *The Mysteries of Nature and Art* Newton had discovered the elements of experimentation and practical skills – lessons he would never forget but would employ both as an orthodox scholar and in his role as alchemist. But here were texts by greater writers and natural philosophers. We do not know for sure the contents of the library, but it is safe to assume a scholar such as Dr Clark would have collected the works of the great names of the past and perhaps the more controversial figures of the day, and it is likely that Newton now first discovered Francis Bacon and René Descartes, Aristotle and Plato, acquiring a fuller and more useful education than he could possibly have gained within the narrow confines of the school curriculum.

Word of Isaac's truancy soon reached two different parties involved in the argument over his future. Hannah heard of her son's antics through the complaints of the servant, and Henry Stokes discovered how his ex-pupil was showing admirable determination not to fall under his mother's yoke. Stokes had tried to dissuade Hannah from taking Newton away from school but had been unsuccessful. Now, hearing how Isaac was doing everything he could to foil his mother's efforts, Stokes decided to try again.

Initially, nothing changed. Despite the irritation caused by her son's behaviour, Hannah would not listen to suggestions that he should pursue an academic career and desert the farm. To be fair, Hannah was herself poorly educated and could not have appreciated the world of learning that Isaac took to so naturally. To her, the only thing that mattered was the management of the estate: it was the source of their prosperity, and she could not understand what her son could possibly gain by attending university. She had already lost two husbands and was expected to maintain a farm, run a household and look after three young children. She could not bear to lose Isaac too.

But, after Stokes appealed to her a second time, she realised she could hold Isaac back no longer. (Her decision was no doubt sweetened by Stokes's offer to remit the standard charge of forty shillings paid to the school by the parents of all boys who came from beyond the town.[26])

Stokes then talked to William Ayscough (who had probably

influenced Hannah's change of heart and was himself a graduate of Trinity College, Cambridge), and probably to Humphrey Babington, a relative of the Clarks and a fellow at Cambridge University. Together they smoothed the way for Isaac's admission, and by the autumn of 1660 the young man was back in Grantham preparing for Cambridge.

Helped by those around him who understood his desire to learn, Isaac now, for the first time, found himself completely content. Throughout his childhood and teenage years he had constantly been pulled in different directions. At school he clashed with the teaching tradition on the one hand and his contemporaries on the other. He eventually found his true nature not from the comfort of others or through the small accomplishments of orthodoxy, but in the discovery of a larger world beyond the confines of his upbringing. By 1660 he had passed the threshold and entered the world in which he would flourish.

CHAPTER 2

The Changing View of Matter and Energy

If God created the world, where was he before the Creation? ...
Know that the world is uncreated, as time itself is, without beginning
and end. **Mahapurana** (India, ninth century)

What is matter, and how does it move? These are questions that have occupied the thoughts of physicists from ancient times to the present day, and they were fundamental queries for Isaac Newton.

Our modern view is based upon the rather exotic world of quantum theory, but for most everyday purposes the way in which we manipulate matter and energy relies upon rules and systems discovered between Newton's lifetime and the present century. For many historians of science, Newton's ideas about how matter behaves and how energies and forces operate can be seen as a watershed in the development of physics. Indeed, some perceive his work as making possible the Industrial Revolution. Newton provided a focus: he was an individual scientist who drew together the many threads that led from ancient times to his fathering of modern empirical science (a study based upon mathematical analysis as well as experimental evidence). Behind Newton lay some 2,000 years of changing ideas about the nature of the universe; his great achievement was to clarify and to bring together the individual breakthroughs of men like Galileo, Descartes and Kepler and to produce a general overview – a set of laws and rules that has given modern physics a definite structure.

The ancient Greeks were the first to record their ideas about the nature of matter, and we know of several different schools of reasoning. The two most important for our purposes are the teachings of

Aristotle and the ideas of a rival theory – the atomic hypothesis of Democritus.

The Greek philosophy that prevailed up to Newton's time was that traditionally attributed to Aristotle – the notion of the four elements: earth, water, air and fire. The alternative was the ideas of Democritus, born some seventy-five years before Aristotle, in 460 BC, who taught that matter is made up of tiny invisible parts, or atoms. Because Aristotle and Plato both largely disapproved of Democritus's atomic theory, however, it was almost completely ignored from Aristotle's day until its partial revival during the seventeenth century.

The first person to formulate the idea of the four elements was actually a Sicilian philosopher named Empedocles, some half-century before Aristotle's birth, but the idea was refined and made popular by Aristotle. It is thought that the concept first arose from watching the action of burning. For example, when green wood is burned, the fire is visible by its own light, the smoke vanishes into air, water boils from the wood, and the remaining ashes are clearly earth-like. This gave rise to the idea that everything in the universe is composed of different proportions of these four fundamental elements – an idea which became the foundation of Aristotle's work in natural philosophy that was handed down to future generations.

Aristotle was born in 384 BC at Stagira in Chalcidice. The son of the physician to Philip, King of Macedon, he later became the pupil of Plato and, in middle age, the teacher of Alexander the Great. He wrote a collection of tracts that were not only influential in his own time but whose rediscovery in an incomplete form by European scholars during the thirteenth century heralded a return to learning and the earliest emergence of the Renaissance. Those most relevant to his thoughts on natural philosophy (what by the eighteenth century had become known as physics) were *On Generation and Corruption* and *Physical Discourse*, which concentrated upon ideas concerning matter, form, motion, time and the heavenly and earthly realms.

To Aristotle, the earthly realm was composed of a blend of the four elements which, if left to settle, would form layers: water falling through air (or air moving up through water, as do bubbles), solid earth falling through water and air, and fire existing in the top layer because it moves up through air. Using this model, Aristotle would have explained the fall of an apple as being due to the earthy and watery parts of the solid apple trying to find their natural place in the universe, falling through air to reach the ground. As well as popularising the idea of the four elements, Aristotle also pioneered the concept of the Unmoved Mover – the name he gave to the omnipotent being who maintained the movement of the heavens, keeping the Sun and the planets travelling around the Earth.

Aristotle's work was encyclopedic in range, and he wrote on almost all subjects known at the time, covering logic, philosophy, biology, astronomy and physics. His strongest subjects were logic and, of the sciences, biology; his weakest was physics. Most significant for how Aristotle arrived at many of his scientific ideas was his creation of syllogistic logic: the principle that a conclusion can be reached as a logical consequence of two preceding premises. An example of this is the collection of statements 'All elephants are animals; all animals are living things; therefore all elephants are living things.'

Syllogisms are powerful tools in the study of logic, and were used as a fundamental mathematical procedure until the nineteenth century, when they were superseded by more versatile ideas, but their use is a rather superficial way to conduct science, because syllogistic logic does not contain an element of experiment: syllogisms consist merely of two statements and a conclusion based upon superficial observation or deductive reasoning.

Plato, Aristotle's teacher (and the man who established the school at the Academy in Athens which lasted nine centuries), actively disliked experiment and so it was never established as a guiding principle for Greek natural philosophy. Instead, Aristotle and the generations of Greek thinkers who followed him created a rigid set of rules based upon syllogistic logic only, producing a distorted picture of reality. But, because of Aristotle's stature, this limited approach became endowed with an aura of infallibility which persisted until the beginning of the modern era. The historian Charles Singer has said of this unfortunate process:

> The whole theory of science was so interpreted, and the whole of logic was so constructed, as to lead up to the ideal of demonstrative science [i.e. conclusions reached through reasoning alone], which in its turn rested on a false analogy which assimilated it to the dialectics of proof. Does not this mistake go far

to account for the neglect of experience and the unprogress-
iveness of science for nearly 2,000 years after Aristotle?[1]

In the same vein, the writer and historian Sir William Dampier
pointed out that:

Aristotle, while dealing skilfully with the theory of the passage
from particular instances to general propositions, in practice
often failed lamentably. Taking the available facts, he would
rush at once to the wildest generalisations. Naturally he failed.
Enough facts were not available, and there was no adequate
scientific background into which they could be fitted.[2]

The modern scientific method involves reasoning *and* experiment.
To give a simple example: early on in a scientific investigation an
idea is postulated – often based upon an inspired insight. This is then
developed into a tentative hypothesis by means of pure reasoning –
a process called the inductive method. The practical consequences
of this hypothesis must then be deduced mathematically and the
idea is tested experimentally. If there are discrepancies between the
hypothesis and the experimental results or observations, the hypothesis must be altered and the experiments be repeated until there is
either agreement between reasoning and observation or the original
idea is discarded. If the reasoning and the practical verification
eventually agree, the hypothesis is promoted to the status of a theory.
This can then be used to attempt to explain a more generalised
scenario than the original concept and may hold for many years.
But, crucially, it is still never considered to be the only theory that
could fit the facts, and good science allows for new ideas to be
introduced that may destroy the old theory or demand radical
changes.*

---

* The influential philosopher of science Sir Karl Popper is critical of this system
and instead supports the idea of tackling science by attempting to show that
ideas are false rather than finding evidence to show that a hypothesis is correct.
In this way, concepts that survive attempts to disprove them can be promoted
to the status of theories. Although this philosophy is recognised and supported
by many contemporary scientists, what has become known as the 'scientific
method' described here has held sway since Newton's time.

Aristotle's dominance left no room for alternative ideas. Democritus, the father of the atomic theory, believed that 'According to convention there is a sweet and a bitter, a hot and a cold, and according to convention there is colour. In truth there are atoms and void.' Aristotle dismissed this notion by relying upon syllogisms that were founded upon inadequate knowledge. For example, he claimed that, if the atomic theory were true, matter would be heavy by nature and nothing would be light enough in itself to rise. A large mass of air or fire would then be heavier than a small mass of earth or water, so the earth or water would not sink (or the air and fire rise) and therefore the elements would not find their natural positions. This argument illustrates how Aristotle was not approaching the problem in the way a modern objective scientist would – he was not able to consider questioning his own cherished beliefs even when presented with a strong alternative theory.

Aristotle's dogma became almost a religion among his followers, and his teachings were passed on to future generations virtually unquestioned, misguiding future thinkers and leading science along a partially blind alley for several hundred years without interruption.

By the time of Aristotle's death, in 322 BC, the Egyptian city of Alexandria was about to emerge as the intellectual centre of the world. At its heart was the great library which is said to have contained all human knowledge in an estimated 400,000 volumes and scrolls. From Alexandria, learning spread eastward with the conquests of Alexander the Great and west into Europe, where Greek philosophy, science and literature acted as the foundation for Roman culture. This was especially true of science: the Roman era could boast many great intellects – Pliny, who lived during the first century AD and wrote a thirty-seven-volume treatise, *Naturalis Historia*, and Plutarch, a thinker of the following generation, to name only two. But these men did little original science and concentrated on refining and clarifying Greek teachings passed on to them.

Of the Greek science that survived through to the early Roman era, the work of Aristotle, Plato, Archimedes and Pythagoras was best preserved, although the ideas of Democritus were championed by the Roman philosopher Lucretius in his poem *De Natura Rerum*. By the time Roman power was melting away and the library at Alexandria was decimated at the hands of the Christian bishop Theophilus around AD 390 (it was later sacked a second time by the

Arabs during the seventh century), Aristotle's work was becoming unfashionable.

The reason for this lies in a shift from pure intellectual inquiry to a distrust of any learning beyond theological exegesis: this plunged most of civilisation into what has become known as the Dark Ages. In this era, as the Roman Empire was in rapid decline, education and learning became dominated by religious fanaticism. The disciples of this new movement, the Stoics, believed in the supreme importance of pure spirit over material existence and therefore shunned learning about the physical world as an end in itself. To them, Aristotle's work was too mechanistic, too embedded in physical reality.* Instead, the musings of Plato held much greater relevance and were perfectly in tune with the new obsession with religious meaning.

Plato had taught an anthropocentric view of reality in which everything was created and carefully controlled by a supreme being who held the interests of humanity paramount. For Plato, the movements of the planets were there simply to enable the marking of time, and he viewed the cosmos as a living organism with a body, a soul and reason. He also saw numerical relevance and meaning in all natural processes, and because of this he placed great importance upon mathematics. However, he abhorred experimental science, which, according to one historian, he 'roundly condemned as either impious or a base mechanical art'.[3]

There is no clear point at which the Dark Ages ended in Europe. Learning in some form had been kept alive in the monasteries, but the interest of the Christian fathers had lain in mysticism and religious relevance rather than practical or theoretical science. The Arabs, who had made great strides in the understanding of alchemy, mathematics and astronomy throughout the period, maintained an interest in pure science, and as this knowledge filtered gradually into Europe the shadow of ignorance lifted. But it was a slow process, taking three or four hundred years.

Sometime between 1200 and 1225, Aristotle's works, which had

---

* The origins of this school of thought date to Zeno of Citrium, who was born around 336 BC. He proposed an austere ethical philosophy based upon the duty of the individual to preserve dignity and reason. Zeno taught that each human being shares in a Reason or a divine *logos* that orders the cosmos.

been saved in part by the Arabs and amalgamated with their own ideas, were rediscovered by European intellectuals and translated into Latin. From this point on, Aristotle's science returned to favour and took over from Platonic mysticism, gradually fusing with Christian theology.

Although this development may be viewed as an improvement upon the Dark Age mistrust of science and the Stoics' preoccupation with spirituality, it created a new obsession – a marriage of Aristotelian natural philosophy with Christian dogma. This meant that any attack upon Aristotle's science was also seen as an attack upon Christianity. Together, the two doctrines formed a powerful alliance and created a world-view that was taught by rote almost unchallenged in every university in Europe for almost half a millennium, from the thirteenth to the seventeenth century.

These twinned beliefs produced a self-contained picture of the universe: God created the world as described in the Scriptures and guided all actions. All movement was not only set in motion by God but was supervised by divine power. The Church's doctrine of divine omnipotence thus dovetailed perfectly with Aristotle's belief in the Unmoved Mover – that no movement was possible unless initiated by an unseen hand. All matter consisted of the four elements and was not divisible into atoms as Democritus had proposed. To Aristotle, every material object was an individual complete entity, created by God and composed of a particular combination of the four elements. Each object possessed certain distinct and observable qualities, such as heaviness, colour, smell, coolness. These were seen as *solely* intrinsic aspects or properties of the object, and their observed nature had nothing to do with the perception of the observer.

To the thirteenth-century mind, the notion that properties of an object such as smell, taste or texture were partly open to interpretation in the mind of the observer would have been totally alien. Every property of an object was intrinsic and the same for all observers. Furthermore, because Aristotle had rejected atomism, the concept that matter was composed of tiny, indivisible elements would have been equally foreign to most people of the time. And, because Aristotelian ideas were now bound up inextricably with religion, any philosopher who openly challenged any aspect of accepted scientific ideology put his life in danger.

Yet, despite the severe limitations this placed upon the development of scientific inquiry, the Middle Ages did produce a collection of notable and original thinkers who contributed to a gradual reawakening of rationality. Together, these men led the way to the Renaissance and the full flowering of innovative science that followed.

Still wrapped up in the need to marry natural philosophy with theology, the thinkers of this period – who became known as the Scholastics, the most famous of whom were St Thomas Aquinas and Albertus Magnus – stuck to the traditional Aristotelian line, shunning experiment. However, they did champion the search for truth outside the limited realm of pure theology. Although they maintained a firm belief that man was the central object of Creation and that the universe was designed for man by God, they had progressed to the idea that the study of Nature and the physical world could lead to greater theological enlightenment. It was not until the deaths of Aquinas and Albertus Magnus (towards the end of the thirteenth century, some seventy-five years after Aristotle had been reintroduced into Europe) that the work of the great Oxford scholar Roger Bacon began to erode the restrictions of Scholasticism.

In some ways Bacon was a man born ahead of his time. Although he subscribed to many traditional beliefs of the Scholastics, he was the first to see the usefulness of experiment and he composed three far-sighted tracts – *Opus Majus*, *Opus Minor* and *Opus Tertium* – which outline his philosophy and his experimental techniques in a range of disciplines. This effort established Bacon's reputation for posterity, but did little for him during his lifetime. Viewing his work as anti-Establishment and its anti-Aristotelian elements as subversive, Jerome of Ascoli, General of the Franciscans (later Pope Nicholas IV), imprisoned him for life as a heretic.

The scientific renaissance that followed Bacon's time marks a change in philosophical beliefs every bit as significant as that in the arts. Men such as Leonardo da Vinci, who approached science from a practical standpoint, foreshadowed many of the ideas of Galileo, Kepler and Newton, but did not write up their discoveries in any coherent form. The best we have is Leonardo's collection of notebooks, which indicate his studies and philosophies. In one sense, Leonardo was all experiment and represented the opposite extreme to the Greeks.

Leonardo held an opposing view of motion to Aristotle. Aristotle claimed that nothing moved unless it was made to do so by God, the Unmoved Mover. Leonardo suggests the exact opposite, writing in his notebook, 'Nothing perceptible by the senses is able to move itself . . . every body has a weight in the direction of the movement.'[4] In other words, matter has an innate tendency to move in a certain direction unless stopped. This anticipates the notion of inertia first postulated by Galileo some half-century later and eventually formalised by Newton.

Galileo, who was born in 1564 (about forty years after Leonardo's death), is regarded by historians of science as the greatest thinker in the realm of motion and matter up to Newton's time. It is generally agreed that his practical demonstrations paved the way for Newton's own blend of experimental verification and mathematical integrity.

Galileo's work in this area was revolutionary because he was the first to devise repeatable experiments that showed that Aristotle's ideas were quite wrong. He is probably most famous for his use of the telescope, which destroyed the traditional ideas of how the solar system is constructed (see Chapter 4), but equally important for the progress of science was his work in what became known as the science of dynamics.

Aristotle held that bodies were either intrinsically light or heavy and they fell at different velocities because of their innate tendency to seek their natural places. In 1590 the Flemish philosopher Simon Stevin had shown that light and heavy objects falling through a vacuum reach the ground simultaneously. Galileo repeated this experiment the following year (although probably not from the Leaning Tower of Pisa as tradition had it) using a cannonball and a musket-ball and showed that the two fall at equal speed if the resistance of air is ignored.

More importantly, Galileo suspected from this experiment that a falling body moves with a speed proportional to the time it has been falling. But, because the balls fall too quickly for the eye to measure their actual speed, he could not formulate a mathematical relationship between the speed of descent and the time it took. In order to find this relationship, he needed to conduct an experiment in which the speed of descent could be measured.

He quickly established that, ignoring friction, an object rolling down an inclined plane acquires the same speed as it would if it was falling vertically through the same distance. This enabled him to construct a series of experiments in which he let balls roll along inclined planes and measured the time of their journey and their speeds. This confirmed that the speed of a falling object does indeed increase with the time of the fall.

In a variation on this experiment, he allowed a ball to roll down an inclined plane and roll up another. In a further test, he allowed the ball to travel on beyond the slope along a horizontal path, where it continued steadily until slowed and eventually stopped by friction.

It was these experiments that convinced Galileo that Aristotle's idea of the Unmoved Mover was false. Objects do not move because they are constantly being pushed or pulled: rather, they possess inertia – an innate tendency to move unless stopped.

This was a revolutionary notion, but his views on other questions concerning matter and energy also entitle Galileo to be seen as the first of the modernists. He rejected Aristotle's idea of the four elements and subscribed to Democritus's atomic theory at least three decades before it began to make a reappearance in the schemes of Europe's leading thinkers, though he was unable to prove it. He also flew in the face of Aristotle's insistence that objects possess integrally all the properties we sense when we observe them, declaring:

> I feel myself impelled by necessity, as soon as I conceive a piece of matter or corporal substance, of conceiving that in its nature it is bounded and figured by such and such a figure, that in relation to others it is large or small, that it is in this or that place, in this or that time, that it is in motion or remains at rest, that it touches or does not touch another body, that it is single, few or many; in short by no imagination can a body be separated from such conditions. But that it must be white or red, bitter or sweet, sounding or mute, of a pleasant or unpleasant odour, I do not perceive my mind forced to acknowledge it accompanied by such conditions; so if the senses were not the escorts perhaps the reason or the imagination by itself would never have arrived at them. Hence I think that those tastes, odours, colours etc. on the side of the object in which they exist, are nothing else but mere names, but hold their residence solely in the sensitive body; so that if the animal were moved, every such quality would be abolished and annihilated.[5]

So, contrary to Aristotle, Galileo states categorically that there are two distinct qualities of bodies. The first may be considered *primary qualities*, which are inseparable from and fundamental to the nature of the object in question – what twentieth-century scientists would ascribe to the atomic structure and chemical nature of an object. The others are *secondary qualities*, which are interpreted by the senses of the observer.

These revolutionary notions of Galileo's – ideas which have perhaps been swamped by his more famous discoveries in astronomy and dynamics – greatly influenced the French philosopher René Descartes, who for a time informed Newton's thinking on the subject of matter and the nature of the physical universe.

Descartes is most famous today for two developments – Cartesian coordinates, which still play a key role in mathematics, and dualism, a philosophy which proposes a sharp distinction between body and soul, matter and spirit. According to Cartesian dualism, the spirit is personal and nebulous, and matter must therefore be impersonal and concrete.

In Descartes's image of the universe, matter is immersed in an unseen, immeasurable medium called the ether. God endowed the universe with movement at the beginning of time and allows it to run spontaneously but in accordance with his will. Because in this scheme matter fills all of space, there can be no such phenomenon as a vacuum and all motion is produced by matter impressing on other matter within the medium of the ether. Descartes expressed this in his famous theory of vortices, in which he pictured movement, such as the fall of a stone to the earth, as being like the movement of a feather or a straw caught in an eddy or a whirlpool.

Descartes rejected mysticism and the occult in his writings and visualised the universe as a machine. Every action involving matter was purely mechanistic, and matter had no contact with spirit. To Descartes, all animals – including humans – were also mere machines. Humans had a spiritual aspect, a soul, but this had no link with our physical selves.

These ideas were highly controversial. On a scientific level, Descartes's concepts were unverifiable and he did not contrive experiments to support his theories. On a superficial level, his vortex theory did not clash with the doctrines of Galileo, in that it did not contradict experimental evidence. Galileo had shown that, because of inertia, all movement continued until stopped, and Descartes proposed that the universe had been set in motion by God. The two ideas were not incompatible: if we assume the Creator set things in motion, they would continue until stopped by, say, the intervention of mortals.

But the most radical aspect of Cartesian philosophy was that it implied to many that, once the universe had been set in motion, God was no longer needed. The Creator had been effectively demoted from 'Supreme Good' to 'First Cause'. Naturally this was a view hotly disputed by theologians and the majority of philosophers, many of whom had been brought up on Aristotle and still thought along the same lines as the Scholastics of the thirteenth century.

Descartes died when Newton was eight years old, but his philosophies were becoming immensely fashionable as Newton entered university and extended his reading beyond the curriculum. Because it contained material referring to his disputed theories of divine function in a mechanical universe, Descartes's most famous book, *Discourse on the Method* (published in 1637), was unpopular with the ecclesiastical authorities, but his theories were discussed openly in the more liberal universities of Europe and began to spread.

As Descartes's theories of dualism became known, three other philosophers were helping to create the intellectual scene to which Newton would add his own unique ideas.

Pierre Gassendi, who was a Catholic priest and a close contemporary of Descartes, revived the work of Democritus and proposed an atomic theory in which matter was composed of tiny indivisible parts. Unlike Descartes, Gassendi did not attempt to describe a mechanistic universe in which all action on a fundamental level occurred by way of vortices – a theory which for many people marginalised God. Instead, he envisaged a universe composed of Democritus's atoms presided over by an all-pervading Creator. Gassendi's outlook has been dubbed 'Christianised atomism', because it maintains an omnipotent and omnipresent role for God. This was more acceptable than Descartes's model to men like Newton, who sought a mechanistic model for the universe but could not countenance any diminishing of the Creator's position.

Another great innovator of the time was Robert Boyle, today seen as the supreme experimentalist of his era. Boyle believed in practical analysis and was more concerned with *how* a phenomenon occurred rather than *why* it happened.

Boyle tried to unite elements of Descartes's philosophy of a mechanistic universe with the revived atomic theory, but he did not subscribe to the contention that God had no role in the physical world after initiating primal movement. Like Gassendi, he held that God's 'general concourse' was continually needed to keep the mechanical universe working. But a greater contribution to the study of matter and energy was his demonstration of the fallacy of Aristotle's notion of the four elements.

In one of these displays, Boyle illustrated how fire could not be considered a basic element and that Aristotle's claim that fire could resolve things into their elements was false. He demonstrated that, contrary to Aristotle's belief, gold can withstand fire and can also be alloyed with other metals and then recovered in its original form, suggesting the existence of unalterable 'corpuscles' of gold. He also showed that even when fire did break down materials it required different degrees of heat and different time periods to succeed, and more often than not it produced new substances that were also complex. Finally, he showed that some materials could not be reduced by fire alone.

The last of the major seventeenth-century figures who greatly influenced Newton's intellectual development was Francis Bacon. Bacon was not solely a philosopher. He was Lord Chancellor under James I, and was an essayist and moral philosopher who wrote widely about the way he thought science should be conducted. In his *The Advancement of Learning* (1605), *The New Organon* (1620) and especially *The New Atlantis* (1627), he criticised the blind pursuit of Aristotelian philosophy and the rote-learning system of the universities. And, most importantly, he was the first to formulate what has become known as the experimental or *inductive* method. It was Bacon who, some time before Descartes dismissed magic and superstition, argued that scientific discipline should be guided and inspired by religious motivations. In *The Advancement of Learning* he wrote:

> To conclude therefore, let no man out of weak deceit of sobriety, or an ill-applied moderation, think or maintain, that a man can search far or be too well studied in the book of God's word, or

in the book of God's works; divinity or philosophy; but rather
let men endeavour an endless progress or proficiency in both.[6]

Although he would have agreed with Descartes's dismissal of metaphysics, Bacon objected to scientific ideas being driven purely by
philosophy and the deductive reasoning employed by Aristotle, which
Descartes did not completely shake off. In effect, Bacon was the first
to conceive of a 'Christian Technocracy'. Quoting Daniel in the Old
Testament, that 'many shall run to and fro, and knowledge shall be
increased', he envisaged a science driven by religion, guided by strict
logical rules and experimental verification (almost as modern scientists perceive it) and aimed at enlightenment and practical applicabil-
ity. Although Cartesianism provided Newton with a platform of
reasoning about a mechanical philosophy which in turn led to the
Industrial Revolution, it was Bacon's scientific method, which was
readily adopted by generations of natural philosophers, including
Newton, that provided the *modus operandi* for the Scientific Revolution.

So, by the middle of the seventeenth century, as Newton was
preparing to enter the academic world, natural philosophy was in a
state of flux. The old notions of Aristotle still provided the traditional
backbone of university study in the areas of logic, astronomy and
natural philosophy, but this was primarily because of an old school
of influential academics. Gradually, radical ideas from the Continent
were eroding the Greek philosopher's supreme position. According
to one historian of science, 'From being a realm of substances in
qualitative and teleological relations, the world of nature had definitely become a realm of bodies moving mechanically in space and
time.'[7]

It was within this climate of change that Newton entered university
in 1661 and took the first steps towards finding his own path through
the shifting philosophies of the time and establishing his own views.

CHAPTER 3

Academia

[Truth is] the offspring of silence and unbroken meditation.
ISAAC NEWTON[1]

C ambridge during the 1660s was far from a Utopia of academic purity. It was both academically backward and a
dangerous place in which to live. The buildings were over-
crowded and huddled together along filthy streets which at night were
unlit and during the day teemed with merchants, beggars, unschooled
children and gowned students. An anonymous visitor to the town
described it as:

> so abominably dirty that Old Street, in the middle of the winter's
> thaw, or Bartholomew's Fair, after a shower of rain, could not
> have more occasion for a scavenger than the miry streets of this
> famous corporation, and most of them so very narrow that
> should two wheelbarrows meet in the largest of their thorough-
> fares they are enough to make a stop for half an hour before
> they can well clear themselves of one another to make room for
> passengers. The buildings in many parts of the town are so little
> and so low that they looked more like huts for pygmies than
> houses for men.[2]

Covering little more than half a square mile, Cambridge had a
population of about 8,000 including almost 3,000 students, graduates
and university staff. Students could easily find themselves at risk –
their souls in jeopardy from the attentions of prostitutes and inn-
keepers (a danger made much of by the hypocritical masters), and
their physical safety threatened by ubiquitous thieves and murderers.
In a letter to his mother written in 1664, one John Strype, a young

student in his first year at the university, describes graphically the social climate in the town:

> We have hereabouts most intolerable robbing: never by reports so much. I have heard within two or three days of six or seven robberies hereabouts committed: whereof two or three killed. No longer than last Sabbath, a mile off, a man knocked on the head. Lately a scholar of Peter House had both his ears cut off, because he told the thieves, after he had delivered some money to them, that he would give them leave to inflict any punishment upon him, if he had a farthing more: but they searching him, found, it seems, 20s. more: so they took him at his word, and inflicted the cheater's punishment upon him.[3]

Such incidents were not attributable solely to the perceived wealth of the students, nor was it simply that students were easy targets for thieves; there had been bad feeling between town and gown for centuries. Although town considerably outnumbered gown, the lives of the townsfolk were dominated by an autocratic university governing body that often behaved in corrupt and self-interested fashion. Most of the town's tradesmen relied upon the university for their livelihoods, and many resented the draconian powers of the Vice-Chancellor. His sphere of influence was by no means restricted to university property or the student body: he was, in all but name, a feudal lord who controlled all forms of commerce and oversaw all legal and financial matters within the town. A royal charter drawn up in 1600 stipulated that Cambridge was allowed a mayor, bailiffs and burgesses and that the civic authorities could have and use their own seal. But the final clause of the charter specified that 'Nothing in this charter shall prejudice or impede the privileges, liberties and profits of the Chancellor, Masters and Scholars of the University.'[4]

Little over a year before Newton arrived in the town, the Mayor had been humiliated by the Vice-Chancellor and made to apologise for apparently overstepping his authority. In his written recantation, he was forced to make it clear exactly who was boss:

> Whereas I, Edward Chapman, Mayor of the Town of Cambridge, did upon the XXVIth day of February 1660 by error send my warrant for releasing of William Land, John Devole

and James Delamot out of the Tolbooth Gaol, to which they had been committed by the then Vice-Chancellor, Dr Ferne, I therefore, in satisfaction to the University, hereby acknowledge the error and do promise not to do or to my power suffer anything hereafter to be done that may anyways infringe the liberties or privileges of this University to my knowledge. In witness whereof I have set my hand the second day of March in the year of our Lord God 1660.[5]

Amazingly, little changed until the late Victorian era, when both the power of the university over the town and the limitations placed upon the freedom of students were gradually eroded. In Newton's student days – and until long after Darwin attended the university during the late 1820s – the activities of the students were monitored by the university police, the proctors. Students were forbidden to associate with tradesmen, to drink in taverns, to have dealings with prostitutes and to break an evening curfew. Although many of these rules were frequently broken by the students and their enforcement was lax, examples were made.

Outside the city walls, England had changed and was continuing to change, but little of this was reflected in the attitude of the university authorities or in the antiquated curriculum taught. The peaceful restoration of the monarchy in 1660 had brought with it a nationwide climate of renewal. Cromwell's Protestant Commonwealth had died with him in 1658, and, although the country would remain suspicious of the Catholic leanings of the house of Stuart, Newton entered Cambridge in 1661 in a new age of religious tolerance and political stability.

This radically altered the broader character of society, and fellows of the university who had fallen victim to Puritan purges were reinstated (although not to the exclusion of former Roundhead sympathisers). Yet the university authorities maintained a hold over the town hardly changed since medieval times. Since the reign of Henry VIII, the King had the legal right to shut down any college in the realm and claim its possessions. Consequently the university remained loyal to the Crown, and as a monarchist institution it was rooted in tradition and notions of a glorious past.

To Isaac Newton – a country boy who had never visited a town larger than Grantham – Cambridge was Avalon. He left Woolsthorpe on the second or third day of June 1661 and set out along the Great North Road on the fifty-mile trip to the town that would be his home almost without a break for the next thirty-five years. *En route*, he broke his journey first at Sewstern, where he took his first look at a piece of land bequeathed to him in Barnabas Smith's will (the annual income from which would pass to him after his twenty-first birthday), and then at Stilton on the approach to the Great Fens, a day's ride from Cambridge.

According to Stukeley, on Newton's last day under Stokes's tutelage the proud headmaster made his prize student stand in front of the school while he delivered a speech praising the boy and, with tears in his eyes, urged Newton's fellow pupils to follow his academic example. Apparently the other boys were as moved as their headmaster. More believable is Stukeley's admission that the farm hands and servants at the manor were glad to see Newton leave home and 'rejoiced at parting with him, declaring, he was fit for nothing but the "Versity"'.[6]

Hannah, however, had ensured that her son would not be allowed fully to escape the mundanity of 'real life' and the hardships he may have thought he was leaving behind.

When Newton enrolled at Trinity College, on 5 June 1661, he entered the college on the lowest rung of the social ladder, as a subsizar (becoming a sizar after he had matriculated at Trinity a month after his arrival). Subsizars and sizars were little more than servants who paid their way by emptying the bedpans and cleaning the rooms of the more privileged students. These included the élite – fellow-commoners, young men from noble families, and pensioners (usually the sons of wealthy businessmen).

The exact form that sizarship took for Newton remains unclear. Traditionally, sizars waited on other students, but there was another type who worked solely for one fellow, invariably their tutor. It has always been supposed that Newton's sizarship was of the first type, and this may be true, but there is evidence to suggest that he was in fact sizar to Humphrey Babington, brother of Mrs Clark, the Grantham apothecary's wife, and fellow of Trinity.

It may even have been that Newton was only able to attend the university thanks to Babington taking him on as his personal servant. Babington had himself been a Cambridge student. As a royalist sympathiser, he had been sacked from his fellowship under the Puritan

purge of the Commonwealth years but was reinstated with the Restoration. After Newton's death, Ayscough family tradition had it that
'the pecuniary aid of some neighbouring gentleman'[7] had enabled
Newton to study at Cambridge.

If Newton was Babington's sizar his duties would have been particularly easy, because his master was in college for only a few weeks
a year and would have demanded little of him. What is clear is that
the conflict of interests between Isaac's mother and those who saw
scholarly potential in the young man did not end when Hannah
complied reluctantly with the wishes of Babington, Ayscough and
Stokes. Newton's academic fees at the university were in the region
of £10 to £15 per year, and he was given an allowance of a further
£10. Both of these expenses were met by Hannah. But, considering
she commanded a very comfortable annual income of around £700,
it is evident that she wanted deliberately to make life hard for her
son at Cambridge.

Sizarship was bad enough for those who could afford nothing
better, and the failure rate of sizars was naturally much higher than
that of the more privileged pensioners and fellow-commoners. But
for Newton the shame of having to empty the chamber-pots of rich
contemporaries, or the stigma of running errands for his tutor, must
have weighed heavily.

Although he may have had an easier time of it than most subsizars,
Newton was still, in the eyes of the college and his contemporaries,
on the lowest rung of the social ladder. As a consequence, he
would have been treated with contempt by those in superior social
positions or else ignored by the sons of the wealthy who considered
the university a playground – a place in which to waste a few
years before accepting undemanding roles in the upper reaches of
society.

Aside from making him even more determined to create an impression, this new humiliation did little for the positive aspects of Newton's personality as a youth. It fuelled the flames of his insecurity
and led to a desire to improve his social status and to sever further
the links with his family, to leap at any chance of social improvement.
If Hannah had imagined that by deliberately making life difficult for
her son he might be persuaded to give up notions of an academic
life and return to the family farm, she clearly did not know him. If
her actions created anything positive it was to convince him he had

to break away from Woolsthorpe, to turn even further in upon himself and to excel within his vocation.

The academic pattern at Cambridge had been set by the Eliza-bethan Statutes of 1571, which not only dictated the manner of dress and conduct of students and academic staff but also determined the structure of degree courses. To obtain a BA, all students had to reside in the university for a minimum of twelve terms of tuition (four years) and to attend all public lectures given by the members of the college faculty. There was really only one course. The first year covered rhetoric, the art of eloquent oral and written communication, encompassing classical history, geography, art, scripture and litera-ture. Also, by the end of their first year students were expected to be fluent in Latin, Greek and Hebrew.

For a time, Newton became a conscientious and dedicated student, but, initially at least, he neither shone nor attracted the attention of his masters. In fact, he was all but invisible. Like most of his fellow students, he had little intellectual guidance. Upon his arrival, he was assigned a tutor who was both his teacher and a surrogate parent – one Benjamin Pulleyn, of whom little is known except that he entered Trinity in 1650 as a sizar and rose to the position of Regius Professor of Greek, a seat he occupied for twelve years. Pulleyn was a lax tutor in an academically sterile university. Known as a pupil-monger – he took on as many students as possible, to bolster his meagre income – he did almost nothing to help Newton, who was just one of over fifty undergraduates in his care.

Within weeks of his arrival, Newton had cut himself off from the other sizars and, following the pattern of life at school in Grantham, he began a very lonely first year at the university. It is significant that not one anecdote of Newton's earliest period at Trinity has been passed on to us from fellow students. There is no record of a personal relationship with any other student even in the most vague terms, except that he appears to have detested his room-mate. We only know this from two 'confessions' which appear in the Fitzwilliam Notebook. The first of these is 'Using Wilford's towel to spare my own'; the other involves Newton owning up to the sin of 'Deceiving my chamberfellow of the knowledge of him that took him for a sot'.[8] From the first of these we can glean that Newton's first room-mate was the otherwise unknown Francis Wilford, who appears in the *Alumni Cantabrigienses* as a pensioner admitted to the college on the same day as Newton. It is also clear that Newton did little to endear himself even to the unfortunate Wilford; small wonder his first year was a lonely one.

Apart from the frustration his mother had caused him, Newton had two other problems during his early days at the university. The first was his age. Almost nineteen that first autumn, he was two years older than the average student. Although some have suggested that this may have been to his advantage academically, in terms of helping him to mix with the other students it could only have been a hindrance. The second and more serious difficulty, and one which was to remain with him throughout his academic life, was his Puritan faith. The teaching of the era centred around the great universities was supported and sustained by the orthodoxy of the Anglican Church. And, although the Restoration had heralded a religious tolerance that would remain a central pillar of British society, Newton was expected to subscribe officially to the tenets of the Anglican Church and to keep his Puritan beliefs to himself.

But, in spite of the potential problems offered by his religious leanings and the extra barrier they created between him and other, orthodox, students, his Puritan ethics also fuelled his drive to learn and focused his thoughts and energies. The distress his mother had caused him early in life had left Newton damaged and emotionally impotent. Puritanism offered him a world with strict emotional and sensual limits in which he did not have to find excuses for his inability to love – a world in which the twin pillars of God and Knowledge (the search for which was a God-given responsibility) could replace most other needs. With Puritanism and the thirst for understanding as his guides, he could at least attempt to shun sex, ignore any lingering desire to marry or to have a family, and keep in check his material ambitions and social goals.

In his first academic year, at least, Newton was preoccupied with sin and with the slightest let-up in religious observance – an obsession which led later in that year to the purchase of the notebooks in which he confessed his misdeeds, past and present. Although he lightened up a little and enjoyed the odd ale and game of cards in a tavern later in his postgraduate days, during his first few months in Cambridge, outside his lectures and tutorials, Newton existed in a permanent state of isolation – lonely, disorientated and trying to feel his way into an alien world of new-found but largely scorned freedom.

His was not the Puritanism of the political extremist (of which there were still many following the turbulent days of civil war and regicide); nor was he the Puritan of the Victorian caricature – the solemn kill-joy who saw debauchery and evil in all the doings of his fellow man. Newton was of the type that elevated the principles of hard work and dedication to learning as the highest hopes of humanity. He believed that the acquisition of knowledge and the unravelling of Nature's truths were to the greater glory of God. But to his contemporaries he must have appeared a flashing beacon of misanthropy.

If he professed indifference towards almost every other student he encountered, they must have been even more dismissive of him. He could suffer this, and indeed appeared to care little what his fellow students thought of him. An example of his high-mindedness comes from the oldest letter in Newton's hand, written to a sick friend around 1661:

Loving friend,
   It is commonly reported that you are sick. Truly I am sorry
for that. But I am much more sorry that you got your sickness
(for that they say too) by drinking too much. I earnestly
desire you first to repent of your having been drunk & then to
seek to recover your health. And if it please God that you
ever be well again then have a care to live healthfully &
soberly for time to come. This will be very well pleasing to all
your friends, especially to
   Your very loving friend.

                                                               I. N.[9]

During his early days at university it was not just his pious detachment from everyday pleasures that so alienated Newton: he did little to encourage others to like him. An example of this was his decision to become a money-lender.

It is easy to imagine Newton at the age of nineteen or twenty growing to accept that he could not mix easily with the other students. He had also been left to his own devices to supplement Hannah's allowance, and by this time he was certainly showing an active interest in money. Indeed, one of his repeated confessions in the list of sins of the Fitzwilliam Notebook is that of paying too much attention to money: 'Setting my heart on money more than God', as he put

it. This was followed by several incidents of 'relapse'.[10]   Being the
meticulous record-keeper for which he was later renowned, Newton
noted every transaction in another notebook he purchased at Trinity:

	£	s.	d.
Lent Pollard	0	2	0
Lent Bigg	0	7	6
Lent Pollard	0	1	0
Lent Agatha	0	1	0
Lent Andrews	0	11	5
Lent Oliver	0	1	0
Lent Wilford	0	6	0
Lent Gosh	1	0	0

His Puritan caution showed in the fact that he never lent more than
£1 to a debtor, and when he did deal with such large sums his
nervousness shows in a note beside this transaction: 'to be paid on
Friday'.[11]

Newton was never a big-time loan shark, but by the end of his
second year business was flourishing and he kept it up until he
became a man of independent means two years later. Quite how he
started in business is unclear. Bearing in mind his own precarious
financial position when he first arrived at Trinity, one can only
assume he took a risk by making a short-term loan and then began
to realise the potential of the venture.

A short time later, things began to improve on other fronts. Eighteen months after arriving at Trinity, Newton managed to change
room-mate. John Wickins, the son of the Master of Manchester
Grammar School, entered Trinity as a pensioner early in 1663 and
met Newton towards the end of his first term. Sadly, aside from a
few comments about Newton's hypochondria and brief descriptions
of his work patterns, Wickins, who lived with Newton for over two
decades (until he gave up his fellowship in 1683), left little record
of their close association.[12] The most detailed recollection that he
passed on to his son Nicholas in old age was a brief description of
his first meeting with Newton, in 1663:

My Father's intimacy with him came by mere accident. Father's
first chamber-fellow being very disagreeable to him he retired

one day into the walks where he found Mr Newton solitary &
dejected. Upon entering into discourse they found their cause
of retirement the same & thereupon agreed to shake off their
present disorderly companions & chum together, which they did
as soon as conveniently they could & so continued as long as
my father stayed at college.[13]

Wickins's reticence in discussing what must have been one of
the most important relationships of his life is odd. He and Newton
separated in 1683 under a cloud, and, despite Wickins living for
another thirty-six years, the two men never met again.

So, who exactly was John Wickins, and what was the nature of his
relationship with Newton? From the story of their introduction, it is
clear they must have been of similar temperament. Both were
unhappy with their 'disorderly companions' and each quickly saw a
kindred spirit in the other. Their staying room-mates for the next
twenty years (including a move in 1673 to rooms in Great Court)
is evidence of their closeness.

Wickins also became Newton's assistant. He regularly transcribed
experiment notes and helped set up apparatus and monitor investigations. Their rooms became a live-in laboratory, at first strewn
with documents and simple home-made optical instruments but later
crowded with furnaces and bottles of chemicals. Wickins eventually
became a clergyman, married and had a family. Shortly after his
departure, Newton sent him a parcel of Bibles to be distributed to
his flock in the village of Stoke Edith, near Monmouth. The only
other correspondence occurred some thirty years later, when Wickins
wrote to ask his erstwhile room-mate for a further donation of Bibles
and attempted to start a friendly exchange. Newton duly sent the
Bibles but brushed off any subtle overtures of Wickins by ending his
letter with the rather curt 'I am glad to hear of your good health, &
wish it may long continue, I remain . . . Newton'.[14]

For all the attempts that have been made to find clues in the
meagre correspondence between the two men, the strongest evidence
for an acrimonious break lies in the fact that Wickins neither wrote
a word about Newton nor related more than a scrap of anecdote
about their time together. When, soon after Newton's death in 1727,
Robert Smith, Plumian Professor of Natural History at Cambridge,
wrote to Nicholas Wickins requesting information about his father,

he was told that John Wickins had long considered collecting together everything in his possession related to Newton but had done nothing about it. This would not have been a difficult task, because all Nicholas Wickins could pass on to Smith were three short letters transcribed into a notebook, five other notes concerning mundane financial matters, and the anecdote describing their first meeting.

Fortunately, much more is known of Newton's academic life as an undergraduate. As at other great seats of learning throughout Europe, the curriculum at Cambridge was based almost exclusively upon the teachings of the Greek masters – especially the ideas of Aristotle, with which Newton would already have been familiar from his reading at the Clarks'. Throughout his first year he attended his lectures conscientiously, but he was already beginning to question the validity of classical ideas.

Like many of the more conscientious students, he had been following the latest philosophical developments and was reading 'fashionable' philosophers, such as Descartes and Galileo, whose works were gradually becoming available in England. As a result, sometime in early 1663, Newton underwent a radical change of approach. During a lecture, while making meticulous notes on Aristotle's teachings, mid-page he stopped abruptly. Then, after leaving dozens of pages blank, he wrote at the top of a fresh page, 'Quaestiones Quaedam Philosophicae' – 'Some Problems in Philosophy'. Beneath this he wrote, 'Amicus Plato, amicus Aristoteles magis amica veritas' – 'I am a friend of Plato, I am a friend of Aristotle, but truth is my greater friend.'[15]

This collection of 'Quaestiones' – or the Philosophical Notebook, as it is sometimes called – marks the point at which Newton stepped away from tradition and began to question what he was taught. He began by creating forty-five headings in the notebook – topics concerning the nature of the universe which he would attempt to investigate and answer. These included 'Of Water and Salt', 'Attraction Magnetical', 'Of the Sun Stars & Planets & Comets' and 'Of Gravity & Levity'. In some cases nothing has been written under the heading, but elsewhere there is a paragraph or two of neatly written text, while some headings are followed by lengthy discourses.

Like his schoolboy exercise books, these undergraduate notebooks contain questions and attempts at answers taken from the works of well-known natural philosophers. In many places the arguments are then dissected and questioned further. Sometimes a section of text is followed by a piece composed by Newton in which he seems to be addressing the quoted author and asking him questions directly or drawing attention to things that do not appear clear. In this way, Descartes and Boyle come under scrutiny along with the antiquated philosophies of Aristotle.

An example is a piece under the heading 'Of Water and Salt' which involves an early hypothesis to explain the ebb and flow of the sea, later explained by Newton in Proposition XXIV of Book II of the *Principia*, first published some twenty-four years later:

> To discover whether the Moon pressing the atmosphere causes the flux of the sea, take a tube of about 30 inches filled with quicksilver or else take a tube with water which is so much longer than 30 inches as the quicksilver is weightier than water & the top being stopped the liquor will sink 3 or 4 inches below it leaving a vacuum (perhaps). Then, as the air is more or less pressed without by the Moon so will the water rise or fall as it does in a weatherglass by heat or cold.[16]

At this stage of his career Newton could offer no explanation for this, but he analyses it in terms of what others say. Can the movement of the quicksilver be explained by the theories of Aristotle, who would declare that the substance is merely trying to find its place in the world? Or is Descartes closer to the truth: is the rise and fall of the surface of the quicksilver due to the movement of particles and ether bearing down upon it, creating vortices within the liquid?

Elsewhere there are speculations based upon thought experiments. Under the heading 'Of Gravity & Levity' Newton wrote:

> Try to discover whether the weight of a body may be altered by heat or cold, by dilatation or condensation, beating, powdering, transferring to several places or several heights, or placing a hot or heavy body over it or under it or by magnetism, whether lead or its dust spread abroad, whether a plate flat-ways or edgeways is heaviest.[17]

Although these inquiries seem to us to have obvious answers (why, for example, should an object weigh different amounts if it is laid flat or edgeways?), no one before Newton had recorded their efforts to verify these things. Rather than accepting tradition, Newton wanted to clarify such matters for himself.

He acquired some of these notions from books available in the extensive library at Trinity College, which contained works by the great natural philosophers of the day. Here could be found texts by Descartes, Boyle, Thomas More, Hobbes, Copernicus, Tycho Brahe and Galileo (with the exception of Galileo's two most important works, *Dialogue Concerning the Two Chief World Systems* and *Dialogue Concerning Two New Sciences*, both of which appear to have been too risqué for the conservative thinkers who authorised the purchase of books for the library). The problem for Newton was not the range of material to be found in Cambridge university libraries: it was that students could use the libraries only at special times, and then only when supervised by a tutor. From what we know of Newton's tutor, Pulleyn, who was usually unavailable and quite uninterested in natural philosophy, it would seem most likely that Newton gained access to these all-important works through the agency of another fellow, almost certainly Humphrey Babington.

Although at first inspired and influenced by Descartes, Newton quickly rejected the Frenchman's mechanical theory as a concept that denied the omnipotence of the Creator. He was able to accept Pierre Gassendi's Christianised atomism, but even this was with reservations. In the 'Quaestiones' he wrote:

Of Atoms

It remains therefore that the first matter must be atoms and that matter may be so small as to be indiscernible. The excellent Dr More [the Cambridge fellow Henry More] in his book of the soul's immortality has proved this beyond all controversy, yet I shall use one argument to show that it *cannot be divisible in infinitum* & that is this: Nothing can be divided into more parts than it can possibly be constituted of. But matter (i.e. finite) cannot be constituted of infinite parts.[18]

Newton is here using logic to dispel the possibility of anyone taking the atomic theory too far. Matter being a finite thing, it cannot, he reasoned, be divided forever into infinitesimally small parts. (If Newton sounds overconfident here and seems to be treating the issue with the same overzealousness that Aristotle might have employed, we can perhaps put it down to his relative youth. These were, after all, musings in a private notebook.)

The key influence in guiding Newton towards a view of the universe that maintained a supreme role for the Creator was the Cambridge philosopher Henry More, a man who was interested in all areas of natural philosophy and mysticism and a leading member of the group of fellows known as the Cambridge Platonists.

Born a gentleman, More had gained the finest education at Eton and was elected a fellow of Christ's College in 1639. Believing in the pursuit of knowledge as a means of exalting God, and upholding the Scholastics' edict 'Understand so that you may believe, believe so that you may understand', More declined all offers of ecclesiastical positions and even the mastership of Christ's College in order to lead an academic life unhindered by other responsibilities.

He and the other Cambridge Platonists believed that the world was permeated by spirit, which More termed the 'Spirit of Nature'. This esoteric 'force', he believed, mediated between God – who controlled all actions, all purpose and all outcomes – and a purely mechanical universe – the mundane physical world in which we live and conduct our lives.

As a young scholar, More had shared many of Descartes's ideas and had initially seen Cartesian philosophy as a means of reconciling theology and natural philosophy; but gradually he had turned away from this view, later becoming its vehement opponent. At the root of More's ideology, and of his influence on Newton, was an amalgamation of atomism and Christian Platonism. Plato had believed in the notion of *spirit*, an essence within all things, manifest in man as the soul, but also an extension of God, a force at work in Nature, guiding the universe. In Descartes's philosophy there appeared to be no continuing need for God. Descartes never intended this interpretation and was himself a devout Christian, but to More, and later to Newton, the mechanisms and ideas portrayed in Descartes's *Discourse on the Method* and *Principles of Philosophy* could easily be interpreted as atheistic. In More's universe, matter was guided by spirit, manipulated by God entirely at his discretion.

More, who was born in Grantham, had tutored Joseph Clark (the brother of Clark the apothecary) at Cambridge. He visited Grantham occasionally, and it is possible that he may thus have met Newton several years before the young man entered university. It is clear from entries in his philosophical notebook that Newton came under More's influence quite early in his university career. As well as the mention of a text by 'the excellent Dr More' in his notes on atomism, Newton has listed headings clearly influenced by More's main areas of interest, such as 'Of the Creation', 'Of the Soule' and 'Of God'.[19] These may have been prompted by Newton's natural curiosity for things spiritual, but it is also likely that they stemmed from reading More's most important book, *The Immortality of the Soul*, to which Newton had referred in the earlier entry 'Of Atoms'.

More's influence upon Newton extended beyond the inspiration provided by his writings.[20] Newton's ideas and loyalties changed so radically within such a short period of time during his second year at university that the influence of at least one academic guide is likely. Having been Clark's tutor and an associate of Babington, More probably talked to Newton on a number of occasions during the young man's final years in Grantham and Woolsthorpe and may even have singled him out upon his entry into the university. He was another father-figure within the academic and social network forming around the serious-minded and inquisitive young man. Although Babington was more of a practical guide (and almost certainly provided access to his private library), he was in Cambridge only rarely. More provided a greater and more lasting intellectual foundation.

But, if More's influence was strong, to the modern mind he seems to have offered a confusing philosophy. To us, atomism is the foundation of modern physics, but the seminal work of Rutherford, who first postulated the existence of smaller particles within the atom, early in the twentieth century, led to the oddities of quantum theory. From this derives indeterminism, as expressed in Heisenberg's Uncertainty Principle, leading to theories of unpredictability and a philosophical viewpoint far removed from the image of a universe manipulated by a benign God. Though some have managed to visualise and have faith in a strange marriage between quantum theory and God, mainstream modern atomism could not be further removed from More's idea of a personal, all-pervading deity. Yet, to More – naturally unaware of where it would one day lead – atomism was a way of proving the actions of an omnipresent Creator, a confirmation of the Testaments; because, as Newton had underscored in his notebook, matter could be divided only to a finite degree, and the resulting fundamental particles must have been created and guided by a divine hand.

If More offered a theoretical foundation which combined natural philosophy with theology, from reading Galileo and Bacon contemporaneously Newton had also learned how to construct a working system with which to verify his ideas. By the summer of 1664 he was able to state in his notebook that 'The nature of things is more securely and naturally deduced from their operations one upon another than upon the senses. And when by the former experiments we have found the nature of bodies . . . we may more clearly find the nature of the senses.'[21] What he means by this is that scientists cannot simply trust what they observe with their senses, but need to experiment before attempting to deduce the nature of the universe and the objects that fill it – that there may be more going on than we know from the information our senses give us by superficial observation.

Newton's first experiments, begun during the summer of 1664, were probably his investigations into the nature of light. Years later these appeared in the *Opticks*, first published in 1704.

His earliest interest in light began when he bought a glass prism at the Stourbridge fair, held on a piece of land beside the river about a mile from the centre of Cambridge. Amid stages for the jugglers and clowns, minstrels and children's games, dancers and actors stood stalls selling all manner of oddities – trinkets from exotic travels in Bohemia, potions and elixirs, and toys. It was from such a stall that Newton purchased his prism.

According to Conduitt, 'In August 1665, Sir I. bought a prism at Stourbridge fair to try some experiments upon Descartes's books of colours.'[22] However, on this occasion Conduitt got his dates wrong and Newton actually acquired the prism on his visit to the fair in 1664, not 1665. Plague prevented the holding of the fair in 1665 – a fact documented in the diary of Alderman (later Mayor) Samuel Newton (no relation): 'On the first of September, a proclamation was posted prohibiting Stourbridge fair on account of the great plague in London.'[23] It is also agreed by most authorities that, because of the plague, Newton had left Cambridge to return to Woolsthorpe before the beginning of August 1665, and was absent from the university for most of the next two years.[24]

The prevailing hypothesis of light at the time was that of Descartes.

He believed that light was a 'pressure' transmitted through the transparent medium of the ether. Sight, he claimed, was due to this pressure impinging upon the optic nerve.

Newton was acquainted with this hypothesis and had already made notes on the subject in his philosophical notebook. But it is likely that, in keeping with his support for atomism, by the summer of 1664 he was beginning to doubt the accuracy of Descartes's explanation. He was already thinking that light might be corpuscular, and by imagining light to be particle-like he was more readily able to explain phenomena such as reflection, refraction, and optical and chromatic distortions. Writing to Henry Oldenburg, the Secretary of the Royal Society, some eight years later, Newton described his earliest experiments with the prism:

> I procured me a triangular glass-prism, to try therewith the celebrated *Phenomena of Colours*. And in order thereto having darkened my chamber, and made a small hole in my window-shuts, to let in a convenient quantity of the sun's light, I placed my prism at its entrance, that it might be thereby refracted to the opposite wall. It was at first a very pleasing divertissement, to view the vivid and intense colours produced thereby . . .[25]

With the prism he was able to demonstrate how white light is composed of a range of component colours and how it can be split into the colours of the spectrum, with blue light, at one end of the spectrum, being bent (or refracted) more markedly than red light, at the other end. Furthermore, he was able to judge – correctly – that the colour of an object depends upon which part of the spectrum is absorbed by it and which part reflected. 'Hence redness, yellowness etc.,' he wrote, 'are made in bodies by stopping the slowly moved rays without much hindering of the motion of the swifter rays, & blue, green & purple by diminishing the motion of the swifter rays & not of the slower.'[26]

In short, an object will look red if the other colours (what Newton refers to as 'the slowly moved rays') are absorbed by it more than is red light. The red will then be reflected back much more than the other colours. In the same way, an object will appear blue because it reflects blue more than the other colours (those that Newton called 'the swifter rays'). The ability to absorb or reflect different parts of the spectrum depends on the nature of the object and produces the rich diversity of colour that we observe in the universe.*

Following this discovery, Newton copied out an extract from Descartes's *Dioptricks* and wrote after it:

Of Light

Light cannot be by pressure for we should see in the night as well or better than in the day we should see a bright light above us because we are pressed downwards . . . there could be no refraction since the same matter cannot press 2 ways, the Sun could not be quite eclipsed, the Moon & planets would shine like suns. A man going or running would see in the night . . .[27]

Throughout the summer and autumn of 1664 Newton conducted further experiments and observed diffraction through feathers and different fabrics held up to the light: 'A feather or a black riband put between my eye and the setting Sun makes glorious colours,' he observed.[28] But then his fervour for experiment seemed to take him over. Within days he performed two experiments that left him almost totally blind.

The first near-catastrophe was when he looked directly at the Sun for too long, with the intention of observing coloured rings and spots before the eyes – a practice he repeated over and over again. In a letter to his friend the political philosopher John Locke, written a quarter of a century later, in 1691, he describes the experience:

I looked a very little while upon the Sun in a looking glass with my right eye and then turned my eyes into a dark corner of my chamber & winked to observe the impression made & the circles of colours which encompassed it & how they decayed by degrees & at last vanished . . . And now in a few hours' time I had brought my eyes to such a pass that I could look upon no bright object . . . but I saw the Sun before me, so that I could neither

---

* This argument ignores the fact that the brain interprets blends of colours to create the impression of, say, green or violet. This interpretation of colour is called the 'physiological colour'. The colour discussed here is the 'physical' colour.

write nor read but to recover the use of my eyes shut myself up in my chamber made dark for three days together & used all means to direct my imagination from the Sun.[29]

This mishap could be put down to Newton's ignorance of the true danger involved, but what is the explanation for the following experiment?

I took a bodkin [from the illustration accompanying this entry in the notebook, astonishingly, this appears to be a small dagger similar to an envelope knife], and put it between my eye and the bone as near to the backside of my eye as I could: & pressing my eye with the end of it (so as to make the curvature in my eye) there appeared several white, dark and coloured circles. Which circles were plainest when I continued to rub my eye with the point of the bodkin, but if I held my eye and the bodkin still though I continued to press my eye with it yet the circles would grow faint often disappear until I resumed them by moving my eye or the bodkin.[30]

Youthful enthusiasm and dedication are one thing, but most people would agree that sticking a blade into one's own eye goes far beyond the call of duty. As a result, by nearly causing permanent blindness, he came close to destroying his scientific career almost before it had begun.

In spite of these set-backs, Newton was learning rapidly from his experiments. The synthesis of Baconian method, innate talent and theoretical rigour was almost complete, but one crucial element was still missing.

There is general disagreement regarding the timing and even the exact method by which Newton acquired the mathematical knowledge that transformed his approach. No mathematics appear in the Philosophical Notebook, but early in 1664, and before his optical experiments, he began to make mathematical notes in what he called the 'Waste Book', the barely used notebook of his stepfather, Barnabas Smith.[31] By late that summer he was already familiar with the most complex mathematical ideas of the times, gleaned largely from major texts of the period, including John Wallis's *Arithmetica Infinitorum* (1655) and Descartes's *Geometry*.[32]

Until his entry into Cambridge University, Newton's mathematical knowledge had been limited to simple arithmetic, perhaps some algebra and a little trigonometry. But it is a mark of his genius that during the course of only two years he taught himself advanced mathematics and developed the calculus.

From letters and a collection of private papers which a few privileged disciples were allowed to rake through towards the end of his life, it is clear that Newton approached mathematics with the same autodidactic fervour he had shown as an adolescent pursuing his quest for knowledge at the Clarks' home. But this time, in his enthusiasm, he skipped the fundamentals. The French mathematician Abraham Demoivre made the most intense study of Newton's earliest mathematical work, and in a memorandum he wrote in 1727 he gives us an account of Newton's stumbling, maverick approach:

In '63 [Newton] being at Stourbridge fair brought a book of astrology to see what there was in it. Read it 'til he came to a figure of the heavens which he could not understand for want of being acquainted with trigonometry. Bought a book of trigonometry, but was not able to understand the demonstrations. Got Euclid to fit himself for understanding the ground of trigonometry. Read only the titles of the propositions, which he found so easy to understand that he wondered how anybody would amuse themselves to write any demonstrations of them.[33]

Whether or not Newton had any official help with mathematics towards the end of his second year at Cambridge is difficult to ascertain. In March 1664 Isaac Barrow began a series of mathematical lectures as part of his duties as the first Lucasian Professor – a position he had accepted that winter. We know that Barrow and Newton were acquainted closely a few years after this, and that Barrow surrendered his chair to Newton in 1669, but it is by no means certain that Newton attended Barrow's mathematics lectures. According to statutes laid down by the King, these lectures were for fellow-commoners only. This may have prevented Newton; however, such rules were flaunted openly and, being the sort of student he was, Newton may have worked his way in despite his lowly social position within the university.

For the future advancement of science, his efforts at teaching himself advanced mathematics were of the utmost significance. Without an understanding of algebra, Newton could not have developed the calculus, and without that he could not have manipulated and communicated his physics – calculus provided the formal structure needed to turn his notions of gravity from concept to hard science.

In 1664 such grand designs were some way ahead; more pressing were the demands of the university. Although he had been working consistently hard, his efforts had been exerted almost entirely outside the curriculum. Like Darwin, Einstein, Hawking and many other great scientists after him, Newton found himself ill-prepared for the various exams he needed to pass in order to continue as a student.

Having realised that his charge was more interested in mathematics and the latest philosophical ideas from Europe than in the official curriculum, Newton's tutor, Benjamin Pulleyn, referred him to Isaac Barrow for his scholarship appraisal. Newton was required to pass an examination in April 1664 which would make him an undergraduate scholar, allowing him to sit for his BA the following spring. Pulleyn presumed that Barrow would be the most useful fellow to access the young man's talents. Unfortunately, Barrow decided to quiz Newton on Euclid. This could have spelled disaster, because Newton had paid little attention to simple Euclidean theorems *en route* to more advanced mathematics. Conduitt tells us:

When he stood to be scholar of the house his tutor sent him to Dr Barrow then mathematical professor to be examined, the Dr examined him in Euclid which Sir I. had neglected and knew little or nothing of, never asked him about Descartes' Geometry which he was master of. Sir I. was too modest to mention it himself & Dr Barrow could not imagine that one could have read the book without first [being] master of Euclid, so that Dr Barrow conceived then but an indifferent opinion of him but however he was made scholar of the house.[34]

Having been made aware of his deficiency, true to form, Newton immediately went back to the basics of mathematics and quickly absorbed Euclidean geometry and simple algebraic theorems. His dedication is evident from the fact that the most dog-eared and tatty book in Newton's library was *Euclidis Elementorum* by Isaac Barrow.

Newton may have made up for his mistakes, but, viewing the situation dispassionately, it is clear that he must have received help in convincing the fellows of his true worth. If the interview with Barrow had indeed gone as badly as Conduitt reported, it must have created a poor initial impression and Newton's supporters must have brought their influence to bear in order to salvage the young man's career. Humphrey Babington was rising high in the college hierarchy (becoming a senior fellow in 1667), and he enjoyed the King's favour. The well-documented fact that Newton visited him frequently during the plague years spent in Woolsthorpe shows that the two men remained in contact throughout Newton's early years in Cambridge. Having helped to get him into Trinity, Babington would not have wanted him to flunk his scholarship. He almost certainly realised the young man's potential and may have appreciated his disenchantment with the outdated university curriculum.

Even though he brushed up his Euclid, Newton clearly did little in the way of formal study for the BA examinations the following spring. As a result, he did graduate – but in an undistinguished manner. According to Stukeley, 'when Sir Is. stood for his Bachelor of Arts degree, he was put in second posing, or lost his groats, as they call it,* which is looked upon as disgraceful'.[35]

In a larger historical perspective, the fact that in the spring of 1665 Newton graduated with a mere second-class BA is laughable, but in the pantheon of scientific greats this is not so unusual. Robert Darwin had to remove his son Charles from medical studies in Edinburgh because it was clear he would make nothing of his time there; Albert Einstein scraped through his degree and then found it almost impossible to find a job; and Stephen Hawking, who was unpopular with the Oxford University authorities because he spent more time on the river than in lecture theatres, was awarded a first only to ensure that he did his PhD in Cambridge. But, for the twenty-two-year-old Newton, graduation, whatever the grade, was enough to secure his future at the university. Setting an example for his scientific heirs, he had long since decided that his vocation was to unravel the laws

---

* 'Lost his groats' refers to the tradition whereby a student gives nine small coins, or groats, to an examining official. If the student does well, the coins are returned; if not, they are 'lost'.

governing God's universe; passing exams was merely a means to an end and was conducted with the minimum of effort. He now had official sanction to pursue his true goal, but even he, with the arrogance of youth and a single-minded determination, could not have realised just how soon would come his first successes *en route* to his dream.

CHAPTER 4

Astronomy and Mathematics Before Newton

In every piece there is a number – maybe several numbers, but if so there is also a base-number, and that is the true one. That is something that affects us all, and links us all together.

ARVO PÄRT (composer)[1]

Number and pattern have always held a fascination, and the true origins of mathematics and astronomy are certainly ancient. The earliest form of organised mathematics, in which numbers were meaningfully manipulated and patterns recorded, is credited to the Babylonians of around 4000 BC, who recorded star patterns and named constellations. They had also developed a surprisingly advanced set of mathematical rules, including a sophisticated method of counting – a skill employed by the record-keeper, the farmer and the architect. It is thought that the last of these professions may also have employed simple forms of algebra and geometry.

Modern research, such as John North's work on ancient stone circles, has demonstrated that the ancient Britons must also have possessed some knowledge of geometry in order to build such structures as Stonehenge, started about 3500 BC,[2] and the ancient Egyptians had highly developed mathematical and engineering skills which they employed in the building of the Great Pyramid at Giza some 1,000 years later. In these ancient civilisations, mathematics and astronomy were blended together intimately and had rich associations with mysticism and the occult. Astronomy and astrology were viewed as one and the same, and mathematics gained an almost spiritual status as a tool for the astrologer/astronomer. It was not until Greek times that mathematics and, to a lesser extent, astronomy were separated from religion and considered worthy of academic attention. While maintaining their spiritual associations, they then gradually became subjects for pure analysis and reasoning.

All mathematics may be viewed as composed of three central subjects: arithmetic, geometry and algebra. As the most immediately useful to a wide range of crafts and professions, arithmetic was the earliest form of mathematics to be developed, and grew to include all forms of number manipulation.

In its simplest form geometry deals with the shapes of things, in either two or three dimensions (although modern mathematicians also deal with multidimensional space – a study still called geometry). This area of mathematics found ready use with the architect and the builder. For the astronomer it was an invaluable tool in the search for patterns in the stars, which in turn fuelled the development of astrology.

Algebra, which was only scantily formulated before the early seventeenth century, is a language in which symbols are assigned to properties of objects. It enables mathematicians to construct equations that describe a situation or the interplay between properties (either real or imaginary) using strict rules that govern what may be done with representative symbols. A simple example would be the equation $s = d/t$. In words this would be 'Speed equals distance travelled divided by time taken'. Further examples would include equations used to find the rate at which water flows through a pipe, how quickly a rocket accelerates from the launch pad, or how efficiently a muscle uses energy from glucose.

Arithmetic and geometry may be considered more everyday than algebra, in that they represent the world and the things we observe *directly*. Algebra is one level of abstraction away from reality, because symbols are used to represent properties, rather than being actual measurements of things. This distinction could account for the fact that arithmetic and geometry were developed into sophisticated tools and used widely very much earlier than algebra.

The Greeks viewed mathematics in a different way to the civilisations that predated them, in that they appear to have been the first to consider pure mathematics – to contemplate mathematical abstractions, rather than using mathematics solely as a tool for constructing religious structures or to develop the mystical arts. Using mathematical skills, the Greeks were able to develop elaborate theories to describe the structure of the observable universe and to postulate ways in which the planets, the Sun and the stars could be arranged in the heavens.

According to most accounts, Anaximander, who lived between 610 and 545 BC, is thought to be responsible for the first development of what became the *geocentric* view of the universe – the concept that the Earth lies at the centre of the universe. Before then, the Earth was believed to be a floor with a solid base of limitless depth.[3]

Anaximander reached his conclusions by astronomical observation, believing that the visible sky was a dome or half of a complete sphere. But it was not until the fourth century BC, when Greek explorers began to travel further afield, that this idea began to be widely accepted. An indication of the rapid progress that was made during this period is that by the third century BC, around 300 years after Anaximander, Greek astronomy had progressed to the point where Eratosthenes, a contemporary of Archimedes, was able to estimate the circumference of the Earth, putting it at 24,000 miles (only 800 miles short of the modern measurement). He was also able to calculate the distance between the Earth and the Sun, assigning it a figure of 92 million miles (a little over 1 per cent out from the modern value of 93 million miles).

This progress in astronomical knowledge was due largely to the development of geometry between the lifetimes of Anaximander and Eratosthenes. Many advances derived from a strong need for practical mathematical tools for use by land surveyors and farmers – 'rules of thumb' and practical guidelines. Such developments helped philosophers and mathematicians to derive axioms and general principles that led to further discoveries. The first great mathematician to work in this way was Pythagoras, a man most people remember from school maths lessons as the creator of a theorem concerning right-angled triangles.

In fact Pythagoras, who was born at Samos shortly after Anaximander's death, derived much more than a single geometric relationship: he was the most important figure in formulating the whole basis of pure mathematics. His school was pseudo-mystical, in that he and his followers believed that the universe had been designed around hidden numeric relations and that its entire structure and the complex interplay of the four elements (later popularised by Aristotle) were governed by mathematical patterns. He and his followers discovered the mathematical relationship between sounds, using vibrating strings, and originated the concept of the 'music of the spheres' – the idea that the ratios observed between notes on the musical scale could be mirrored in the distances of the planets from the Earth.*

Fortunately, many of Pythagoras's ideas were preserved by another great mathematician, who lived two centuries later, Euclid of Alexandria – the man most commonly perceived as the father of modern geometry. Although Euclid was an original thinker and added much to the knowledge of geometry, his greatest contribution was to collect earlier work, especially that of the Pythagorean school, and to rationalise it into a collection of books he produced around 300 BC. These have survived to the present day and formed the basis of all geometry until the middle of the last century. So fundamental is this work to our understanding of mathematics that the three-dimensional space in which we perceive the universe is known as 'Euclidean' space, and it was only during the nineteenth century that mathematicians began to speculate about the possibility of non-Euclidean space – geometry which did not adhere to Euclidean rules.†

Astronomy and mathematics developed little between the waning of Greek culture and the domination of the Arabic intellectual system which began to emerge during the second half of the first millennium AD. The exceptional name from this era is the Alexandrian Ptolemy (c. AD 100–170), who codified the geocentric theory, a concept that remained at the heart of astronomical thinking until the sixteenth century.

Little is known of Ptolemy's life, but he made astronomical observations from Alexandria during the years 127–41 and probably spent most of his life there. Principally a geographer, he wrote a treatise entitled *Almagest* which contained many of his own observations and theories as well as summaries of Graeco-Roman thinkers. He also produced geometric models which he used to predict the positions of

---

* According to legend, Pythagoras realised the mathematical relationship between musical notes after passing a brazier's shop and hearing an anvil struck by different-sized hammers. He then went on to experiment with vibrating strings to derive precise relationships between notes.

† Non-Euclidean geometry was employed by Einstein to enable mathematical verification of his concept of space-time early this century.

the planets, imagining all heavenly bodies to travel in a complex set of circles known as *epicycles*, within the framework of a geocentric system supplied by many of the Greek astronomers.

To the modern mind, the concept of the Earth lying at the centre of the universe is an absurdity, but there were very good reasons why this concept held sway for so long and became so thoroughly ingrained in Western intellectual systems. It was certainly not born out of ignorance on the part of the Greek philosophers and astronomers who created it. These same philosophers could, after all, measure the distance between the Earth and the Sun with an accuracy of 1 per cent. It was more to do with deliberate obfuscation of the facts in order to comply with the Greeks' anthropocentric vision.

This historical interpretation has become fashionable only in this century and has been championed by a number of historians of science, including the eminent writer Arthur Koestler, who popularised it in his influential work *The Sleepwalkers*.[4] The Greeks, like the scholars of Europe in the Middle Ages, were obsessed with the idea that man was the centre of Creation and that consequently the Earth must be at the centre of the universe. To accommodate this dogma they created an incredibly elaborate mechanical system that would account for their observations of the heavens. If there had been no philosophical imperative for the Sun, the Moon and the five known planets to orbit the Earth in perfect circles, then it would have been quite within the powers of the late Greek and Alexandrian astronomers to show that the Earth, along with Mercury, Venus, Mars, Jupiter and Saturn, orbited the Sun, and that the Moon orbited the Earth. They might even have been able to deduce that the orbits were elliptical rather than circular. Instead, in order to account for the observed movements of the known heavens and to satisfy the prevailing philosophy, Ptolemy needed to create a system of forty different 'wheels within wheels' – a crazy pattern of gears or Ferris wheels.

The most difficult problem he faced was how to explain what is called the retrograde movement of some planets: that at certain times of the year planets appear to move backwards against the backdrop of stars from one night to the next. Today we know that this is because planets follow elliptical orbits around the Sun and move at different speeds, so there will be times when the Earth appears to 'overtake' the slower-orbiting outer planets and they appear to travel backwards.

To picture this clearly, imagine the solar system as a motor-racing track with the planets represented by the cars in different lanes. If the cars move at very different speeds, from the viewpoint of a car moving quickly on the inside track a slower car in an outside lane would appear to be moving backwards.

Because it complied with the anthropocentric world-view, Ptolemy's complex, but quite incorrect, system was adopted as the only valid description of the universe and was later sanctioned by Christian theology. But even then there were some who doubted. When he learned of Ptolemy's thousand-year-old system, the thirteenth-century Spanish King Alphonso X, known as Alphonso the Wise, declared, 'If the Almighty had consulted me before embarking upon the Creation, I should have recommended something simpler.'[5]

Koestler believed that 'There is something profoundly distasteful about Ptolemy's universe.'[6] What he meant by this was that, as we learn more about the universe, we find that the underlying rules by which it operates are fundamentally elegant and simple. This is what scientists mean when they talk about the 'beauty' of the mathematics representing universal laws such as those of gravitation or radioactive decay. Although science appears to the uninitiated to be incredibly complex, the laws that govern the behaviour of matter and energy are remarkably simple. In retrospect we can see that any system like Ptolemy's had to be wrong. He and others of his time were trying to squeeze the facts into a false theory in order to fit a belief. They were starting with a rigid conviction and attempting to make the universe suit their dogma. Ptolemy himself wrote, 'We believe that the object which the astronomer must strive to achieve is this: to demonstrate that all the phenomena in the sky are produced by uniform and circular motions.'[7]

Like Aristotelian dogma, Ptolemy's system survived the Renaissance, but it was gradually eroded by a growing awareness that came from exploration of our own world. As European explorers circumnavigated the globe and discovered America and Australia, it was gradually accepted that the Earth was spherical (a fact known to the Greeks), but their voyages also offered opportunities for observation of the heavens from the perspective of the southern hemisphere and other regions never before visited. Exploration also expanded trade and supplied increased wealth, and learning grew exponentially. At the same time, mathematical and astronomical techniques were improving, and the traditional notions of Ptolemy, Aristotle and Plato were challenged on an intellectual as well as a practical and observational level.

Mathematics had been refined greatly by the Arabs throughout the period of the Dark Ages in Europe. Based in cities such as Isfahan and Baghdad, Arabic mathematicians merged material gained from Alexandria, India and China. Preoccupied with astrology, they were greatly taken with Ptolemy's system, preserving the *Almagest* after the sacking of the library of Alexandria and passing it on to the West around the thirteenth century.

During this period, Arabic mathematicians refined Greek geometry and developed algebra significantly, and when Europe finally emerged from the Dark Ages the Arabic system of numerals was adopted along with the Arabs' place-valued decimal system – the system gives values to numbers according to their positions on either side of the decimal point, increasing by factors of ten from right to left and decreasing by factors of ten from left to right.

All of these advances helped to re-establish analytical astronomy in the West. Although there remained a strong tradition of interest in astrology, and many of the alchemists and magicians who travelled around Europe well into the seventeenth century also earned money from practising astrology, the Christian Church did not officially recognise the art. Because learning had been sustained by a unification of science and theology, many of those interested in astronomy and mathematics were monks and theologians and they could not be seen to be dabbling in astrology. Astronomy, however, could be justified as a purely intellectual pursuit, as a component of worship. Indeed, the man who led the re-evaluation of the Ptolemic system was a priest: the Polish canon Nicolas Koppernigk, known to posterity as Nicolas Copernicus, who wrote a book called *On the Revolutions of the Heavenly Spheres*, first published in 1543.

The simple image of Copernicus upturning the established, fourteen-hundred-year-old ideas of Ptolemy almost overnight and revolutionising our thinking about the structure of the universe is quite false. Copernicus's work was revolutionary, but, except in one respect, it did not offer a completely new system. Copernicus's model of the solar system was actually more complex than Ptolemy's and comprised forty-eight circles or wheels within wheels to account for the observed facts (eight more than in Ptolemy's model). The one vital and controversial aspect of Copernicus's work that challenged accepted thinking was his belief that the Earth did not lie at the centre of the universe.

Having said that, Copernicus did not place the Sun at the centre of the system either, but accounted for astronomical observations by keeping most of Ptolemy's epicycles and having the Sun, along with the Earth and the known planets, orbiting a point (close to the Sun) which he claimed to be the true centre of the universe.

Copernicus suppressed his findings for over thirty years, and, through fear of religious persecution, did not allow his book to be published until he was on his deathbed. In fact he began his treatise boldly, by asserting that the Sun lies at the centre of the universe, but then appeared to change his mind. After the first few pages, he went on to complicate his theory more and more with unnecessary refinements, finally placing the Sun slightly off-centre. This prevarication made the entire work almost unreadable and frequently contradictory. Perhaps it was because of this that, despite containing what became one of the most influential theories in the history of science, in commercial terms *On the Revolutions of the Heavenly Spheres* was one of the least successful books ever written.

Running to 212 sheets in small folio, the heart of *Revolutions* is contained in the first twenty pages, in which Copernicus outlines the central tenets of his theory, stating, 'in the midst of all dwells the Sun . . . Sitting on the royal throne, he rules the family of planets which turn around him . . . We thus find in this arrangement an admirable harmony of the world.'[8]

The central tenets of Copernicus's theory were revolutionary for the time in which he lived, and they bear comparison with the modern view far better than Ptolemy's geocentric picture. Copernicus states that the Sun lies stationary at the centre of a finite universe bounded by the fixed stars. The Earth and all the planets orbit the Sun in circles; the Moon revolves around the Earth. The apparent daily rotation of the firmament about the Earth is not because it is at the centre of the universe but because the planet revolves on its axis. The apparent annual motion of the Sun around the Earth in Ptolemy's description is actually due to the passage of the Earth around the Sun. Finally, Copernicus could explain the bug-bear of the Greeks' system – the apparent retrograde motion of certain planets – by describing how the planets all orbit the Sun at different speeds. He was even able to explain the slight irregularities of the seasons as being due to the Earth 'wobbling' on its axis.

It is possible that Copernicus confused the written account of his theory deliberately, to ward off the expected attacks of religious orthodoxy. But it is equally likely that, having come to the irrefutable conclusion that the Earth revolved around the Sun, he could not himself accept the consequences of the simple system he had stated at the beginning of his book. Copernicus was an Aristotelian, rooted in medieval thinking, and had been educated to accept Greek teaching verbatim. In no sense was he the revolutionary figure posterity has painted him as being. Throughout his notebooks there are points where he could have reached far more profound conclusions but missed them because he was strait-jacketed by his religious convictions and his traditional education.

In spite of his fears, Copernicus's *On the Revolutions of the Heavenly Spheres* had little immediate impact upon religious or scientific thinking, and was not included in the Index Prohibitorum (the list of books banned by the Catholic Church) until 1616, seventy-three years after its publication. What did change history and posthumously placed Copernicus at the centre of an ecclesiastical and intellectual storm was the work of his successors, who based their ideas upon his discoveries and combined them with meticulous observation and more refined mathematical knowledge.

This mathematical knowledge had been developing in parallel with advancing ideas in astronomy based upon observation. The Babylonians had developed a simple form of algebra five or six thousand years before Copernicus, but their understanding of the subject had been limited to the use of what mathematicians call linear and quadratic equations. These describe equations of different levels of complexity. A linear equation contains just numbers and symbols which may be added, subtracted, divided or multiplied together. A quadratic equation contains terms (such as x or y) which have been squared (raised to the power of 2). Both of these types of equation are simple to solve and fall well within today's secondary-school curriculum. More difficult, but far more powerful in the hands of the scientist, are solutions to what are called cubic equations – equations which contain terms raised to the power of three (or cubed).

Cubic equations had resisted all attempts at solution until the beginning of the sixteenth century, when two mathematicians could independently lay claim to success: Niccolo Tartaglia and Scipione Ferro. Their methods of solving cubic equations were eventually published in 1545, two years after the death of Copernicus, in a book by Gerolamo Cardano called *Ars Magna* which paved the way for a succession of new algebraic techniques.

The sixteenth century was also a time when international trade and mercantile enterprise flourished and businessmen and economists employed new and more efficient mathematical techniques. Towards the end of the century, in 1585, Simon Stevinus of Bruges created rules for solving equations of higher powers than three, and three decades later, in 1614, the Scottish mathematician John Napier devised the technique of logarithms – an incredibly powerful algebraic and arithmetic tool that opened the door to further rapid advance.

As well as assisting the pure mathematician, all of these methods had a beneficial impact upon the development of astronomy. Application of these techniques was most ably exploited by a man who was himself a mathematician as well as a practising astronomer, Johannes Kepler. Kepler, born in Württemberg, first appeared on the scene in 1596, when, at the age of twenty-four, he published a book called *Cosmographic Mystery*, in which he postulated his earliest model of the solar system and defended Copernicus's ideas.

Although *Cosmographic Mystery* was a promising work for someone so young, it gave no better mechanism for how the planets moved or how the mechanical structure of the solar system was maintained than did Copernicus's complex model. But, a short time after its publication, Kepler was offered an opportunity that was to transform his career and lead to an advancement in astronomy almost as profound as Copernicus's own.

Impressed by the *Cosmographic Mystery*, in 1600 the ageing Tycho Brahe – mathematician to the court of Emperor Rudolph II in Prague – invited Kepler to come to Prague as his assistant. A year later Brahe died, leaving the twenty-nine-year-old Kepler to take over his observatory and inherit his vast collection of astronomical data.

At heart Kepler was a Pythagorean, in that he believed that the universe was an harmonic entity, that number ruled every aspect of

Creation, and that regular simple patterns lay behind all facets of the observable realm. Combined with meticulous observation and mathematical rigour, it was these convictions that led him to his great discoveries.

Using the vast body of data that Tycho Brahe had collected during twenty years of observations, Kepler discovered that there was a minor discrepancy between the observed position of the planet Mars and that calculated using Copernicus's model. He could trust the observational data because Brahe had used a set of newly invented sextants and quadrants that were accurate to between one and four minutes of arc. (A minute of arc is one-sixtieth of an angular degree, or one five-thousand-four-hundredth of a right angle.) The difference between the calculated value (based upon Copernicus's theory) and the observed value for the position of Mars was eight minutes of arc.

Starting from this discrepancy, Kepler came to the conclusion that the orbits of the planets around the Sun were not circular, as Copernicus had proposed, but elliptical, and he went on to prove it by matching precisely his calculations for planetary positions based upon elliptical orbits with accurate observations from Brahe's data. This offered conclusive support for the heliocentric, or sun-centred, model of the solar system, because accurate observation matched values derived from independent calculation. Kepler was then able to formulate three laws (since known as Kepler's Laws) first described in his two great books, *New Astronomy*, published in 1609, and *Harmony of the World*, which appeared in 1619.

The first law is a simple statement of Kepler's discovery – that all the planets travel in paths which are ellipses with the Sun at one focus. The second law states that the area swept out in any orbit by the straight line joining the centres of the Sun and a planet is proportional to the time taken for the orbit. In other words, the area of space an orbit borders is proportional to the length of time the planet takes to orbit the Sun: the further a planet is from the Sun, the larger the area and the longer an orbit will take.

Kepler's third law came some time later, with publication of his *Harmony of the World*. This law describes the mathematical relationship between the distances of the planets from the Sun and the times they take to complete their orbits. Kepler found that the square of the periodic time which a planet takes to describe its orbit is proportional to the cube of the planet's mean distance from the Sun.

Just as Kepler was devising these laws, the telescope was being turned into a powerful tool by the Italian natural philosopher Galileo. The instrument was actually invented by a Dutchman, Hans Lippershey, in 1608, but Galileo's device, designed and built within two years of Lippershey's, was far superior and could magnify up to thirty times – making it powerful enough to distinguish craters on the Moon's surface and to observe a set of moons orbiting the planet Jupiter.

So revolutionary was this invention that (perhaps through fear of the inevitable consequences of such a discovery) many could not contemplate its uses, and several leading political and military figures of the time had to be teased into trying out the instrument.

What Galileo observed immediately confirmed Kepler's laws. The system of moons around Jupiter could be visualised as a model of the solar system, and the moons' revolutions could be measured and compared to calculated values, showing their paths to be elliptical. Furthermore, Aristotle had believed that the heavenly sphere (anything outside of the Earthly realm) was perfect and featureless, yet the telescope showed clearly visible craters on the surface of the Moon.

Gradually, the edifice of Aristotelian ideas and the ancient astronomy of Ptolemy and the Greeks was crumbling and being replaced with accurate observation supported by the clinical precision of mathematics. Together these would eventually become an irresistible force for a major change in the way the universe was perceived.

But the road to empiricism was bumpy and dangerous. The sixteenth and early seventeenth centuries were a period of huge ecclesiastical upheaval throughout Europe that included the worse excesses of the witch-hunts and the terror of the Inquisition. Kepler's own mother, who lived in the small town of Leonberg that was sympathetic to Catholic activists, was accused of being a witch in 1615 and suffered over a year in jail and faced torture several times before being acquitted.

Kepler and Galileo corresponded, and the German astronomer sought the Italian's public support for his theory, asking him openly to support what he had accepted in private. But Galileo was unable to do this. Living as he did in the most volatile religious environment in Europe, and keeping only one step ahead of the ecclesiastical authorities, who were fearful that this new science would undermine their authority, Galileo could not risk such an endorsement. Kepler was lucky to be living and working in northern Europe, where, for the most part, he experienced far greater religious tolerance.

Soon after the invention of the telescope, the powerful cleric Cardinal Bellarmine began to undermine Galileo's work. In 1616 he pronounced that the Copernican system was 'false and altogether opposed to Holy Scripture'.[9] Galileo was then famously persecuted for his support of the heliocentric model of the universe and spent his final years under house arrest. He managed to avoid the punishment of the heretic only by denouncing his convictions and what he knew to be fundamental truth before the Inquisition in 1633, just nine years before Newton's birth.

But, once started, nothing could stop the flow of progress. By the early part of the seventeenth century the stage was set for another series of revolutionary advances in mathematics that would pave the way for the work of Newton and the triumph of change from a geocentric to a mechanistic and heliocentric viewpoint – from Aristotelian guesswork to empiricism, observational precision and mathematical rigour. The man responsible for this last development before Newton was to take up the baton and reach the law of gravity and the development of the calculus was the French philosopher René Descartes.

Descartes's great mathematical breakthrough was the realisation that the equation was not the only way in which mathematical terms could be related. During the 1630s he devised the idea of constructing coordinates to represent pairs of numbers relating to algebraic terms (usually x and y). These came to be known as Cartesian coordinates and opened up the vast range of possibilities offered by the drawing of graphs – lines and curves bordered by axes. The technical name for this branch of mathematics is analytical geometry, and it first appeared in an appendix called 'Geometry' tacked on to the end of Descartes's *Discourse on the Method*, which was first published in 1637.

Descartes's technique galvanised the world of mathematics, and within a few years of publication the *Discourse* had influenced the work of mathematicians and astronomers throughout Europe. Such men as the English mathematicians William Wallace and Isaac

Barrow, as well as natural philosophers and mathematicians on the Continent, led by Pierre de Fermat and Christiaan Huygens, used Descartes's findings as a springboard for their own efforts, which began to focus on the properties of the curves that could be drawn using Cartesian coordinates.

A simple example is the graph produced by plotting the distance travelled by a ball dropped from a high tower against the time for which it has fallen. Galileo had shown that the speed of a ball increases with time. If after one second the ball has fallen 16 feet, after two seconds 64 feet and after three seconds 144 feet, clearly it is accelerating. If these values are plotted on a graph with speed on the *y*-axis and time on the *x*-axis a curve is produced.

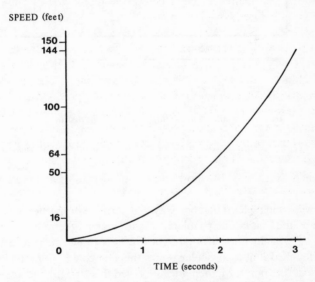

Figure 1. The curve produced by plotting distance against time.

Now it is comparatively easy to calculate properties for straight-line graphs. For example, the area under a straight line can be calculated by simple geometry known to the Babylonians, and the gradient of a straight line (or its steepness) can be found by dividing the change in the values along the *y*-axis by the corresponding values along the *x*-axis. So, if the distance–time graph had been a straight line, the gradient would have given us the speed of the ball (the change in

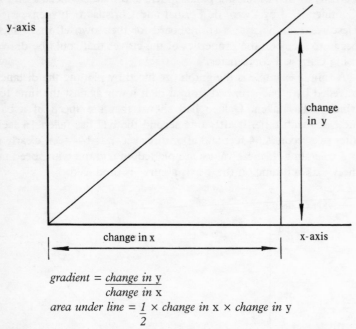

$$gradient = \frac{change\ in\ y}{change\ in\ x}$$

$$area\ under\ line = \frac{1}{2} \times change\ in\ x \times change\ in\ y$$

Figure 2. Calculating the gradient and the area under a straight line.

distance with time). But how can the properties of a curve be calculated?

It was soon realised that one way to determine properties of curves, such as the one in our problem, was to imagine them as constantly shifting straight lines: if a straight line was drawn next to a curve and touched it at a particular point, this line could approximate the curve at that point. Mathematicians called this straight line a *tangent*, and found that they could treat a tangent like any other straight line – they could, for example, find its gradient and therefore work out a value for the speed of the ball at that particular point. But this was still an approximation – and a very limited one at that.

Simple problems concerning objects travelling in circular motion had been studied by earlier generations of philosophers, especially Galileo, but by the 1660s astronomers weaned on Kepler's work were becoming interested in mathematical models to describe the

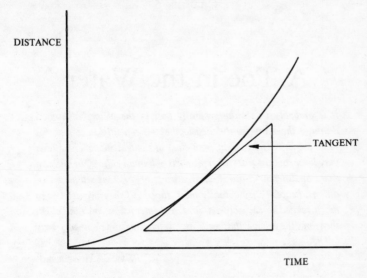

Figure 3. The tangent to the curve.

new celestial mechanics – the mathematics of how the planets maintain their orbits around the Sun. They of course realised that the mathematics of curves could lead to a fuller understanding of planetary motion, but limited solutions such as those offered by drawing tangents were not accurate enough to correlate with increasingly sophisticated methods of gathering observational data. Although the mathematicians and astronomers of Europe were exploring methods of working with curves and some, such as Fermat and the great English polymath Christopher Wren, came to very limited solutions that worked in specific cases, there was a need for general solutions, or methods that could be applied to all situations. Newton gradually became aware of this as he studied the work of his predecessors while an undergraduate student at Cambridge during the early 1660s. By the middle of the decade all the elements were in place for a mathematician of genius to produce the required new mathematics. And, thanks to a series of unpredictable events, Newton was able to find the time and inspiration to do just that.

CHAPTER 5

A Toe in the Water

It is probably true quite generally that in the history of human thinking the most fruitful developments frequently take place at those points where two different lines of thought meet. These lines may have roots in quite different parts of human culture, in different times or different cultural environments or different religious traditions: hence if they actually meet, that is, if they are at least so much related to each other that a real interaction can take place, then one may hope that new and interesting developments may follow.

WERNER HEISENBERG[1]

When, in the spring of 1669, the Trinity fellow Francis Aston was preparing to leave on a European tour, he wrote to his friend Isaac Newton asking for his advice on how best to conduct himself and what to look out for on his travels. This is surprising, since Newton had never travelled abroad and had only recently made his first trip to London. But it illustrates the high esteem in which Newton was held by his colleagues so early in his career, even in connection with matters outside his area of expertise. More significant still is Newton's reply to Aston's letter, for, as well as asking his friend to gather alchemical information for him and to attempt to track down the famous alchemist Giuseppe Francesco Borri, then living in Holland, Newton went on to offer a long list of dos and don'ts as though he were a seasoned globe-trotter. These included the recommendation:

If you be affronted, it is better in a foreign country to pass it by in silence or with jest though with some dishonour than to endeavour revenge; for in the first case your credit is none the worse when you return into England or come into other company that have not heard of the quarrel, but in the second case you may bear the marks of the quarrel while you live, if you outlive it at all.[2]

The reason for this easy confidence is that by the 1660s Newton had already adopted what one of his biographers has called 'a Polonius-like pose'.[3] Even as a boy he had been confident to the point of alienating others, but Newton the man, the twenty-six-year-old fellow of Trinity College, Cambridge, six months away from accepting the Lucasian chair, was already so accomplished that if he had done nothing further with his life he would still have found a significant place in the history of science.

Although his genius was realised by only a handful of associates in Cambridge and he was totally unknown to the scientific community, by 1669 Isaac Newton was in fact the most advanced mathematician of his age, creator of the calculus as well as elucidator of the basic principle behind the inverse-square nature of gravity and the theory of the nature of colours. Within the space of four years he had grown from unnoticed undergraduate to a man on the foothills of greatness. But, while he had been internally fostering these scientific upheavals, catastrophes had befallen the larger, external, world – catastrophes that had even threatened the ivory tower that Newton inhabited at the very heart of academe.

The plague of 1665 was not the first in English history, but coming as it did straight after the Civil War, and taking the lives of almost 100,000 people (some 70,000 of them in London, which then had a population of under half a million), it was seen by many as yet another fulfilment of the prophecies in the Book of Revelation. The fact that it extended into the year 1666, with its numeric similarity to the 'sign of the beast', only made the psychological impact of the catastrophe more poignant. Daniel Defoe reports that 'Some heard voices warning them to be gone, for that there would be such a plague in London, so that the living would not be able to bury the dead.'[4]

Some 300 years earlier the Black Death had killed an estimated 75 million people in Europe – about a third of the population – but, because most people of the seventeenth century could neither read nor write, it is unlikely that any but the educated few would have realised that plague was a relatively common occurrence. Their only likely knowledge of the virulence of such diseases would have come from their grandparents and great-grandparents recounting horror stories of the last major outbreak, forty years earlier, in 1625.

The plague began in London and spread to other parts of the country rapidly during the hot summer of 1665. It was always at its worst in the east of the city, in the districts of Stepney, Shoreditch, Whitechapel and around the crowded streets clustered at the foot of St Paul's. At its height it claimed 10,000 lives a week, and in one day in September 1665 alone, 7,000 victims died. The disease was in fact bubonic plague – a bacterial infection carried by a flea which infested the black rat (*Rattus rattus*). Wherever rats could breed, the disease spread like wildfire. The flea carried the initial infection to humans via a bite. There was no cure and only a slim chance of survival for those unfortunate enough to become infected. Without the benefit of antibiotics, the only means of containing the disease was quarantine.

By the end of the first summer of the Great Plague, after tens of thousands had lost their lives, the quarantine laws which would eventually help to halt the spread of the disease were finally enacted and major towns and cities became citadels where travellers and visitors were entirely unwelcome. It is clear from a number of reports of the spread of the disease that it took some time for the authorities to realise they were facing a major catastrophe, and by the time they did the plague had a grip on London and had been carried to many other parts of the country. Samuel Pepys, the great monitor of the *Zeitgeist*, first mentioned the plague in his diary entry of 30 April 1665, noting 'Great fears of the sickness here in the City, it being said that two or three houses are already shut up.'[5] But it was not until 15 June that he reported, 'The town grows very sickly, and the people to be afeared of it – there dying this week of the plague 112, from 43 the week before.'[6]

Cambridge escaped relatively lightly, and the university fared amazingly well. The first mention of the disease appears in the *Annals of Cambridge* of August 1665, when, we are told, the plague had prevailed and taken the life of one of the town bailiffs, a William Jennings. Things grew worse as the summer progressed, and the Stourbridge fair was cancelled that year (and in 1666) because of the fear of attracting travellers to Cambridge – especially those from the capital. According to the *Annals*, only 413 people died in all the parishes of Cambridge during 1665, and many of these deaths were from natural causes. They then go on to report that during a two-week period in November of that year a total of fifteen deaths from plague were recorded.[7] In the colleges, there was not one case of the disease all year, largely because the majority of students, fellows and staff had left during the early summer, and those few who did stay kept any contact with the townsfolk to an absolute minimum, locking themselves away in their sanctuaries like medieval monks.

The exact date when Newton left Cambridge is unclear. He was certainly there on 23 May, because he paid his tutor Pulleyn £5.[8] He was not in college for most of July and early August (the college was dismissed on 8 August), because he did not claim six and a half weeks worth of commons (food allowance) paid to those who had stayed on to risk plague during the summer. According to most accounts, he left Trinity around the end of June or the beginning of July and did not return, except for a brief spell in early 1666, for almost two years.

He travelled to his mother's home, the manor in Woolsthorpe, where tradition has it that he made his great discoveries concerning gravity and the mathematical breakthroughs that later made him famous. It is in Woolsthorpe, in the orchard next to the house, that the famous apple is supposed to have dislodged itself with impeccable timing and set in motion the development of the theory of gravity. Thereafter, one might assume, the *Principia* was a mere formalising of the great revealed truth. Yet the reality, magnificent though it was in its intellectual depth and its effect upon the course of science, was far more prosaic. The truth is not so much grounded in singular fluke events or any deeply symbolic psychological drives associated with Newton living in the home of his childhood than it is to do with a gradual revelation brought about by concentration and sheer dedication. As Newton himself said when asked how he came upon his great discoveries, 'I keep the subject constantly before me, till the first dawnings open slowly, by little and little, into the full and clear light.'[9]

The apple story is almost certainly a fabrication, or at the very least a highly embroidered version of the truth. Indeed, the very notion, so integral to many early accounts of Newton's life, that there were two special years in his life during which everything was solved

– the so-called *anni mirabiles* of 1665 and 1666 – is an extreme simplification of the facts. Although Newton's achievements during the time he spent in Woolsthorpe sprang from intuition and inspiration and did lead to the great laws that lie as a foundation beneath our technology, they did not appear fully formed and complete. Although the years 1665 and 1666 were truly great ones for Newton's intellectual development, they mark merely the start of his quest. If we are to label Newton's achievements by the calendar, then the true *anni mirabiles* cover more than two decades, from his arrival in Woolsthorpe to the delivery of the *Principia* in 1687, and encapsulate his period of almost single-minded dedication to the practice of alchemy during the 1670s and '80s as well as the gradual transmutation of his intuitive insights into hard science.

Quite how and indeed from where the initial moment of inspiration came remains a mystery, and, despite the anecdotes and varied accounts describing Newton's efforts during 1665 and 1666, we may never know how one of the most important sets of scientific and mathematical discoveries in history was initiated.

The story of the apple has come down to us from a number of sources. First there is William Stukeley. During the spring of 1726, a year before Newton's death, the biographer visited the great scientist in his final home in Kensington. As they walked out into the garden of the house, Newton remarked that it had been on just such an occasion that he had first realised the theory of gravity. Intrigued, Stukeley pursued the matter 'under the shade of some appletrees, only he and myself,' Stukeley recounts. 'Amidst other discourse, he told me, he was just in the same situation, as when formerly, the notion of gravitation came into his mind. It was occasioned by the fall of an apple, as he sat in the contemplative mood.'[10]

Another account comes from Newton's great admirer Voltaire, who made the English scientist famous in France with his *Éléments de la philosophie de Newton* (1736), in the English edition of which he says:

> One day in the year 1666, Newton, having returned to the country and seeing the fruits of a tree fall, fell, according to what his niece, Mrs Conduitt, has told me, into a deep meditation about the cause that thus attracts bodies in the line which, if produced, would pass nearly through the centre of the Earth.[11]

As the passage relates, Voltaire received this story second-hand from Newton's half-niece Catherine Barton, wife of John Conduitt. Voltaire never met Newton.

There is one other contemporary account of note: that of Henry Pemberton, who was the editor of the third edition of the *Principia*, published in 1726. He describes the scene in a similar way: 'The first thoughts, which give rise to his *Principia*, he had, when he returned from Cambridge in 1666 on account of the plague. As he sat alone in the garden, he fell into a speculation on the power of gravity.'[12]

The common factor in all these versions of the story is that they derive directly from Newton himself and we therefore have only his word that the story is true. Perhaps, one day during the summer of 1666, he did sit under a tree and see an apple fall and it was this, combined with a wealth of other factors, that inspired his theory. But it is also quite likely that the apple story was a later fabrication, or at least an exaggeration designed for a specific purpose – almost certainly to suppress the fact that much of the inspiration for the theory of gravity came from his subsequent alchemical work.

On a prosaic level, Newton's work as an alchemist was completely anathema to the traditional world of science and to society in general. But, beyond that, attempting to transmute base metals into gold – a preoccupation of the alchemists – was a capital offence. Even in old age Newton was determined to maintain his duplicity, both to protect himself and to preserve unsullied his hard-won image as the greatest scientist who had ever lived.

So, if these stories are false, how then did Newton arrive at the inverse square law for gravity – the first major development towards the elucidation of universal gravitation, the principle that all masses attract all other masses?

The first step was to use his mathematical studies in order to mould a set of general mathematical principles that he could use to investigate planetary motion. Since his earliest inquiries into basic mathematics, begun two years earlier, during the spring of 1664, he had managed to assimilate the entire canon of known mathematics and then to extend it into totally uncharted waters – 'For in those days I was in the prime of my age of invention,' Newton said of himself sixty years later.[13] He was familiar with the latest work on the mathematics of curves and the principle that tangents can approximate to the curve and allow certain calculations to be managed, but, like many mathematicians of the time, he wanted something more precise. In particular, he was interested in finding the area under a curve (the area between a curve and the x-axis) and a precise value for the curvature (or gradient) of a curve.

Scholars are in general agreement that the greatest influence upon Newton's own thinking about these problems came from his reading Descartes's *Geometry* during the summer of 1664, but others have pointed out that Isaac Barrow had also made some considerable progress with the mathematics of gradients and curves and that Newton may have learned a great deal from both men.

During 1665 and early 1666 Newton worked on these problems and devised a method of finding the exact gradient of a curve by a method which has since become known as *differentiation*. To understand this method we must first recall that a graph is a way of representing a set of values describing a situation. In the last chapter, the example of a ball falling from the tower was used to illustrate how a real situation can be described graphically. Equally, an algebraic equation is another way of describing a situation. In fact a graphical representation and an algebraic description can both represent the same thing. This means that the algebraic and graphical representations are paired, so manipulating equations by a suitable method can lead the mathematician to information about the curves these equations mirror.

Newton's greatest mathematical breakthrough was the realisation that a particular manipulation of a suitable equation could lead to a precise value for the gradient of the curve represented by that equation. This method of manipulation is the essence of differentiation. Another process carried out on a equation (a process since named *integration*) leads the mathematician to the area under a curve represented by that equation. The *calculus* is the overall term for these two processes of differentiation and integration, and together they are powerful tools for the mathematician and the scientist.

Although sometimes placed in his 'Woolsthorpe period', work on this development was actually begun while Newton was still in Cambridge. By his own account, he had begun to develop the calculus as early as February 1665.[14] His first mathematics paper, dealing with a mathematical process called the summation of infinitesimal arcs of curves (a major step towards a full realisation of the techniques involved in the calculus), was composed in May 1665.

Once he had a general method for the calculus, the next step was to apply it to the practical problem of planetary motion – how planets orbit the Sun, and the Moon the Earth, and how mathematical laws can represent these movements.

A thought experiment familiar to natural philosophers was that of the stone on a string. This may be visualised by imagining a stone attached to a string being whirled around the experimenter's head. In this model, one force draws the stone towards the centre of the circular path and another pulls the stone away. The Dutchman Christiaan Huygens called the first of these forces the *centripetal force* and the other the *centrifugal force*. The stone continues to travel in a circle around the experimenter's head because the two forces cancel out. If the string is cut the stone will fly off in a straight line at a tangent to the circle.

Using this as a basis, Newton created a thought experiment to determine a way of calculating the outward or centrifugal force experienced by an object travelling in circular motion. To begin with, he imagined a ball travelling along the four sides of a square inscribed in a circle.

He was able to calculate the force with which the ball struck one of the points of the circle (say, A), and by multiplying this by four (for the sum of the sides of the square) he arrived at a value for the force exerted by the ball in one circuit around the square. But a square is a poor approximation to a circle, and to arrive at closer values for the force the object would exert if it was travelling in circular motion Newton imagined polygons with increasing numbers of sides inscribed in the circle. The more sides the polygon possessed, the closer it came to describing the circle.

Using principles derived from his own recent mathematical developments, he eventually obtained a value for the force exerted by an object completing a single truly circular revolution.

From this calculation he could determine the force with which an object travelling in circular motion pulls away from the centre of the circle and hence the relationship between the force and the size of the circle (or orbit). Applying this to rotation of planets around the Sun, he concluded that 'the endeavour to recede from the Sun will

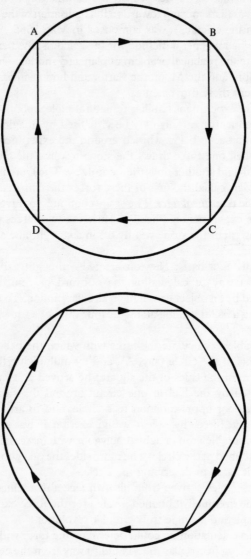

Figures 4 and 5.

be reciprocally as the squares of their distances from the Sun'.[15]★

This means that the distance and the force experienced by a planet receding from the Sun (or the Moon from the Earth) are related by an inverse square relationship. In other words, if planet A orbits the Sun at a certain distance, another planet B (of equal mass) orbiting at twice the distance will experience a receding force one quarter the value of planet A. If another planet of equal mass, planet C, orbits at a distance three times greater than A it will experience a receding force only one ninth the size of the force experienced by planet A.

In keeping with his lifelong habit of working with whatever materials were at hand, Newton began this thought process on the back of an old lease his mother had used some time earlier.[16] Surviving to this day, the piece of parchment shows a muddled collection of jottings and calculations which, although describing work that eventually led to a law of gravitation, at this stage (1666) merely illustrate his contention that planets experience a receding force from the Sun governed by an inverse square relationship. It was only later that Newton was able to equate this receding force with a force pulling planets towards the Sun and to realise that this pulling force is also governed by the same inverse square law.

Yet the thought that there existed an equal and opposite force which countered the receding force could not have been far from his thoughts – not least because of the familiar example of the stone on a string. And indeed, later in life, Newton dated his realisation of a force of gravity (the force pulling a planet towards the Sun) also acting by the inverse square law to the same time as he had conceived how the receding force could be calculated.

Just before he died, Newton wrote of the dawning of the idea in a letter to the Huguenot scholar Pierre Des Maizeaux:

I began to think of gravity extending to the orb of the Moon & having found out how to estimate the force with which a globe

---

★ Like all astronomers of the time, Newton was familiar with Kepler's discoveries that planets followed elliptical paths, but at this stage he worked with circular motion because the mathematics of elliptical motion offered insurmountable hurdles.

revolving within a sphere presses the surface of the sphere from Kepler's rule of the periodical times of the planets being in sesquialterate [3:2] proportion of their distances from the centres of their orbs, I deduced the forces which keep the planets in their orbs must be reciprocally as the squares of their distances from the centres about which they revolve.[17]

It is implicit here that the notion of an inward pulling force was in place at the time when he elucidated the magnitude of the receding force, but there is some debate about when this step really was made. There was a series of calculations made both on the lease document and in other documents written in Latin and almost certainly composed no earlier than 1667. (We know this because Newton never wrote in Latin before his return to Cambridge in 1667.) Together, these documents show a step-by-step development of the idea of both forces operating by the inverse square law. What is certain is that it was sometime before 1667 when Newton applied his method of determining the receding force to the case of the Moon's passage around the Earth.

The calculation was actually a very simple one. In order to work out the strength of the receding force of the Moon by his newly devised method, Newton needed to know the period of one revolution of the satellite. This was readily available from the work of contemporary astronomers: 27 days 8 hours. He also needed to know the distance between the Moon and the Earth, and the best estimate for this at the time was that the distance to the Moon was sixty times the Earth's radius. Unfortunately, however, the best figure then known for the radius of the Earth, based upon a figure calculated by Galileo, was inaccurate – 3,500 miles (over 400 miles too small). Consequently, the figure Newton obtained for the receding force experienced by the Moon was incorrect and did not demonstrate the inverse square relationship accurately.

Disheartened and exhausted by the effort, Newton abandoned the idea for several years, concluding that he had oversimplified the matter and that there must be some other force, perhaps related to Descartes's vortex theory, that could explain planetary dynamics – something he had overlooked. It was not until 1685, when he was preparing the *Principia*, that he finally used a more accurate figure for the Earth's radius (found by a Frenchman, Jean Picard, some years earlier) and consequently found that the inverse square relationship worked perfectly.

It is clear from this succession of calculations alone that Newton did not realise the entire theory of gravity in one flash of inspiration. The Woolsthorpe years provided a foundation, both conceptual and mathematical, upon which during the following twenty years he constructed a detailed theory based on both alchemical knowledge and experimental verification. (See Chapters 7 and 9.) All of these elements were essential. If the mathematics had not been developed during the 1660s, Newton's intuitive grasp of the nature of planetary motion would have remained little more than a good idea. Without his in-depth knowledge of alchemy (which he practised during the 1670s and '80s), he would almost certainly never have expanded the limited notion of planetary motion as he saw it in 1665/6 into the grand concepts of universal gravitation, of attraction and repulsion, and of action at a distance. Finally, if the experimental evidence had not been gathered, then Newton's theories, even if substantiated by mathematics, would not have carried the weight they did in his *Principia*, nor would they have so readily inspired the practical application of mechanics and the laws of motion which led, a century later, to the Industrial Revolution.

In March 1666 Newton returned to Cambridge briefly, but by June the plague again threatened and he was forced to return to Woolsthorpe. During the summer of that year he decided the time was right for him to claim officially his right to gentleman status, and he attended the Herald's visitation at Grantham, making the process legal. For the first time he wrote, 'Isaack Newton of Wolstropp. Gentleman, age 23.'[18]

Although this may have seemed premature, Newton's credentials were quite sound. Certainly, his father could not have taken the title of gentleman, but Isaac junior was not only related to the lower gentry, via the Ayscoughs on his mother's side, he was also in line for a substantial inheritance upon Hannah's death. Above all, he was a scholar – a graduate of Trinity College, Cambridge.

Little is known of this second period in Woolsthorpe, between June 1666 and March the following year. He probably divided his time between studying at home and visiting Babington (who lived close by) and the Clarks in Grantham. It would be fair to assume that Hannah still harboured hopes that his stay would be permanent. But, though Newton could still work efficiently in Woolsthorpe, the rural lifestyle had never suited him and he would have had no intention of remaining away from the university any longer than caution dictated. And by the summer of 1666 one of the factors enabling his return to Cambridge was about to make its mark on history. More than 100 miles away, in London, the Great Fire was about to eradicate the last vestiges of the plague from the capital, and the disease soon began to recede elsewhere.

The university reopened in early 1667 and Newton was able to return to Cambridge and to begin the struggle to obtain both his Master of Arts degree and a guarantee for a future at Trinity – the all-important fellowship.

If fellowships were awarded according to merit, then by any measure Newton would have been worthy of one; but they were not. Indeed, his scientific achievements, even if known to the college authorities, would have had little influence. His success depended more upon available vacancies within the hierarchy and knowing the right people.

The acquisition of a fellowship was of the utmost importance to Newton. Without it, he would have been unable to continue at Cambridge and would perhaps have been forced into obscurity as a farmer or encouraged to accept a rectorship in some isolated rural spot. He had not shone academically, and had not been 'a face' around college; nor was he very wealthy. His tutor, Benjamin Pulleyn, had been helpful in securing the first stage of his pupil's academic ascent and had probably introduced him to Barrow; but, although the Lucasian Professor proved imperative to Newton's later success, he could provide little help in 1667. Again Newton made use of his association with Humphrey Babington, who in 1667 had been promoted to senior fellow – one of eight men who answered directly to the Master and selected new fellows and minor fellows. But, even with Babington's help, Newton might still have foundered if it had not been for a series of serendipitous events. Because of the plague, there had been no fellowship elections during 1665 and 1666. Even so, when Newton returned to Trinity in early 1667 there were only nine positions to be filled from a total complement of some sixty academics.

By chance, that year the number of vacancies had been inflated by several retirements and a death occasioned by events which vividly convey the atmosphere of Restoration Trinity. A senior fellow had been recently removed on the grounds of 'mental aberration' of some unknown variety, and two other fellows had been forced to retire through injuries sustained after falling down the staircase leading to their rooms while in a drunken stupor. A fourth, the poet Abraham Cowley, had died after catching a fever brought on by a night spent sleeping in a field after a bout of heavy drinking. Luckily for Newton, this created a lengthy enough list to give him an opening.

After a scholar had been accepted as a candidate for fellowship, he underwent a succession of gruelling examinations in order to test his suitability. The exams took place during the last days of September, and, despite the continuing distractions of his extracurricular activities, Newton devoted as much time as he could to his preparation after arriving back at Trinity on 25 March.

The examination consisted of three days of oral questioning in the college chapel, followed on the fourth day by a written paper. On 1 October a tolling bell summoned the candidates to learn their fate before the senior fellows: 'by the tolling of the little bell at 8 in the morning the seniors are called & the day after at one o'clock to swear them that are chosen'.[19]

With the fellowship came both responsibilities and privileges. The most important benefit was that Newton now had a job for life and could continue to study at his leisure, pursuing whatever academic route he wished. The college provided a stipend of £2 and a small allowance for ceremonial costumes and academic robes. He was also given a room free of charge, and upon passing his Master of Arts examination the following spring he was accepted as a major fellow and was granted an increase in his stipend to £2.13s. 4d. as well as an improved livery allowance of £1.3s. 4d.*

Newton was clearly delighted by the turn of events. He had pushed back barriers in science and mathematics, had made truly significant discoveries, had declared his social status as a scholar and a

---

* As well as his tiny stipend from Trinity, Newton earned in the region of £80 per annum from his properties and was entitled to a share in surplus annual dividends from college endowments which, at the time, amounted to an estimated £25 per year. He also received an annual allowance from his mother.

gentleman, and had risen through the ranks of the academic élite. Now he knew he was out of reach of the restraining hands of his past. And, for the only time in his entire life, he let his hair down.

For the following year, as he acquired his MA and rose to the rank of major fellow (in March 1668), he led the life of a comfortably placed and successful young man. Uncharacteristically, he visited taverns with Wickins, played bowls, and cast aside the image of the single-minded Puritan, even recording in his notebook a loss of fifteen shillings playing cards.[20] He paid for his and Wickins's rooms to be decorated by a professional painter, and bought new furniture, carpets and pictures and a whole wardrobe of expensive clothes.*

Because Newton was footing most of the bill, he got to choose the colour scheme and the details of the decoration, revealing a new personality trait – an obsession with the colour crimson. New cushions, chairs, bedspreads and curtains were almost all dominated by crimson. He surrounded himself with the colour, and it was a fixation that persisted into old age. In a list of possessions drawn up by Catherine Conduitt after her uncle's death, there are recorded 'a crimson mohair bed complete with case curtains of crimson Harrateen' and, in the dining-room, 'a crimson settee'. Other listed items included crimson drapes and valances in the bedroom, a crimson easy chair, and six crimson cushions in the back parlour of the house.[21]

Why Newton was so struck with the colour we will probably never know, but the obsession went back a long way. As a teenager, in 1659 he had recorded in the Morgan Notebook some three dozen recipes for the formulation of coloured dyes, and the vast majority of these were for different shades of red. An example is 'Take some of the clearest blood of a sheep & put it into a bladder & with a needle prick holes in the bottom of it then hang it up to dry in the sun, & dissolve it in alum water according as you have need.'[22] Newton's optical experiments are also suggestive of his interest in colour, but the reason for his particular interest in crimson is nowhere explained. It would be easy to draw from this predilection conclusions

---

* It appears from college records that he never occupied the rooms first given to him as a minor fellow but, like a number of other fellows of the time, sublet them. According to the Fitzwilliam Notebook, in which he recorded his financial transactions, he received a rent of £1.11s. 0d. for them during 1668.

about his underlying psychological drives – that he was obsessed with the colour of blood, for example – but such ideas contribute little towards a sensible understanding of the man. Rather than revealing any deep insights into his motivations or neuroses, a fascination with crimson probably demonstrates little more than an odd quirk of personality. It is interesting because it shows Newton to have a human side – a weakness for something so materialistic as colourful decorations.

Even more intriguing is the way he was able to let go of his normal self-imposed austerity for a short period and then to snatch it back, never to loosen his grip again. Shortly after this relatively brief, vaguely narcissistic period, he was once again clamming up and turning in upon himself almost as though it had acted as a catharsis, a cleansing. As suddenly as he had shown his face in Cambridge inns early in 1668, he withdrew again into isolated scholarship, and it was now that he began to forge one of the most important relationships of his early academic career.

In many respects, the personality and experiences of Isaac Barrow were the antithesis of Isaac Newton's. Barrow was born into a prosperous merchant family in 1630, entered Trinity College as a pensioner in 1647, and from an early age was seen as a promising academic. The problem with Barrow was not lack of ability in what was even then an intellectually mediocre environment, but his political and religious views: he was both Puritan and royalist.

Over a decade before the Restoration, and in a university dominated by parliamentarians, Barrow's unorthodox and highly vocal political views were unpopular. But for most of the time his academic talents saved him from persecution. He became interested in the works of Bacon, Gassendi, Descartes and others, and, although he did not advance science in the way Newton later did, he was a competent and versatile mathematician and philosopher.

In temperament he was quite different from Newton: he rather enjoyed teaching, and as a young lecturer he was often witty and light-hearted in his approach – at least by the standards of the time. In stark contrast, Newton's assistant Humphrey Newton (who was not related to the scientist) claims he only ever saw his master laugh once: 'It was upon occasion of asking a friend he had lent Euclid to read, what progress he had made in that author, and how he liked him? The friend answered with a question of his own: What use and benefit in life that study would be to him? Upon which Sir Isaac was very merry.'[23]

Unlike Newton, Barrow smoked to excess, was slovenly of appearance, and was also fond of the sound of his own voice: sometime after he became Royal Chaplain to Charles II in 1669, he is said to have delivered a sermon on the subject of charity lasting three and a half hours.

He also travelled widely. In 1655, a few years after obtaining a fellowship at Trinity, he found that his politically unpopular views were threatening to damage his academic career. Securing a travel grant from the college to the tune of £16 per annum for three years, he embarked upon a four-year grand tour of Europe. It was probably during his travels that he first became acquainted with alchemy, an interest he pursued on and off through the rest of his career, passing on his knowledge and contacts to Newton when they became colleagues.

Back in Cambridge and with Charles II taking the throne in 1660, Barrow found the political atmosphere more sympathetic. Stories of his adventures – including besting an argumentative Turk in hand-to-hand combat, and out-arguing a group of Jesuits he had encountered in Paris – had travelled ahead of him, and he was seen at the university as something of an heroic figure. He was quickly appointed Regius Professor of Greek at Trinity, and in 1662 he became the Gresham Professor of Geometry in London, appearing on the list of fellows who first founded the Royal Society (although he did not play an active role there and attended only rarely).

Then, two years later, in 1664, Barrow was offered a new professorship. A fellow of St John's, one of the university's Members of Parliament, Henry Lucas, had succeeded in establishing a new chair at Cambridge with a royal warrant approved by King Charles – the Lucasian Professorship of Mathematics. Lucas wanted the position to go to someone of genuine ability. Barrow, who was perceived as perhaps the most accomplished academic in the college, had little trouble in securing the position – a chair today held by Professor Stephen Hawking.

The professorship came with a handsome remuneration of £100 per year and a minimum of teaching responsibilities, but the holder was not allowed to hold any other academic positions concurrently. Initially, this suited Barrow. He did his best to meet the teaching conditions – an annual set of lectures, and the annual delivery of written copies of at least ten of his lectures to the university library. The fact that on most occasions not a single student appeared for the statutory lectures was probably to do more with the complexity of the material and the low academic standard of the students than with Barrow's delivery. He taught what interested him, mainly mathematics and optics – strong meat for most of the spoilt sons of the wealthy who wanted merely to cruise through their degree course or to pursue a path of least resistance ending with a rectorship in a quiet village, or entry into the family business.

Fortunately for Barrow, the regulations surrounding the position were not monitored strictly. He rarely delivered any form of written lecture, and, demoralised by neglect, he could only bring himself to teach one term in three each year.

After serving as the Royal Chaplain to Charles II, Barrow was briefly Master of Trinity between 1672 and his untimely death in 1677. During his tenure, he almost single-handedly initiated the construction of the Trinity College library, designed by his friend Christopher Wren, and he is seen by posterity as one of the best administrators of the college during the Restoration. Yet his fame stems undeniably from his association with Newton.

They may never have been close friends, but they enjoyed a mutual respect. If there is any truth in the story of Newton having first met Barrow when he was found lacking in basic mathematics, then any misgivings were soon forgotten. Astonishingly, only five and a half years lay between this indifferent first encounter and Barrow handing Newton the Lucasian Professorship. Just as the older man had returned to Cambridge a seasoned and well-travelled hero in 1659, and quickly inspired his superiors, Newton, upon returning from his own intellectual odyssey in 1667, quickly made his mark with Barrow.

There is no doubt that, despite Newton's natural misanthropy, he networked very cleverly – not with natural charm, but solely through the impressive power of his intellect. Babington had been the key – easing Newton's transition from his youth in Woolsthorpe to his fellowship at Trinity – and Henry More had supplied intellectual support. Both Babington and More knew Barrow, so it was a small step for the Lucasian Professor to become Newton's guide through the higher reaches of the Cambridge academic system.

By the spring of 1667 Newton was losing interest in mathematics and becoming increasingly drawn towards alchemy. More, Babington and Barrow are both known to have been interested in the clandestine art, and Newton had read the work of the great experimenter of the era, Robert Boyle, who had already begun to study the subject, attempting to approach it from a scientific perspective. Keen as he had always been to unite disparate fields of study, Newton found a natural source of inspiration in alchemy. The experiments he had conducted as a young boy in the apothecary shop, making dyes and nostrums, were merely a preparation.

Soon after Newton was made a fellow, Barrow became convinced that during his two years in Woolsthorpe the young man had reached conclusions of real scientific relevance. But, at the same time, he became aware of Newton's extreme reluctance to publish his work, and that ideas had to be teased from him. Even at this stage in his career, Newton harboured a deep-rooted suspicion that others would try to steal his discoveries. He also believed strongly that material should be released only after it had been fully developed. Having long disapproved of others confusing *a priori* concepts with verifiable theories, he did not want to be viewed in the same light. An example of this was his distaste for Descartes's belief in the vortex theory, which by the mid-1660s Newton saw as a vague and unverifiable set of ideas, an opinion that led him to say of his own efforts, 'hypotheses non fingo' – 'I do not invent hypotheses.'[24]

This attitude lies at the heart of why twenty years were to pass between Newton's realisation of the basic concept of gravity and the writing of the *Principia*. He never believed in delivering half-baked concepts or proposing ideas that had not been substantiated either by mathematics or by experiment. As a young man in Woolsthorpe, he knew he had discovered something important, but he may even then have realised he was merely scratching the surface of the subject. Barrow, on the other hand, was acutely aware of the need to publish, or at least to communicate with others within the scientific community.

An opportunity to persuade Newton of this soon presented itself. In September 1668 one of Barrow's contacts in London, a fellow of the Royal Society, John Collins, a great facilitator who brought together scientists from all parts of Europe via letters and publications, sent him the latest work of the Danish mathematician Nicholas Mercator. The book was entitled *Logarithmotechnia* and described a new method of calculating logarithms. Barrow passed the book on to Newton, knowing perfectly well that his protégé had gone far beyond this material early in the Woolsthorpe period, perhaps even before leaving Cambridge in 1665.

Understandably, when Newton read Mercator's account he was horrified, and he agreed to write a short paper immediately, outlining what he had achieved along these lines early in his mathematical studies. The result was a short treatise entitled 'De Analysi per Aequationes Infinitas' – 'On Analysis by Infinite Series'. But even then he refused to let Barrow make any attempt to publish it through John Collins or anyone else.

In his first response to Collins, Barrow commented, 'A friend of mine here, that hath a very excellent genius to these things brought me the other day some paper, wherein he hath set down methods of calculating the dimensions of magnitudes like that of Mr Mercator concerning the hyperbola, but very general.'[25] A month later, Barrow appears to have talked Newton into letting him send 'De Analysi' to Collins, but he was forbidden to mention the author by name:

> I send you the paper of my friend I promised, which I presume will give you much satisfaction; I pray having perused them so much as you think good, remand them to me; according to his desire, when I asked him the liberty to impart them to you. And I pray give me notice of your receiving them with your soonest convenience . . . because I am afraid of them; venturing them by post.[26]

Naturally, Collins wanted to publish 'De Analysi', but nothing would convince the young virtuoso to agree. Barrow persisted, however, and gradually and grudgingly Newton at least allowed him to reveal his name to Collins: 'I am glad my friend's paper gave you so much satisfaction. His name is Mr Newton; a fellow of our college, & very young . . . but of an extraordinary genius & proficiency in these things.'[27]

Barrow followed this by passing on to Collins Newton's permission to show the work to the President of the Royal Society and a limited number of interested intellectuals, but neither Collins nor Barrow ever did succeed in persuading Newton to release the work to the printers. In fact 'De Analysi' remained unpublished until 1711 –

some seven years after the *Opticks* and a quarter of a century after the *Principia*.

We might conclude from this incident that Newton was experiencing youthful insecurity, but this does not square with his later attitude of extreme reticence towards publishing, nor does it tally with his overdeveloped sense of self-importance and confidence. His 'De Analysi' was no mere hypothesis, nor was it a controversial piece of work, so he could not claim fears of misinterpretation or of charges of presenting unprovable or half-formed ideas.

The fact that others had discovered similar techniques and that Newton was attracting the interest of intellectuals outside Cambridge must have done much to impress Barrow. During 1668 and 1669 they worked closely together, Newton acting the part of the older man's assistant. Barrow asked him to edit his lectures on his theory of optics, and thanked him with a 'Letter to the Reader' in which he called Newton 'a man of great Learning and sagacity, who revised my copy and noted such things as wanted correction'.[28]

Ironically, Newton knew the ideas contained in Barrow's manuscript to be completely wrong, yet he went along with his older associate's request to help him prepare it for press. Some have seen this as callous on Newton's part, arguing that he should have put Barrow straight on his naïve ideas, but this is quite unrealistic. It is easy to see Newton at this time as the powerful, intellectual giant of his later years, but in fact he was still a young, totally unknown fellow. He had created the beginnings of one of the greatest revolutions in the history of science and had risen to peaks of mathematical awareness no others had yet scaled, but no one save perhaps Barrow and maybe More and Babington was aware of any of this. It is quite unreasonable to suggest that Newton could have quietly informed Barrow that he had completely misunderstood the nature of light and was wasting his time lecturing and writing upon the subject.

There is a further reason why Newton was right to hold his tongue. Barrow, Newton knew, was a very ambitious man and had made no secret that he was thinking of giving up the Lucasian chair far sooner than many would have expected. Even as early as 1668 Barrow had his eye on the mastership of Trinity. He was in his prime intellectually and had achieved a rapid ascent through the academic hierarchy; he saw no reason why he should stop there. He made an attempt to have the conditions of his professorship changed, so he could accept other administrative positions, but failed. When the opportunity arose to become Chaplain to Charles II, Barrow quickly realised that this would offer a faster route to achieving his ultimate goal. Newton was equally quick to realise he would make a perfect successor to the Lucasian Professorship and did everything he could to facilitate this.

Appointment to the Lucasian chair was determined by the executors of Lucas's estate, but they had long grown used to deferring all decisions on academic matters to Barrow. He, of course, recommended Newton, and on 29 October 1669 Isaac Newton, not yet twenty-seven years old, became the second Lucasian Professor of Mathematics at Cambridge – a position he continued to hold for some years after he had left the university in 1696.

It had taken Newton just eight years to make the transition from fresher to Lucasian Professor. And, as well as ascending the academic ladder with astonishing speed, he had embarked upon a set of scientific and mathematical projects that would etch his name in the granite of history. As he stood on the academic pinnacle of his life, other facets of his intellect were beginning to surface. Ahead of him lay years of turmoil – work that would take him to the *Principia* and beyond, to obsession and to the brink of self-destruction. On the verge of acceptance into the scientific community, Isaac Newton was set to embark upon a new, duplicitous, existence. Soon he would become the Cambridge professor who was also 'the last wonder-child' of the Magi.

CHAPTER 6

The Search for the Philosophers' Stone

The changing of bodies into light, and light into bodies, is very comfortable to the course of Nature, which seems delighted with transmutations.

ISAAC NEWTON[1]

On a freezing-cold late-December night in 1666, a black-haired, scruffily dressed stranger appeared at the house of the philosopher Johann-Friedrich Schweitzer (otherwise known as Helvetius) in The Hague, Holland.

Helvetius was a famous and outspoken disparager of magic and alchemy who had published an attack on the English alchemist Kenelm Digby;[2] he was therefore perhaps a little surprised that the stranger, who professed to be a master of the magic arts, had decided to pay him a visit. Nevertheless, he made the visitor welcome and led him into his study, where the stranger showed him an ivory box in which he kept three pieces of stone the colour of brimstone. These, he told Helvetius, were fragments of the legendary philosophers' stone and had the power to transform any base metal into the highest-quality gold. In fact, he declared, as he laid the stones upon one of Helvetius's books, these stones were able to produce over twenty tons of gold.

The philosopher was naturally doubtful and asked the stranger to demonstrate the power of his stones, but he refused. Helvetius next tried to persuade his guest to leave a fragment, so that he could experiment himself, but again the stranger refused; he said that he would return in three weeks and perform a demonstration of transmutation so his doubting host could witness the power of the stones then.

The man left, and Helvetius, being a shrewd fellow, scraped up as many flecks of dust as he could from where the stones had been placed. He then attempted the process the stranger had described with some pieces of lead, but the effort failed – the metal remained ordinary lead. But, true to his word, three weeks later the visitor reappeared and this time rewarded the philosopher with a tiny piece of one of the stones, no larger than a mustard seed. When Helvetius suggested that nothing could be done with so small a fragment, the stranger broke it in half, threw one piece into the fire, and handed back the other half, saying that that was enough to make several ounces of gold.

That night Helvetius and his wife carried out the alchemical process described by the stranger. They melted down some lead coins and threw the tiny grain of stone into the molten metal. When the crucible had cooled they removed it from the fire, and there, at the base of the container, lay a piece of gold which they later found weighed six ounces.

Amazed but still doubtful, they took the nugget to the best gold-smith of the region for his validation. After a brief examination of the metal he declared it to be of the finest quality and offered them fifty florins per ounce for it.

News of the story spread far and wide, and the home of Helvetius rapidly became the centre of the alchemical world, with devotees flocking to The Hague from all corners of Europe. Spinoza, who knew Helvetius well, was initially doubtful of the story and visited the goldsmith to check the story. The goldsmith turned out to be a minter for the Duc d'Orange and clearly knew his trade, and, after seeing the gold for himself, Spinoza, a man not known for his gullibility, was convinced.

This is just one story from the vast canon of alchemical anecdote resulting from over 2,500 years of effort to find the philosophers' stone and an estimated 100,000 books written on the subject; but it is particularly interesting because the protagonist actually appears to succeed in producing what is believed by his contemporaries to be gold. In most other cases the alchemist comes to a sticky end, dying in abject poverty or at the hands of an executioner employed by a disappointed monarch.

Ironically, although Newton was largely responsible for the development of the scientific enlightenment which swept away the common belief in magic and mysticism, he created the origins of empirical science and the modern, 'rational' world in part by immersing himself in these very practices.

Just how far he went with his interest in the occult is still open to debate. We shall see that his fascination with alchemy was a major influence in the development of his ideas about gravity. It is also clear that he was interested in a synthesis of all knowledge and was a devout seeker of some form of unified theory of the principles of the universe. Along with many intellectuals before him, Newton believed that this synthesis – the fabled *prisca sapientia* – had once been in the possession of humankind. As he commented in his notebook:

> So then it was one design of the first institution of the true religion to propose to mankind by the frame of the ancient temples, the study of the frame of the world as the true temple of the great God they worshipped . . . So then the first religion was the most rational of all others till the nations corrupted it. For there is no way (without revelation) to come to the knowledge of a deity but by the frame of nature.[3]

Newton's *raison d'être* was to rediscover this 'frame of nature'. For this reason, on an intellectual level, he considered no avenue of research beyond his probings, no stone unimportant enough to be left unturned, no theory beyond the pale.

A generation after Newton's death, David Hume described this desire to unravel ultimate truth, or to rediscover what had supposedly been lost, when he wrote:

> We are placed in this world, as in great theatre, where the sources and the causes of every event are entirely concealed from us; nor have we either sufficient wisdom to foresee, or power to prevent those ills, with which we are continually threatened. We hang in perpetual suspense between life and death, health and sickness, plenty and want; which are distributed among the human species by secret and unknown causes, whose operation is oft unexpected, and always unaccountable. These *unknown causes*, then, become the constant object of our hopes and fears; and while the passions are kept in perpetual alarm by an anxious

expectation of the events, the imagination is equally employed in forming ideas of these powers, on which we have so entire a dependence.[4]

As a natural philosopher and adventurer of the intellect, Newton would have gone anywhere, done anything, to find the truth; but in practice he could not. He would have sympathised with Roger Bacon's belief that 'Though everything is not permitted, everything is possible', and was restrained by the same limitations as Bacon, half a millennium before him – religious convictions that clashed with the search for pure knowledge wherever it might lead.

Newton learned very early in his search, probably before he left Cambridge during the plague, that the study of alchemy was fraught with difficulties and required great commitment. Yet something must have convinced him of its value. He must have decided from his preliminary reading, and probably from conversations with such men as Babington and More, that alchemy could offer a route to unification. He must have come to the conclusion that, if there was an underlying principle to the occult, then it was this: that, if studied from an intellectual standpoint, the occult could act as a glue, a unifier of fundamental principles; that it could be rationalised and logicalised. Although he could not have visualised it at the time, what he did was to transmute what we now see as the irrational beliefs of the age into a new approach – science.

In order for us fully to appreciate what it was that Newton was trying to do before his furnace and throughout nights made sleepless by obsession and almost demented self-motivation, we should do what he himself did. The first requirement of the adept of alchemy is study – to read everything he can lay his hands on in an effort to assimilate the knowledge of all previous travellers along the path to the philosophers' stone. For several years before he touched a crucible or tended a furnace, Newton buried himself in the literature and the lore of the subject, in order to develop some form of intellectual foundation for the discipline.

It has been claimed that Moses was an alchemist, but this is probably nothing more than an example of those who practise alchemy attempting to add credence to their vocation by giving it a timeless and weighty pedigree. That said, the subject is certainly an ancient one.

The first known alchemical work was probably a book called *Phusika kai Mustika – On Natural and Initiatory Things* – written by one Bolos of Mendes, which contains instructions for the making of dyes and the working of precious metals and gems.* Exactly when this was written is unclear, but it probably dates from around 250 BC. It does not describe transmutations or processes to produce the *elixir vitae* (or elixir of life) – these were later obsessions – and so in many respects it is atypical of the material that alchemists studied in translated form in Europe some fifteen hundred years later.

Alchemy adopted a recognisable form through a gathering of cultures and influences in the Egyptian city of Alexandria, founded by Alexander the Great in the fourth century BC. In this form, which became known as Hellenic alchemy, Greek, Syrian, Persian and Egyptian cultures met and cross-fertilised to produce the foundations of the art as it was handed down to later followers. Between the founding of Alexandria and the decimation of its famous library towards the end of the fourth century AD, some of the basic techniques of alchemy were first devised and popularised in texts credited to Democritus, Isis, Moses, Hermes and many others. These techniques included distillation, which is still used extensively today and stimulated the development of many common pieces of equipment such as the *ambix* or *alembic* (a glass or copper vessel), the *solem* (a distillation tube) and the *tribikos* (a funnel).

The Alexandrian alchemists also understood the importance of controlling the amount of heat applied to a substance during alchemical processes, and appreciated the significance of the various colour changes observed when liquids were mixed and materials were heated. All of these were crucial to the alchemists of Europe, and lay at the heart of much of what Newton and others observed and tried to explain. Indeed, Newton knew a great deal about both ancient

---

* In this discussion I will be referring to the canon of alchemical teaching that emerged in Europe. Eastern alchemy, as practised by the Chinese and the Indians, developed contemporaneously with the earliest emergence of the subject in the Near East.

and modern furnaces, covering several pages of his notebooks with learned discourses on the subject.[5]

Like all European alchemists from the Dark Ages to the beginning of the scientific era and perhaps beyond,* Newton was motivated by a deep-rooted commitment to the notion that alchemical wisdom extended back to ancient times. The hermetic tradition – the body of alchemical knowledge – was believed to have originated in the mists of time and to have been 'given' to humanity through supernatural agents. The very name 'hermetic tradition' derives from the god Hermes, the god of travellers and adventurers, and a mythical figure known as Hermes Trismegistus (Hermes the Thrice Great) is credited with the composition of some of the most important early works of the art. Venerated by alchemists throughout history, it was said of Hermes Trismegistus that he 'saw the totality of things. Having seen, he understood. Having understood, he had the power to reveal and show. And indeed what he knew, he wrote down. What he wrote he mostly hid away, keeping silence rather than speaking out, so that every generation coming into the world had to seek out these things.'[6]

Naturally, it was in the alchemists' interests to further the idea that their art had mysterious, ancient foundations, because it added even greater exclusivity and self-importance to their practices. It was also essential that their techniques be hidden or, as the author of the above passage puts it, 'every generation coming into the world had to seek out these things'. In reality, however, almost all the techniques used by the alchemists of Europe, and indeed the texts they held so sacred, did not date back to Old Testament times and beyond, but originated in Alexandria around 200–300 AD.

The earliest theoretical basis for alchemy in the West stems from Aristotle's notion of the four elements and the concept that one material may, under the correct conditions, be converted, or transmuted into any other by adjusting the proportions of its constituent

---

* It may come as some surprise that there remain believers in alchemy at the end of the twentieth century. Although discredited by the advent of empirical science, alchemy survived the Enlightenment, and has maintained a following, circumventing Victorian rationalism, developments in technology and the atomic theory.

elements. Further to this, Aristotle also believed that each element had its own special characteristics and that these were each related to human emotions and endowments. So, earth is present in black bile, which is associated with melancholy. Water is manifest in phlegm; too much of this produces laziness. Fire is related to the blood and therefore passion, while air is present in yellow bile and linked to anger. According to this hypothesis, any substance can be transmuted into any other by changing the proportions of the four basic elements from which it is made. In this way, it was believed, dull and relatively valueless lead could be changed into precious gold by changing the proportions of fire, earth, air and water within it.

In terms of technique, alchemy was greatly developed by the Arabs following the fall of Alexandria. So much had been lost in the destruction of that city's great library that a mere outline of the subject survived, but from this a slightly altered form of alchemy developed. Early Syriac texts containing the bulk of the seminal works were translated into Arabic and soon spread beyond the Near East. But the most important changes to alchemy came through development not so much of processes or special laboratory methods but of the spiritual and philosophical foundations of the subject.

What later came to be seen as the central pillars of alchemy – the concepts of the philosophers' stone and the elixir of life – came not from the ancient texts supposedly handed down by figures such as Moses: both concepts came from China. The ancient Chinese may actually have been the first alchemists anywhere in the world, and they were certainly the first to try to create a magical material able to transmute matter. They were also the earliest recorded seekers of a potion which could restore life or endow eternal youth. During the fourth century AD they even tested their concoctions on convicted criminals – human guinea-pigs.

Even the fundamental principle of Aristotle's four elements was modified by the Arabs of the fourth century onwards, hundreds of years before it reached Europe. Instead of fire, earth, water and air, metals were believed to be made of just two basic materials – sulphur and mercury. As in Aristotle's hypothesis, it was the proportions of these two basic elements that determined the properties of a substance. If these proportions could be changed, dull lead could be transmuted into coveted and venerated gold.

The surprising thing is that so many learned men, long before

Newton appeared on the scene, could accept such *a priori* assumptions and in many cases give up their lives and careers in pursuit of something so elusive. Even allowing for news (and therefore historical records and tales) travelling only slowly, it seems strange that stories of endless failure, not to mention the sheer randomness of the effort, did not influence enough people to draw a halt to the whole enterprise.

One can argue that during pre-scientific times, in a world seemingly controlled by external supernatural forces – a world in which man appeared a frail creature surviving at the whim of all-powerful Nature – alchemy did not seem so unreasonable. Furthermore, the expected rewards were so great that many may have seen the effort as worthwhile. Perhaps most alchemists were aware of the failures of the past but thought that a defective aspect of technique had caused them. There is also little doubt that human vanity played a key role in the history of alchemy, providing a seemingly endless supply of men (and the occasional woman) willing to attempt the impossible. Many started out with good intentions and believed they could stop when they had proved to their satisfaction that they were chasing moonshine. Unfortunately, most could not – until it was too late. History is littered with the desiccated lives of those who tried to manipulate Nature for wealth beyond the dreams of avarice. Tellingly, there are no comparable tales of anyone who succeeded in any truly significant sense.

With the decline of Rome, western Europe descended into the Dark Ages, and it was not until the eleventh century that learning started to return via the migration of Arabic philosophy and science and the beginnings of trade between East and West. As a result, many of the early alchemical works originating in Alexandria and later modified by the Arabs were translated into Latin and circulated throughout the Continent.

In 1460 Cosimo de Medici, a Florentine duke, had sent emissaries around the world in an effort to track down ancient manuscripts about the hermetic arts. A monk came to him claiming that he had in his possession a work written by none other than Hermes Trismegistus himself and dating from the time of the ancient Egyptians. It was not until 1614 that the manuscript – known as the *Corpus Hermeticum* – was found to be no older than the second or third century AD, but during the intervening period it inspired several generations

of alchemists throughout the Continent and beyond and was probably the single most important factor in the huge growth of interest
in the hermetic tradition and occult practices during the Renaissance.
The manuscript was written from the viewpoint of a seeker of truth
who is led to the wonders of the universe by an omnipotent being;
it began:

> Once upon a time, when I had begun to think about the things
> that are, and my thoughts had soared high aloft, while my bodily
> senses had been put under restraint by sleep – yet not such sleep
> as that of men weighed down by fullness of food or by bodily
> weariness – I thought there came to me a being of vast and
> boundless magnitude, who called me by name, and said to me,
> 'What do you wish to hear and see, and to learn and to come
> to know by thought?' 'Who are you?' I said. 'I,' said he, 'am
> Poimandres, the Mind of the Sovereignty.' 'I would fain learn,'
> said I, 'the things that are, and understand their nature, and get
> knowledge of God.'[7]

Many of the great names of European medieval and Renaissance
philosophy have been associated with alchemy, and during this pre-
scientific era the distinction between what would later be refined into
'science' and what was clearly 'magic' was blurred. It is also clear that
many early technologies and elements of pre-Newtonian scientific
knowledge were intermeshed with some of the more bizarre notions
of the Alexandrian magi. The thirteenth-century philosopher Roger
Bacon, who wrote *Speculum Alchimiae – The Mirror of Alchemy*
– published in 1597, is said to have rediscovered gunpowder, which
was probably first developed by Chinese alchemists over a thousand
years earlier. He also drew designs for a telescope several hundred
years before the Dutch astronomer Hans Lippershey reinvented the
device in 1608.

Other famous adepts who mixed what would now be considered
'respectable' science with magic were two contemporaries of Bacon:
Albertus Magnus and his pupil Thomas Aquinas. As well as
developing notions of natural philosophy such as the aphorism 'Like
seeks like' and theories about the nature of fire which were worthy
for their time, Albertus and Aquinas were also said to have together
used the philosophers' stone and a mysterious *elixir vitae* to develop

automata that could speak and conduct the duties of a domestic servant.[8] True to the spirit of the age, it was not enough that a philosopher should be able to understand how things worked: he had to perform feats of 'magic' to prove himself. Albertus was believed to be able to control the weather and to influence the seasons, to be capable of conjuring up a thunderstorm for his enemies or a sunny day in winter for his friends. Rather than by formulating theories or reasoned hypotheses, it was in this spirit of wonder and a belief in their ability to interact with the forces of Nature that most alchemists worked. Instead of being content to help develop a coherent philosophy or a 'science', they pursued their own fantasies and their own desires (in most cases the production of large amounts of gold), and they helped to defy any attempt to formalise their art by following these dreams in distinctly individual ways – only rarely do alchemical tracts agree upon any method or technique.

For some 500 years, from the twelfth century until Newton's time, Europe was the new centre of the alchemical world – a place where sages and 'wise men' travelled freely from state to state in search of elusive but greatly prized wonders. Many alchemists wrote of their adventures and their experiments, but almost always their recipes were coded so that others could not copy them without first gaining insights into the art and undergoing special initiation. Some spent their entire lives attempting to decode the works of the masters and adding their own interpretation to ideas handed down to them. In some countries and at certain times they were tolerated and even encouraged and financed by the ruling monarch; in other places alchemists and magicians were reviled and their practices deemed illegal.

One alchemist, John Aurelio Augurello, who lived in Italy during the fifteenth century, presented the then Pope, Leo X, with his latest alchemical work, *Crysopeia*, which described the process of making gold. Having dedicated the book to Leo, Augurello was hopeful that the Pope would return the favour with a reward. He did – Leo recalled him to the papal court and with great pomp and ceremony drew from his pocket an empty purse and presented it to the penniless alchemist, saying that, because he was such a great magician and could make gold, he would need a purse to keep it in.

Others, both supporters and opponents, were more extreme. Pope John XXII, himself a practising alchemist, actively encouraged others in the art, whereas Frederick of Wurzburg maintained special gallows for hanging alchemists – and used them frequently.

Most alchemists were either born poor and acquired money temporarily from gullible but wealthy noblemen or successful merchants, or else they were born wealthy and gradually frittered away their inheritance in ill-conceived alchemical experiments, led along the path to ruin by one sacred text or another.

An example of the high-born coming to grief through extreme gullibility and folly is Bernard of Treves. Born in 1406 in Padua, to a wealthy noble family, he managed to squander his entire family fortune by the lifelong pursuit of alchemical pipedreams. Chasing one crazy method after another, duped by a succession of conmen and fraudsters during a lifetime of travel through France, Italy, Germany and Spain, Bernard ended his days in abject poverty on the island of Rhodes an exhausted and broken man of eighty-five.

Alchemists were not usually corrupt or evil, and often succeeded only in destroying their own lives or in disappointing men who had plenty of cash to spare. And, in spite of alchemy's reputation, we should not forget that much of value did emerge from the unremitting efforts of some alchemists. To the 'serious' alchemist – particularly those with academic credentials, who pursued the art not only for personal gain but from a desire to find fundamental truth – many of the legends and anecdotes surrounding the practice were simply an embarrassment. The famous alchemist and writer Elias Ashmole said in his widely read and influential book *Theatrum Chemicum Britannicum*:

> it is not less absurd, than strange, to see how some men . . . will not forebear to rank true magicians with conjurors, necromancers, and witches . . . who insolently intrude themselves into magic, as if swine should enter into a fair and delicate garden, and (being in league with the Devil) make use of his assistance in their works, to counterfeit and corrupt the admirable wisdom of the magi between whom there is as large a difference as between angels and devils.[9]

Although for some time alchemy was actually outlawed in Britain, this did not stop those who were truly dedicated, and the art flourished in England as much as it did on the Continent. The English anti-alchemy law was a rather vague affair. Passed by Henry IV in 1404, it stated that the making of silver or gold was a felony. The reason was the same as that for witch-hunts: the Church feared it would lose control over the souls of the people (and therefore the purses of its flock); the secular authorities were afraid that if an alchemist should succeed it would disturb the status quo irretrievably.

Whether or not alchemists were tolerated in England was as much to do with the reigning monarch as it was in other countries of Europe. Both Henry VIII and Elizabeth I were involved intimately with the practices of alchemists. Henry gave four successive patents to chemists, monks, mass-priests and others to search for the philosophers' stone – in order, so he claimed, to assist the Crown in the payment of debts if they should succeed. George Ripley, who was a great influence upon Newton's alchemical work, lived in England during the 1470s and dedicated his famous book *The Compound of Alchymy; or the Twelve Gates leading to the Discovery of the Philosophers' Stone* to the reigning monarch of the time, Edward IV. Perhaps even more influential was John Dee, one of the most famous of all alchemists, and so close to Queen Elizabeth I that she actually funded many of his efforts.

Dee has become a legendary figure in the history of the occult, but so much has been written of his bizarre adventures and exploits that it is often forgotten that at the foundations of his philosophy was alchemy, and that much of what he did was driven by a search for the philosophers' stone and the elixir of eternal life.

Dee began his career as a scholar at Cambridge, where he was enrolled at the age of fifteen in 1542, but he soon became entranced by studies of magic and the occult which almost got him expelled from the university and even brought him within an ace of arrest. Things got so bad that he was forced to flee to Europe for a short time, and he made a living as a soothsayer and astrologer there before returning in 1551 to an England more sympathetic to alchemists during the reign of the boy-king Edward VI. For a time, Dee was favoured at court, but this happy state of affairs did not last long and he soon found his career as an astrologer and seer on a roller-coaster ride as monarchs came and went. Edward was succeeded by the Catholic, anti-occultist Mary and then by Elizabeth, who supported and encouraged him.

With Elizabeth's patronage Dee settled into a comfortable

existence, setting up a laboratory at his home in Mortlake which the Queen was said to have visited on several occasions. He was once again welcomed at court, and became not only a favourite but the Queen's personal astrologer and confidant.

Then, it appears, he overreached himself. In 1582 he claimed to have been visited by an angel who gave him a crystal which, he was told, would enable him to converse with spirits and to foretell whatever future events he turned his mind to. This would not have been a problem were it not that around the same time Dee fell in with a shady figure called Edward Kelly, who eventually persuaded him to travel with him to Europe.

Along the way they kept body and soul together by practising astrology and fortune-telling, but at times they almost starved to death through squandering anything they earned on alchemical experiments. On one occasion they narrowly escaped execution at the hands of a chagrined nobleman.

Eventually Dee and Kelly fell out and Dee returned to England, where he continued to try to contact the spirit world with his crystal. He died in 1608, little more than three decades before Newton's birth, not in great poverty but nevertheless another example of a wasted talent washed up on the shore of unattainable dreams.*

Although he died over half a century before Newton took to alchemy, Dee's ideas had a considerable influence upon Newton's own alchemical thinking, primarily because the two men shared an interest in the philosophies of a group called the Rosicrucians. In recent years a number of scholars have suggested that Dee was the founder of the Rosicrucians, and, according to at least one writer on the occult, Frances A. Yates in her book *The Rosicrucian Enlightenment*, Newton was extremely interested in Rosicrucian ideology.[10]

The origin of the Rosicrucian movement is not known for certain, but evidence suggests that during his travels during the 1580s and 1590s Dee inspired many scholars to explore the occult and the mystical arts.[11] One of the most influential was the hermetic philos-

---

* Even in death Dee was unfortunate, his work being slandered and marginalised by a politically minded Royal Society. Among other contributions to science, he was the first editor of the English translation of Euclid's *Elements* and wrote on navigation and calendar reform.

opher Dr Henricus Khunrath of Hamburg, who wrote a book widely read among the alchemical community called *The Amphitheatre of Eternal Wisdom*, published in Hanover in 1609. This, it is believed, had some influence upon what became known as the Rosicrucian Manifestos – short pamphlets laying out the fundamental principles of the society.

The Rosicrucian movement was a genuine secret society, in that no one but the selected leaders of the group knew who was and who was not a member. It caused a great flurry of excitement in France and Germany for a decade or more early in the seventeenth century, but ultimately it achieved little except for the influence it had upon the philosophies of men like Newton. Even the Victorian anti-alchemy writer Charles Mackay was able to acknowledge the contribution made by the Rosicrucian movement in cleaning up the alchemists' act. In his famous denigration of the occult, *The Alchymists: From Memoirs of Extraordinary Popular Delusions*, he says of the Rosicrucians, 'Before their time, alchymy was but a grovelling delusion; and theirs is the merit of having spiritualised and refined it.'[12]

The Rosicrucians had been discussed and written about in underground circles throughout France and Germany by supporters and enemies alike for almost two decades when, at the beginning of March 1623, the citizens of Paris awoke to find walls covered with notices and placards dotted around the city proclaiming the arrival of the movement in their midst. The announcement declared:

We the deputies of the principal college of the brethren of the Rose-cross have taken up our abode, visible and invisible, in this city, by the grace of the Most High, towards whom are turned the hearts of the just. We show and teach without books or signs, and speak all sorts of languages in the countries where we dwell, to draw mankind, our fellows, from error and from death.[13]

The Rosicrucians believed they were of a superior race to the humdrum run of humanity, that they could make themselves invisible and were able to converse with spirits and angels. They claimed to be almost immortal – or at least, through the agency of the legendary *elixir vitae*, to be able to decide when the time was right for them to die. They were, they stated, in possession of the philosophers' stone, with which they could produce almost infinite wealth and with it control the governments of the world. Deeply religious and avowedly Christian, they were anti-Catholic and highly politicised.

By the time Newton was studying and practising alchemy the Rosicrucian movement had had its day,* but many of the men whom he admired and whose works he studied had been intensely interested in the movement – men like Elias Ashmole and the famous German alchemist and writer Michael Maier (who may have even been involved in the foundation of the Rosicrucians). Newton would have been interested in their alchemical pretensions and their claims of supernatural powers. More prosaically, but just as importantly, he would also have been sympathetic towards their vehemently anti-papal stance.

To the Rosicrucians, religion was a central theme around which all else revolved – a principle with which Newton would have empathised. They believed that secret ancient knowledge could be tapped only by the virtuous, and that the mystical knowledge, the *prisca sapientia*, was fundamental to human enlightenment – again, beliefs shared by Newton.

For all their crankiness, the Rosicrucians could be considered forward-thinking and believed in the power of what would soon after their day evolve into science. Perhaps this is what the usually sceptical Mackay saw as their 'merit'. They were not merely seekers of wealth or vain searchers for eternal youth, but believed for instance in the idea that the most fruitful approach to learning was a combination of natural law and the intellectual power of mathematics – undoubtedly another aspect of the movement which would have appealed to Newton. Frances A. Yates has said of them, 'To the genuine Rosicrucian, the religious side of the movement was always the most important. The Rosicrucian attempted to penetrate the deep levels of religious experience through which his personal religious experience, within his own confessional affiliation, was revived and strengthened.'[14]

Yet the Rosicrucians were just one of many influences upon New-

---

* According to some scholars of the occult, the Rosicrucians may still exist in transmuted form. Some argue that they continue to influence world politics in a clandestine fashion, but ultimately more divisively than their largely ineffectual ancestors.

ton's alchemical learning and researches. At the time of his death his library contained 169 books on alchemy and chemistry, including works by some of the most important names in the history of the subject, and it has been said that Newton possessed the finest and most extensive collection of alchemical texts ever accumulated up to his day.* Among these books was a copy of the Rosicrucian Manifestos published in *The Fame and Confession of the Fraternity R.C.*, a 1652 English translation by the alchemist Thomas Vaughan which was heavily annotated by Newton himself. He also read two important books about the Rosicrucian movement by Michael Maier, *Themis Aurea* and *Symbola Aureae Mensae Duodecim*, and made extensive notes on all three works.[16]

In all, Newton possessed nine works by Maier, eight by the celebrated Spanish alchemist Raymund Lull (sometimes named Raymond Lulli), who was a contemporary of Roger Bacon, and four volumes by a peer of Bernard of Treves, a Benedictine monk named Basilius Valentinus. Along with these were works by Thomas Vaughan under his pen-name of Eugenius Philalethes, texts by George Ripley and the Polish adept Michael Sendivogius, and, as a centre-piece to the collection, one of Newton's first and most used purchases of alchemical literature – Elias Ashmole's six-volume *Theatrum Chemicum Britannicum*. This chronicled, among other arcane information, the work of some of the greatest alchemists from the Arab tradition, including the medical alchemist Paracelsus and perhaps the most important of all followers of the hermetic tradition – Cornelius Agrippa.

Although famous in his day and known throughout Europe as a philosopher of almost peerless ability, Agrippa is another example of a man of talent misled by impossible dreams who never managed to realise his hypothetical powers. Born in Cologne in 1486, by the age of twenty he was already an accomplished scholar whose services

---

* It is important to remember that, at the time, the distinction between the two disciplines was blurred. According to John Harrison's *The Library of Isaac Newton*, 138 volumes are considered 'pure alchemy', while 31 others are categorised as 'chemistry' texts.[15] Newton had many more alchemical volumes during his Cambridge years, but sold them or gave them away after moving to London in 1696.

were highly sought after by several monarchs and heads of state throughout the Continent. He travelled widely, working in turn for Emperor Maximillian, King Francis I and Margaret of Austria and, along the way, turning down a generous invitation to join the court of Henry VIII. Although the tales that went before him suggested he could turn iron bars into gold at a glance, fly unaided and converse with angels, unsurprisingly, he demonstrated none of these powers and appears to have sustained a livelihood through medical quackery and the art of bluff, all fuelled by a voracious ego. He wrote a number of books, some of which appeared in Newton's library in translated form or as part of collections. His most important book was the *Vanity and Nothingness of Human Knowledge*. Other works considered various means of transmutation, and it is these that probably interested Newton most when he began his own alchemical researches.

Paracelsus is a name which has become almost synonymous with early medical practice. A near-contemporary of Agrippa, he was born near Zurich in 1493 and believed, like many alchemists, that the alchemical process was a microcosmic representation of the Creation. What made him unusual was his interest in applying alchemy to medicine. Alchemy's special work, he said, was:

> To make arcana [a celestial power Paracelsus believed to be contained in metals], and direct these to disease . . . The physician must judge the nature of medicine according to the stars . . . Since medicine is worthless save in so far as it is from heaven, it is necessary that it shall be derived from heaven . . . Know, therefore, that it is arcana alone which are strength and virtues. They are, moreover, volatile substances, without bodies; they are a chaos, clear, pellucid, and in the power of a star.[17]

Following many an alchemist in a stereotypical fixation with finding the unattainable and achieving the impossible, he travelled Europe in search of the secrets of the ancients, squandering much of his talent and any money he earned along the way. Like most of his fellow seekers, he died in poverty, discredited by the intellectual Establishment.

By comparison, Michael Maier and many of the alchemists of the seventeenth century – Thomas Vaughan, Robert Fludd, Elias

Ashmole and others – were far more realistic in their approach and ideals.

Maier, who was born in Germany in 1566, was an academic who for many years held a respectable position as physician to Emperor Rudolph II. After the death of the Emperor in 1612, he began to travel through Europe, forming a network of contacts with other alchemists and philosophers. He visited England and for a time was a close associate of the English alchemist and Rosicrucian Robert Fludd (whose writings were yet another important source for Newton). Later, between 1614 and 1620, he embarked on writing a series of books which proved to be highly influential in the alchemical world. These included a book of alchemical emblems or symbols called *Atalanta Fugiens*, which contained esoteric text relating to a unification of alchemy with rationalism and orthodox religion; it is seen by some as one of the earliest models for the ethos of the Royal Society. As well as his two linked volumes concerning the Rosicrucian Manifestos, *Silentium post Clamores* and *Themis Aurea*, Maier also wrote a book which would have been of particular significance to Newton, called *Septimana Philosophica*, a treatise on the hermetic origins of alchemy. Although much of Maier's writing is arcane in the extreme and he, perhaps more than any other adept, was deeply involved with the ancient roots of the art, he was also a highly intellectual alchemist and was probably the most powerful influence upon Newton's thinking on the subject.

Yet, given that alchemy was steeped in subjectivity and often deliberate obfuscation, why was Newton so keen to research the subject?

The answer is complex, and Newton's reasoning on this matter has only gradually come to be understood since his death. Early biographers such as Stukeley were probably unaware of the facts, but by the middle of the nineteenth century, when David Brewster wrote his acclaimed biography,[18] Newton's secret writings had been unearthed and were available to serious scholars. Brewster, however, chose to brush them aside, embarrassed for the man long regarded as the great figurehead of science.

Nevertheless, as an old man speaking with unusual frankness to John Conduitt, Newton said, 'They who search after the Philosophers' Stone [are] by their own rules obliged to a strict & religious life. That study [is] fruitful of experiments.'[19] That last sentence might imply that Newton was aware of the important role the subject had played in the elucidation of his theories. Sadly, we will never know for sure whether he fully realised this.

All studies of Newton's life and character agree that he was not a man overly interested in money – or at least not in the procurement of gold simply for its own sake. He was keenly aware of the value of money, and tried to turn a profit wherever he could, but avarice went completely against the grain of his religious feelings and money was not as important to him as the unearthing of universal truth. As Maynard Keynes pointed out in his famous speech for the tercentenary celebrations of Newton's birth, 'He regarded the universe as a cryptogram set by the Almighty.'[20]

Even so, the notion that Newton waded through dubious texts and obscure alchemical poetry in order to get to 'the truth' must surely be only part of the answer. Alchemical literature is obscure deliberately. Alchemists made a virtue of elusiveness in the same way that all brotherhoods, sects and even modern professions enshroud their work in jargon in order to obstruct outsiders, to make what they do appear exclusive and erudite, even when it is not. Newton believed a close study of the alchemical tradition was essential before embarking upon his own practical researches; but why would he even attempt to find the philosophers' stone, given that the alchemical tradition is so illogical, the obvious fact that no single alchemist has succeeded throughout history so clear?

The answer is to be found not by trying to equate Newton's intellectual stature with the largely unscientific nature of alchemy, but, instead, within his personality. There is no doubt that alchemy is a compelling, alluring concept – albeit one that has caused many promising careers to be wasted, many fine intellects to be devastated by the pursuit of an impossible goal. Its fascination may account for Newton's continued enthusiasm once he began, but there were further reasons for his passion.

It is clear from many of his own writings that he was convinced that the ancients had once held the key to all knowledge and that this had been dissipated into the arcane philosophies. He was also worried that someone else would achieve the seemingly impossible and discover the philosophers' stone before him. His ego could simply never allow that to happen until he had exhausted all avenues of investigation and was eventually forced to give up or to lose his mind in the pursuit.

Although Henry IV's law against 'multiplyers' was not revoked until 1689, and the Enlightenment was close at hand, the late seventeenth century saw a blossoming interest in alchemy. Several figures besides Newton were fascinated with the art, the most famous being Robert Boyle. Although Boyle and Newton were friends and colleagues within the Royal Society, sharing views on alchemy and orthodox science, they were also rivals. Newton acknowledged the older man as a mentor and an important influence (Boyle was fifteen at the time of Newton's birth and was seen as the country's leading scientist by the time the younger man appeared on the scientific stage), but Newton was never one to defer to anybody over anything.

Boyle was vocal in his fascination with alchemy, and sometimes sailed close to the wind with his papers and published arguments. Like Newton, he spent many years researching the art, but he differed in that he was not afraid to publish at least some of his more 'scientific' findings – which he went to great lengths to call 'chemistry' rather than 'alchemy'. With a return to tolerance after the Restoration, Newton must have been anxious of rivalry from all sides (including the Continent, where pseudo-scientists abounded).

On the surface, alchemy is a highly practical subject, and it is for this aspect of the art that posterity should be most grateful – for the development of laboratory equipment and techniques still used by the modern chemist. Without alchemy, refinement techniques to produce modern drugs, to purify water in Third World countries and to synthesise plastics and other modern materials might well have taken much longer to be developed. Ignoring the influence it had upon Newton's theoretical concepts and the dissemination of pre-scientific knowledge into Renaissance Europe, almost by accident alchemy aided the arrival of the Industrial Revolution and the development of technology as much as any 'orthodox' science.

The alchemist began by mixing in a mortar three substances – a metal ore (usually impure iron), another metal (often lead or mercury) and an acid of organic origin (most typically citric acid from fruit or vegetables). He ground these together for anything up to six months, to ensure complete mixing. This blend was then heated very carefully in a crucible, the temperature being allowed to rise very slowly until it reached an optimum at which it was maintained for around ten days. This was a dangerous process that produced toxic fumes, and many an alchemist working in cramped, unventilated rooms succumbed to poisoning from mercury vapour; others went slowly mad.*

After the allotted period of heating was completed, the material in the crucible was removed and dissolved in an acid. For many generations alchemists experimented with different types of solvent, and in this way nitric, sulphuric and ethanoic acid were all discovered (possibly by the Arabs of the fourth and fifth centuries). This dissolving process had to be conducted under polarised light – light which vibrates in only one plane – and to produce what they believed to be polarised light alchemists used sunlight reflected by a mirror or worked solely by moonlight.

After the material had been successfully dissolved, the next step was to evaporate the solvent and reconstitute the material – to distil it. Successive generations of alchemists refined the equipment and developed improved distillation techniques based upon the earliest methods practised by the magi of Alexandria almost 2,000 years ago. Today, no chemical laboratory would be complete without distillation apparatus, alcohol could not be produced in large quantities without a still, and the same equipment on a much grander scale lies at the heart of the oil refinery that allows crude oil to be separated into its components.

This distillation process was the most delicate and time-consuming stage of the whole operation, and it often took the alchemist years to complete it to his satisfaction. It was also another highly dangerous phase – the laboratory fire was never allowed to go out, and claimed many lives through the centuries.

If the experimenter was not consumed by flames and the material was not lost through poor control, then the alchemist could move on to the next stage – a step most clearly linked with mysticism. According to most alchemical texts, the moment when distillation should be stopped was determined by 'a sign'. No two alchemical manuals agreed upon what constituted this sign, however, and the

---

* A number of commentators have suggested some of the more bizarre aspects of alchemy and the reason why so many alchemists went off the rails had their origins in the fumes the alchemists inhaled (from mercury in particular). Some historians also account for Newton's nervous breakdowns (see Chapter 10) by suggesting that he suffered mercury poisoning as a result of his experiments.

poor alchemist simply had to wait until he deemed it the most propitious moment to stop the distillation and to move on to the next stage.

The material was then removed from the distillation equipment and an oxidising agent was added. This was usually potassium nitrate – a substance certainly known to the ancient Chinese and quite possibly to the Alexandrians. However, combining this with sulphur from the metal ore and carbon from the organic acid gave the alchemist quite literally an explosive mixture – gunpowder. It was probably by reaching this stage that Roger Bacon discovered gunpowder in the thirteenth century, and many an alchemist who survived poisoning and fire ended his days by going up with his laboratory.

Those who managed to master all these stages of the complex and time-consuming process were then able to continue to the final stages. The mixture was sealed in a special container – 'hermetically' (from Hermes) – and was warmed carefully. After cooling the material, a white solid was sometimes observed. This was known as the White Stone, and was capable, it was claimed, of transmuting base metals into silver. The most ambitious stage – producing a red solid called the Red Rose by an elaborate process of warming, cooling and purifying the distillate – could lead eventually (if all the signs were auspicious) to the production of the ultimate substance, the philosophers' stone itself, the fabled material that could transmute any substance into pure gold.

All of these seemingly practical stages in the process were described in the literature allegorically, enveloped in a mystical language with secret, esoteric meaning. So the blending of the original ingredients and their fusion via the use of heat was described as 'setting the two dragons at war with one another'. In this way, it was believed, the male and the female elements of the substances, symbolised by a king and a queen, were released and then recombined or 'married'. This was the concept behind one of the most famous of all alchemical books, the allegorical romance *The Chemical Wedding*, which, on one level, has been interpreted as describing the transmutation process.[21]

The steps to be taken by the alchemist were always described in the most obscure terms, and allegorical imagery was also used to encode the secret methods. One description, attributed to a second-century female alchemist called Kleopatra, begins:

Take from the four elements the arsenic which is highest and lowest, the white and the red, the male and the female in equal balance, so that they may be joined to one another. For just as the bird warms her eggs with her heat and brings them to their appointed term, so yourselves warm your composition and bring it to its appointed term. And when you've borne it out and caused it to drink of the divine waters in the Sun and in the heated places, cook it upon a gentle fire with the virginal milk, keeping it from the smoke. Then shut the ingredients up in Hades and stir carefully until the preparation becomes thicker and does not run from the fire. Then remove it from the fire; and when the soul and spirit are unified and become one, project upon the body of silver and you will have gold such as the treasuries of kings do not contain.[22]

Other texts were so indistinct and symbolic that they lost all concrete or quantitative content, becoming almost visionary:

In brief, my friend, build a temple of one stone, resembling white lead, alabaster, with no beginning nor end in its construction. It should have within it a source of pure water, sparkling like the Sun. Observe carefully on which side of the temple the entrance lies, and, taking a sword in your hand, seek out the entrance, for the opening is narrow indeed. A serpent sleeps at the entrance, guarding the temple. Seize him and sacrifice him; skin him, and taking his flesh and bones, dismember him. Then reunite his parts with his bones at the entrance to the temple, and, making a step of him, climb up and enter; you will find here that which you seek. The priest, this man of copper, whom you will see seated in the spring, mustering his colour, should not be thought of as a man of copper, for he has changed the colour of his nature and become a man of silver. If you wish, you will soon have him as a man of gold.[23]

It is clear that what the alchemist was engaged in was an odd combination of modern chemistry (the elements of the art which have survived as science) with a strong element of mysticism and spirituality which, to the twentieth-century mind, appears irrational.

To us it seems odd that the esoteric aspect of the alchemist's efforts and beliefs were to him the most important.

Of course, the unsuccessful alchemist often used the more abstract aspects of his work to explain away failure – the astrological conditions were not precisely in his favour, or some malevolent external influence was disturbing the great work – but the spiritual element of the experiment was in fact the key to the true alchemist's philosophy. It is this which has led to the suggestion that, for many alchemists, it was the practical process that was in fact the allegory and their search was really for the elixir or the philosophers' stone within *them*: that, by conducting a seemingly mundane set of tasks, they were following a path to enlightenment – allowing *themselves* to be transmuted into 'gold'. This is why the alchemist placed such importance on 'purity of spirit' and spent long years in preparation for the task of transmutation before so much as touching a crucible.

It was Carl Gustav Jung who brought this idea into twentieth-century thinking. He was fascinated with the psychology of alchemy and came to the conclusion that alchemical emblems bore a close relationship to dream imagery – an observation which eventually led him to the concept of the collective unconscious. According to this concept, at a deep level of the subconscious mind the psyche of an individual merges with the collective psyche of humankind, so that all individuals share a common heritage of symbols or images that Jung named *archetypes*. These, he believed, manifest themselves in dreams and subconsciously affect waking thought patterns.

An avid student of his own dreams, Jung analysed their meaning in relation to alchemy:

Before I discovered alchemy, I had a series of dreams which repeatedly dealt with the same theme. Beside my house stood another, that is to say, another wing or annex, which was strange to me. Each time I would wonder in my dream why I did not know this house, although it had apparently always been there. Finally came a dream in which I reached the other wing. I discovered there a wonderful library, dating largely from the sixteenth and seventeenth centuries. Large, fat folio volumes, bound in pigskin, stood along the walls. Among them were a number of books embellished with copper engravings of a strange character, and illustrations containing curious symbols such as I had never seen before. At the time I did not know to what they referred; only much later did I recognise them as alchemical symbols. In the dream I was conscious only of the fascination exerted by them and by the entire library.[24]

The alchemists, Jung reasoned, were inadvertently tapping into the collective unconscious. This led them to believe that they were following a spiritual path to enlightenment when they were actually liberating their subconscious minds through the use of ritual. This is not far removed from the mental process exploited by evangelical preachers or faith healers, or from the ecstasy experienced by ritualistic voodoo dancers, screamers and chanters, and even dancers at raves. Jung said of the alchemical process:

> The alchemical stone ... symbolises something that can never be lost or dissolved, something eternal that some alchemists compared to the mystical experience of God within one's own soul. It usually takes prolonged suffering to burn away all the superfluous psychic elements concealing the stone. But some profound inner experience of the Self does occur to most people at least once in a lifetime. From the psychological standpoint, a genuinely religious attitude consists of an effort to discover this unique experience and gradually to keep in tune with it (it is relevant that the stone is itself something permanent), so that the Self becomes an inner partner towards whom one's attention is continually turned.[25]

To the alchemist, the most important aspect of the art was participation of the individual experimenter in the process of transmutation. The genuine alchemist was absolutely firm in his belief that the emotional and spiritual state of the individual experimenter was involved intimately with the success or failure of the experiment. And it is this concept, more than any other, which distinguishes alchemy from the orthodox chemistry that superseded it. The alchemist placed inordinate importance upon this element of his work, and for many sceptics it was this which pushed the subject into the realms of magic and left it for ever beyond the boundaries of 'science'.

Although alchemy had more than its fair share of cheats and tricksters throughout its long history, the genuine devotee (rather than those who sought gold merely to satisfy their own greed) believed the adept had to be pure of soul – this was an absolutely essential feature of the art. By appealing indirectly to his vanity, this was yet another aspect of alchemy that attracted Newton.

It was a prerequisite of the practice that the alchemist had to spend many years not only in studying the literature but in preparing himself spiritually for the great task ahead, and Newton certainly considered himself worthy of the challenge. On a number of occasions he referred to the spiritual dimension of alchemy. In one of his alchemical notes he quotes Hermes Trismegistus:

> Yet I had this art and science by the sole inspiration of God who has vouchsafed to reveal it to his servant. Who gives those that know how to use their reason the means of knowing the truth, but is never the cause that any man follows error & falsehood.[26]

And again, in a note entitled 'Observations of the Matter in the Glass', he states clearly how important he considers the virtuous purpose of his work and the reasons for doing it:

> so be it far from me to make myself a name or otherwise to use it excessively farther than for competent necessities for myself, but specially for thy honour & glory & maintenance of thy truth, & to the good of the poor fatherless, the poor widows & other thy distressed members here on Earth.[27]

For the modern scientist it is the element of personal involvement that principally casts doubt upon the relevance and practical use of alchemy as an intellectual process. Enthusiasts of alchemy claim that there are many parallels between modern physics and the traditions of alchemy; they refer especially to what they see as the anthropomorphic dimension of some of the latest ideas at the forefront of quantum mechanics. But these claims are quite unjustified.

Quantum theory's suggestion that the experimenter plays a role in the experiment may appear to be a link with the beliefs of the alchemist, but there is no direct comparison between this and the idea of the alchemist influencing the contents of his crucible.

Quantum theory is a precise, mathematical science based upon fundamental concepts which show rigorous consistency and relate closely and cohesively to other scientific disciplines. Most importantly, *quantum theory works*, for without it we would have no lasers, television, satellite communications, CD-players or other modern technological devices. The practicality of quantum theory is unquestionable; the usefulness of alchemy in extending the frontiers of science is non-existent. The laws of modern physics can be demonstrated by *repeatable* experiments. Although the language of science is undecipherable to the uninitiated, it is a common, strict language – consistent and communicable. Unlike the ancient alchemist, modern physicists do not hide behind a façade of mystical code and they work independently of religious feeling or emotional character.

If we are to appraise alchemy in the cold, clear light of the late twentieth century, we should acclaim it for giving the world some useful tools and techniques still used in modified form today, and, most significantly, for inspiring a train of thought in at least one great philosopher of the seventeenth century. With that we should be content.

CHAPTER 7

The Sorcerer's Apprentice

God is subtle but he is not malicious.
ALBERT EINSTEIN[1]

Continue east about four hundred yards from the end of Oxford Street, take a turn into Museum Street and then, at the end, right into Great Russell Street, and you quickly come to the gate of the British Museum. Dodging the tourists, walk up the front steps and through the bag-checkers into the main reception. To your left is the main staircase, a massive two-tier, stone ensemble as wide as a house. At the top stands the Roman gallery, and as you pass through the rooms you move across the centuries – Early Medieval, Late Medieval, Middle Ages, Rooms 41, 42, 43 . . . In Room 46 you will find the Tudor collection, and there, in the far left exhibition cabinet, lies something quite extraordinary. Walk up to it and crouch down. On a glass shelf at waist height, settled on a plastic cradle, is the crystal ball of the alchemist and philosopher John Dee. No more than three inches in diameter and perfectly smooth, it looks black from some angles, translucent from others. Behind it, a few inches away, stands an electronic thermometer and humidity-sensor, its digital display flashing and a red LCD blinking discreetly.

After withdrawing your gaze from the ball's depths and retracing your steps, back from the museum to the tarmac beyond, the Nikons and the Sony camcorders, as you wander through the exhaust fumes of Oxford Street and the twilight of the late twentieth century you have straddled the divide between myth and science, metaphysics and physics, from magic to mechanic, from John Dee to Isaac Newton.

Newton did not distinguish between these things so readily as we do now. For twenty-seven years, from 1669 until he left for London in 1696, Newton pursued a vast collection of themes both scientific and alchemical, as well as subjects as seemingly diverse as biblical chronology, numerology, history and mythology. Newton seems to have led a double life during the 1670s – the scientific inquirer, moulder of a new approach to science, a revolutionary genius and far-sighted revealer of truth, coexisted in covert harmony with the occultist, the seeker of the ancient flame of wisdom and arcane knowledge. He juggled the responsibilities of his position in the academic world with his clandestine and totally heretical ideas, keeping hidden his unorthodox and socially unacceptable religious views.

Newton's earliest interest in alchemy stemmed from the Grantham apothecary shop, and he learned from the kindly Mr Clark a great deal about the apothecary's art. From his schooldays, Newton loved to create strange concoctions, not simply from curiosity but also because he was a hypochondriac (a facet of his character that grew more pronounced as he grew older). Wickins was witness to some of his methods in Cambridge: 'He sometimes suspected himself to be inclining to a consumption, & the medicine he made use of was the Lacatellus Balsam which, when he had composed himself, he would now & then melt in quantity about a quarter of a pint & so drink it.'[2] This, it turns out, was a delicious combination of turpentine, rose-water, beeswax, olive oil, sack and red sandalwood. Newton considered it so powerful that, as well as drinking it, he often applied it externally, finding it useful for 'green wounds' and 'the bite of a mad dog'.[3] He also wrote in praise of the curative powers of opium.[4] Isaac Barrow was keen on the drug and probably introduced him to it.

In his boyhood fascination with odd draughts, we may also see the early rationalist. He kept meticulous records of his experiments and noted any recipes he came across in the apothecary's books or specialised treatments that Clark himself had heard about or contrived. Later, sometime during the early 1660s, he became interested in what was considered 'conventional chemistry', as opposed to alchemy. The difference between the two disciplines was subtle. Chemists and alchemists dealt with the same compounds, even used the same apparatus and shared inherited knowledge; what lay between them was approach and intent.

Chemistry was considered by many as rather vulgar – a gross and unsubtle pursuit. Alchemists considered themselves superior to chemists principally because of its practitioners' belief that alchemy required an interaction between experiment and experimenter and that to be a successful alchemist one had to be pure of spirit, to 'receive' understanding by divine transmission. The fifteenth-century alchemist Thomas Norton said of this talent, 'A most wonderful majesty and archimajesty is the tincture of sacred alchemy, the marvellous science of the secret philosophy, the singular gift bestowed upon men through the grace of Almighty God – which men have never discovered through the labour of their own hands, but only by revelation and the teaching of others.'[5]

By becoming spiritually involved with a 'divine' process, the alchemist believed he could share in its divinity. Whereas alchemists were in pursuit of mythical goals and wrapped their work in mysticism and occult meaning, 'chemists' worked in a manner which passed as empirical (at least for the time) and, most importantly, they were not usually motivated by greed or the pursuit of dreams as were many charlatan alchemists. Instead, chemists were considered to be little better than craftsmen or tradespeople. Apothecaries, dye-makers, tanners, brewers and distillers all practised chemistry within their narrow specialisms, and their techniques formed collectively the canon of chemical knowledge.

Alchemy evolved into chemistry as practised later by rationalists and post-Industrial Revolution 'scientists', but Newton was travelling in the opposite direction. First he absorbed the limited arena of conventional chemistry, then he moved on to what he considered the more exciting realm of alchemy – a subject which must, to his eyes, have held almost limitless potential.

During Newton's life and until the work of Lavoisier, Dalton, Priestley and others in the late eighteenth century, the intellectual (as opposed to the motivational) foundations of chemistry and alchemy overlapped. Those who tried to get to the root of chemistry inevitably found themselves faced with a wall of alchemical lore, and those who wished to acquire the philosophers' stone either followed a confusing path through myth and magic or else they built on a very limited chemical tradition. So 'chemists' such as Robert Boyle were as involved in alchemical pursuits as much as they were with chemistry experiments, and many alchemists inadvertently developed concepts and practical skills later requisitioned by the chemist.

Newton began his studies in alchemy and chemistry around 1667. In typical fashion, he compiled a glossary, or dictionary, which included chemical names, apparatus and a list of terms, in all about 7,000 words in length, with each term and name having a brief definition.[6] The dictionary is almost exclusively 'chemical' in content, and in this single document is to be found almost the entire chemical knowledge of the time. But even then, shortly after returning to Cambridge and some two years before acquiring the Lucasian Professorship, he was aware of the need for secrecy. In the dictionary there are a number of sensitive terms which fit into the alchemical school rather than formal chemistry – words such as *Anima*, meaning the innermost aspect of the personality, and *Elixar* [*sic*], meaning the elixir, the mythical panacea or cure-all sought by many alchemists. Strikingly, all of these entries are left undefined.*

As well as cataloguing the foundations of the discipline, Newton later kept to his usual technique of absorbing as much information about the subject as he could before venturing into the realm of experiment himself. His mentor in this new field of study was Robert Boyle, at the time acknowledged as the greatest living authority in the field of chemistry.

As an undergraduate, Newton had made notes on Boyle's work, and he would probably have cited him along with Descartes as one of his intellectual forebears. But Boyle's influence upon Newton continued and grew as Descartes's waned.

Although Robert Boyle did not revolutionise chemistry to the extent that Newton upturned physics, comparisons may be drawn between the ways in which the two men used alchemy. Boyle practised alchemy as well as chemistry and utilised many of the esoteric aspects of the former to push forward the theoretical limits of the latter. What he laid down as formative concepts, Lavoisier, Priestley and others later confirmed by experiment and put to practical use. Boyle was an atomist who believed that the shapes and natures of individual atoms created differences in chemical behaviour. He proselytised the concept of basic elements that could be combined into groups (molecules) which could be broken down again and rearranged – all hypotheses later integral to the chemical revolution of the late eighteenth and early nineteenth century.

---

* These omissions may of course have been simply due to his not having discovered suitable definitions at this stage.

Born the fourteenth child and youngest son of the first Earl of Cork in 1627, Boyle led the life of a wealthy aristocrat. He was educated by a tutor at home and then at Eton and showed an early flair for languages, becoming fluent in both French and Latin by the age of eight. As a teenager he travelled Europe with an older brother, Francis, and a tutor, making a grand tour of Italy and France which lasted from 1638 until soon after his seventeenth birthday. In Italy he discovered the work of the recently deceased Galileo (who had died in 1642) and became interested in science. Then, when his father died suddenly, he and his brother were called back to England and Robert lived quietly in Dorset, where he practised alchemy and dabbled in a range of scientific and pseudo-scientific experiments for the best part of a decade. In 1654 he moved to Oxford, where he conducted experiments with a vacuum pump designed and built for him by Robert Hooke, and he became one of the founders of the 'Invisible College' – the earliest form of the Royal Society.

Newton and Boyle met publicly for the first time in 1675,* and their publicly recorded correspondence began a year later, when Newton became interested in Boyle's 'Of the Incalescence of Quicksilver with Gold', published in the Royal Society journal, *Philosophical Transactions*.

Boyle had begun his alchemical experiments some twenty years before Newton, in 1646, and, as well as writing a collection of secret texts, he had already published several books. The most important and influential was *The Sceptical Chymist*, published in 1661, in which he laid the cornerstone of many of his theoretical ideas in chemistry and made a clear and definitive distinction between chemistry and alchemy. Although he was a practising alchemist, enamoured of the art, he did not see it as influencing his chemical work. For his own reasons, and perhaps not merely for the sake of clarity, Boyle insisted on separating the two fields and treating each differently. He published his chemical discoveries freely, but kept most (though not all) of his alchemical findings to himself and his close circle of friends and associates (including Newton).

---

* They may have met earlier than 1675 at secret meetings of alchemists at Ragley in Warwickshire, the home of Viscountess Conway, under the aegis of Henry More.

Boyle and Newton had very different approaches to public awareness of alchemy. Although Boyle did not shout his alchemical discoveries from the rooftops, he was not nearly so secretive about his
work as Newton. Indeed, 'Of the Incalescence of Quicksilver with
Gold' dealt with alchemical themes, and when Newton heard that
Boyle wanted to make it available to a wider readership he railed
against the idea. In a letter sent to Henry Oldenburg, the Secretary
of the Royal Society, with the explicit instructions 'keep this letter
to yourself', Newton strongly recommended that Boyle should not
communicate his alchemical findings to the world:

> It may possibly be an inlet to something more noble, not to be
> communicated without immense damage to the world if there
> should be any verity in the hermetic writers, therefore I question
> not but the great wisdom of the noble author will sway him to
> high silence till he shall be resolved of what consequence the
> thing may be either by his own experience, or the judgement of
> some other . . . that is of a true hermetic philosopher . . . there
> being other things beside the transmutation of metals (if those
> great pretenders brag not which none but they understand).[7]

After they became friends, Newton often suggested that Boyle
should disguise his authorship of papers, and tried to spread his own
insecurities and paranoia to the older man. Boyle seems to have
ignored this advice. He was a wealthy man, and from an old and
powerful family: perhaps this insulated him from fears of exposure
as a magician. Like Newton, he was a devout Christian, but he had
refused to take holy orders (he despised the taking of oaths in general), and because of this, in 1665, he had denied himself the prov-
ostship of Eton.

Boyle was not stimulated by the same intense desire to rediscover
'lost secrets'. He was a far more extrovert and socially well-adjusted
man who pursued knowledge for the sake of it from a position of
extreme comfort and financial security. In some respects he antici-
pated the stereotype of the 'gentleman scientist' which became
familiar in Georgian and early Victorian times through men like
Erasmus Darwin and his more famous grandson Charles. It is easy
to see why Newton was so careful about concealing his own interest

in alchemy – he was terrified of losing the things he had struggled to acquire. Boyle could afford to be carefree.

Newton began to gather his research materials and to buy laboratory apparatus during a succession of trips to London. During the first of these, sometime during late 1669, he was introduced to a network of alchemists and specialist book-dealers who could obtain important texts from all over Europe.

He would take the stagecoach at the Rose public house in Cambridge and end his journey at the Swan tavern in Gray's Inn Lane, Holborn. From there it was only a short trip to Little Britain, an area famed as the literary centre of the capital.

His most useful contact was one William Cooper, a man of impeccable reputation but also a dealer in illegal and highly sought-after manuscripts. Through Cooper, Newton acquired some of the most important alchemical tracts available at the time anywhere in Europe, including the *Theatrum Chemicum* by Lazarus Zetzner in six volumes and a collection by Eirenaeus Philalethes called *Ripley Reviv'd*, based upon the works of the fifteenth-century English alchemist George Ripley, that was only later published in England (in 1678). The bookseller obtained all of these through the underground network of alchemists and experimenters in London and beyond.*

On this first trip to London Newton bought and himself took back to Cambridge a set of chemicals (costing £2) and ordered two furnaces to be delivered to his rooms in Trinity. One of these was made of tin and cost seven shillings; the other, of iron, was priced at eight shillings. These provided Newton with the basic materials for setting up his own laboratory. The furnaces arrived early in the New Year and were duly installed in the rooms he shared with the ever-tolerant John Wickins. But, more importantly, by making contact with Cooper and other dealers in the capital, Newton had also tapped into a wholly secret society of alchemists that had flourished for centuries and secretly pervaded all strata of society, extending tendrils into many diverse areas of intellectual life.

The original founders of the Royal Society, a group that evolved

---

* Eirenaeus Philalethes has been shown only in this century to be the London alchemist George Starkey, who was a friend of the scholar Samuel Hartlib and practised alchemy until the 1660s.

into the very epitome of empirical, 'conventional' science, may in their early days have called themselves the 'Invisible College', but the real invisible college was the network of nameless adepts who kept alight the alchemical flame. These men operated covertly within the same city as the Royal Society – London – and included many of their number.

The London alchemists of the mid seventeenth century had centred around Samuel Hartlib. Hartlib had arrived in England from Polish Prussia in 1625 and had immediately begun to infiltrate the existing network of alchemists and magicians via contacts already established between the English and their European counterparts during the past century. The Hartlib Circle, as the group became known, grew in importance during the 1640s and '50s and included Sir Kenelm Digby (author of *A Choice of Rare Secrets and Experiments in Philosophy*, published by William Cooper in 1682), George Starkey (Eirenaeus Philalethes), Thomas Vaughan (Eugenius Philalethes), Hartlib's son-in-law Frederick Clodius, and Robert Boyle.

The Hartlib Circle aimed to rationalise alchemy by marrying alchemical lore with the intellectual framework of mechanical philosophy; they may thus be thought of as providing the link between vague, personalised medieval alchemical practice and the foundations of empirical chemistry. Alchemy was recognised by most practitioners as an amalgamation of the spiritual and the practical, but the spiritual side had usually predominated. Europe was littered with laboratories where alchemists had slaved over the flames in attempts to realise their dreams, but, until Hartlib and his associates, very little attempt had been made to develop a rationalised alchemical system.

The most famous member of the Hartlib Circle was not Hartlib himself (he was more a facilitator) but Robert Boyle. As well as being almost certainly the ablest experimenter, he was also the most accomplished natural philosopher of the group, a man who represented the ideals and philosophies of the circle in their most intellectual form. His *The Origine of Formes and Qualities*, published in 1666, was a distillation of all the Hartlib Circle had achieved during the previous three decades and provided Newton with an intellectual framework for his own alchemical explorations.

Another key figure in the movement, and the most likely man to have linked the Hartlibians and Newton, was Henry More. When he was not in Cambridge, More spent as much time as he could at the home of his pupil and close friend Anne Finch, the Viscountess Conway. She provided More with rooms at her country estate of Ragley, in Warwickshire, where he could host meetings to discuss and experiment. Ragley soon became a focal point for the Hartlib Circle and also for a growing body of intellectuals interested in alchemy who remained on the periphery of the group. Gatherings of such thinkers continued to be popular long after Hartlib's death in 1662, and may have been frequented by Newton during his early alchemical career.

Sadly, there is almost no Newton correspondence from the early 1670s referring to anything alchemical,* but it is known that he left Cambridge for extended periods on a number of occasions during those years, and his appearance at gatherings at Ragley would have been quite in keeping with his yearning for discovery and his known association with Henry More.

The alchemists of the period referred to one another by code-names. Thus within the surviving correspondence of this group we have a mysterious 'Mr Petty', a 'Mr Gassend' and, most intriguingly, references from Newton to a 'Mr F'. Mr Petty was probably Sir William Petty, one of the founder members of the Royal Society, professor of anatomy at Oxford, inventor, cartographer and quite possibly an alchemist.

It now seems likely that Mr F. was Ezekiel Foxcroft, a fellow of King's College, Cambridge, between 1652 and 1675. He was a close friend of More and was described by contemporaries as 'being interested in chemistry'.[8] The only surviving correspondence between him and Newton is an alchemical manuscript entitled 'Manna', copied in an unknown hand, at the top of which Newton has written, 'Here follows several notes & different readings collected out of a MS communicated to Mr F. by W.S. 1670, & by Mr F. to me in 1675.'[9]

Further into the document, Newton gives us a rare insight into his own feelings about alchemy and his approach to the art:

For alchemy does not trade with metals as ignorant vulgars think, which error has made them distress that noble science; but she has also material veins of whose nature God created

* However, this is an unusual gap in the sequence of his letters, suggesting that there was once a collection of correspondence that has since been lost.

handmaidens to conceive & bring forth its creatures . . . This philosophy is not of that kind which tends to vanity & deceit but rather to profit & to edification inducing first the knowledge of God & secondly the way to find out true medicines in the creatures . . . the scope is to glorify God in his wonderful works, to teach a man how to live well . . . This philosophy both speculative & active is not only to be found in the volume of nature but also in the sacred scriptures, as in Genesis, Job, Psalms, Isaiah & others. In the knowledge of this philosophy God made Solomon the greatest philosopher in the world.[10]

Newton may have contributed texts to the network, although this has never been established beyond doubt. In a perverse way, he would have been more relaxed about submitting alchemical conclusions to the scrutiny of his peers within the secret society than he was about offering his 'scientific' papers to the Royal Society. This is almost certainly because of the covert nature of the process. Newton guarded his privacy jealously and could not stand being challenged over his ideas. Years later he disliked dealing with anyone but Henry Oldenburg in preparation for any of his publications through the Royal Society. Most significantly, as an alchemist, he too could hide behind a pseudonym. Revealingly, his was 'Jeova Sanctus Unus' – One Holy God – based upon an anagram of the Latinised version of his name, Isaacus Neuutonus.

So, armed with his knowledge of the mysteries, his early, tentative links with the alchemical underworld established and a study filled with furnaces, laboratory equipment and chemicals, what did Newton actually set out to do?

To the modern eye, his earliest experiments seem decidedly prosaic, but he was feeling his way into the subject, following the clearly defined path laid down by his predecessors – in particular Robert Boyle. Yet, from the earliest experiments, Newton's methodical approach distinguished him from almost all of the thousands of alchemists who predated him. From the start he applied his practised methods of note-taking, meticulous attention to detail and a genius for observation.

He began by opening a laboratory notebook.[11] The earliest experiments are almost certainly from 1669 and, in keeping with the

alchemical tradition, they are concerned with the nature of metals.*
In particular, he was trying, like many before him, to find a way of
'opening' metals – to make them change their characteristics.

Of all materials, mercury was the most important to the alchemist,
for, by tradition, a substance called 'philosophers' mercury' (sometimes 'our mercury') was the vehicle by which metals could be transmuted. According to Charles Nicholl in his book *The Chemical
Theatre*:

> There are two directions in which alchemical mercury leads us.
> On the one hand, it is a complex elaboration of a chemical
> substance, its various qualities referring back to the properties
> of quicksilver, or 'common mercury' . . . But there is another
> direction entirely, away from chemical matter. Mercury is not,
> finally, a substance, or even many substances: it is a process . . .
> All these [alchemical writings] point to one crucial idea: that
> transformation is something intrinsic and contained inside mat-
> ter . . . Each stage of this self-devouring, self-generating process
> bears the name 'mercury'. Mercury, in short, is alchemy itself.[12]

To the modern chemist this has little concrete meaning, as mercury
is merely another element in the periodic table. But the alchemist,
working long before the periodic table was elucidated and the atomic
structure of elements unravelled, believed that certain elements possessed mystical powers simply because they had visibly impressive
chemical and physical properties.

Of the seven metals used by the alchemist (gold, silver, iron, tin,

---

* A word about dating Newton's manuscripts: Newton very rarely dated any
of his alchemical work, and the best indicator for the date of a manuscript is
the style and size of his handwriting. The scholar B. J. Dobbs has divided
Newton's handwriting into six periods. 1. 'Very early', 1667–69: the handwriting here is almost microscopic, perpendicular and meticulous. 2. 'Middle early',
1670–75: the writing is fuller and more rounded. 3. 'Late early', 1675–80:
bolder and more rounded still. 4. 'Bad ink', 1680–81: a period during which
Newton used poor-quality ink which has rendered his writing now almost
illegible. 5. 'Middle confident', 1682–92: far larger, sloping and more confident.
6. 'Late', 1693–1727: small again, but different in nature to the 'Very early'
period.

mercury, lead and copper), mercury was the only one which existed in the liquid state at room temperature. When each of the other six metals was heated individually, they too liquefied, which alone implied to the medieval mind that there was a 'mercurial principle' in each of the metals and that therefore there must be something special about mercury. The adept Elias Ashmole described the substance as 'that universal and all-piercing spirit, the one operative virtue and immortal seed of worldly things, that God in the beginning infused into the chaos, which is everywhere active and still flows through the world in all kinds of things by universal extension'.[13]

Mercury was viewed as nothing less than the first matter, the *prima materia* of metals. The *prima materia* was the holy spirit of alchemy, the essence that lay at the heart of matter, the spirit pervading the material, and it had to be released from 'dead', inert, metals via transmutation. The concept is an ancient one and had been expressed in Taoist philosophy some 2,300 years before Newton's day:

> There is a thing confusedly formed,
> Born before heaven and earth.
> Silent and void
> It stands alone and does not change,
> Goes round and does not weary.
> It is capable of being the mother of the world.[14]

The concept of the *prima materia* was also analysed by Jung during his extensive study of the psychological meaning of alchemy. He concluded that 'The basis of the opus, the *prima materia*, is one of the most famous secrets of alchemy. This is hardly surprising since it represents the unknown substance that carries the projection of the autonomous psychic content.'[15] And, in Newton's era, the adept Thomas Vaughan waxed lyrical on the subject:

First, that the first matter of the stone is the very same with the first matter of all things. Secondly, that in this matter all the essential principles, or ingredients of the elixir, are already shut up by Nature, and that we must not presume to add anything to this matter, but what we have formerly drawn from it; for the stone excludes all extractions, but what distil immediately from its own crystalline universal minera. Thirdly, and lastly, that the philosophers have their peculiar secret metals, quite different from the metals of the vulgar, for where they name mercury they mind not quicksilver, where Saturn, not lead, where Venus and Mars, not copper and iron.[16]

Small wonder, then, that Newton's first experiments involved an attempt to produce this apparently most useful of substances, philosophers' mercury.

He took as his guide Boyle, who had already conducted experiments with mercury, but Newton pushed the principles further, applied them with far greater care, and spread his net much wider.

Boyle's instructions for the production of philosophers' mercury involved dissolving common mercury in nitric acid followed by the gradual addition of lead filings. Newton followed the instructions precisely and produced a white precipitate, a colourless solution and a silver substance at the bottom of the crucible. But, upon testing, he found that the silver metal was just the mercury he had added originally (released from the solution by the addition of lead, which is more reactive with the nitric acid). There was nothing stunning in that discovery.

He then tried adding different metals to the solution of quicksilver in nitric acid. With tin he again produced the silver metal and a colourless solution, and with copper a blue solution along with the silver metal. Quickly, he realised that all he was doing was obtaining purified mercury (the silver metal) and dissolved compounds of the metals he was adding. It was all very disappointing. After that, he gave up following the traditional line of reasoning and tried using heat to alter the product.

Newton later told Conduitt that John Wickins helped him. Wickins was physically stronger and assisted in moving 'kettles' and manning the furnace as well as transcribing experiment notes. In a rare comment upon his life with Newton, Wickins tells us of his room-mate's 'forgetfulness of food, when intent upon his studies; and of his rising in a pleasant manner with the satisfaction of having found out some proposition without any concern for or seeming want of his night's sleep which he was sensible he has lost thereby'.[17] Sadly, Wickins recalled nothing of what Newton was actually doing, and in any case he may not have understood what he saw.

With this second set of experiments, Newton was again

disappointed, producing a precipitate along with purified but perfectly ordinary mercury. (How Newton or any other alchemist could
have distinguished between 'common' mercury and what he saw as
philosophers' mercury is impossible to tell. We can only assume that
he tested it to see if it exhibited the powers ascribed to it by alchemical
tradition.)

After spending several months on these experiments, Newton
abandoned Boyle's methods and turned to his own, this time using
antimony as his starting material.

The element antimony is found in a natural ore called stibnite
(Sb_2S_3). The alchemist was interested in the substance because, when
purified, antimony appeared to have an affinity with gold (and some
other metals), forming a type of amalgam which they called a *regulus*
(from the Latin for 'petty king').

The reguli – crystalline compounds with striking arrangements of
radiating shards – were produced by blending purified antimony with
various reducing agents – materials which changed the oxidation
state of the antimony (and therefore some of its chemical properties).* Initially, the exciting thing about reguli for the alchemist was
their appearance: in particular, the regulus produced by processing
antimony with iron produced a star-like crystal configuration which
they dubbed the *Star Regulus of Antimony* or the *Regulus of Mars*.

Newton would have learned about the reguli from the fifteenth-
century alchemist Basilius Valentinus, who wrote at length about the
Star Regulus of Antimony in his tract ***The Triumphal Chariot of
Antimony***, from which Newton copied sections into his notebook
(Keynes MS 64). Valentinus said of the substance:

Many have esteemed the Signed Star of Antimony very highly,
and spared neither labour nor expense to bring about its preparation. But very few have ever succeeded in realising their
wishes. Some have thought that this Star is the true substance
of the Philosophers' Stone. But this is a mistaken notion, and

---

* The oxidation state of a substance is linked to the number of electrons
available for reaction or 'spaces' available to receive electrons in an atom during
chemical reactions, and consequently plays a part in the way elements react
together and the types of compounds they may form.

those who entertain it stray far afield from the straight and royal road, and torment themselves with breaking rocks on which the eagles and the wild goats have fixed their abode. This Star is not so precious as to contain the Great Stone; but yet there is hidden in it a wonderful medicine.[18]

Newton described his method of producing this regulus in one of his earliest essays on alchemical practice, in which he said:

These rules in general should be observed. 1st that the fire be quick. 2dly that the crucible be thoroughly heated before any-thing be put in; 3dly that metals be put in successively according to their degree of fusibility – iron, copper, antimony, tin, lead. 4thly that they stand some time after fusion before they be poured off accordingly to the quantity of regulus they yield – iron, copper, tin, lead. 5thly that no salt be thrown on, unless upon iron to keep it from hardening ... 6 that if you would have the saltpetre flow without too great a heat, you may quicken it by throwing a little more saltpetre mixed with 1/8 or 1/16 of charcoal finely powdered.[19]

We know from a letter Newton wrote to Oldenburg in January 1672 that he had produced this mysterious substance sometime before the end of 1670.★ Although he may have been pleased with the result, he also realised it was little more than a step on the way to the dreamed-of philosophers' stone and not an end in itself.

The Star Regulus of Antimony is of particular interest in any effort to link Newton's alchemical researches with his elucidation of the law of universal gravitation. As the name implies, the regulus does look like a star, and its radiating shard-like crystals (Plate 9) may be imagined as lines of light radiating from a starlike centre. But the crystal may just as easily be visualised as representing shards or lines of light pointing inwards – a star at the centre with lines of light, *or force*, travelling *towards* its centre.

★ In this letter Newton proposed a number of uses for the substance. One application was as a material for making mirrors to improve a reflecting tele-scope he was constructing at the time.

Newton, we should recall, probably first produced the Star Regulus of Antimony sometime during 1670, only four years after deriving a mathematical relationship describing the receding force involved in planetary motion. We know that the development of this simple relationship into the full-blown theory of gravity as it appeared in the *Principia* of 1687 came about via a collection of influences, experimental verification and inspiration during the 1670s and 1680s. The creation of the Star Regulus was probably one step along this road – a subconscious contribution to the slow process of realisation firstly of attraction and then of universal gravitation.

By the mid-1670s Newton had been experimenting with alchemical processes for at least five years and, after composing thousands of words of experimental notes, brief conclusions and huge piles of recipes – both original and transcribed from the alchemical canon – he felt ready to write a substantial document describing his findings. This was the 'Clavis', 'the Key', dated around 1675 – a document about 1,200 words in length and the distillation of his first five years of research.[20]

Some doubt has been cast upon the originality of the 'Clavis', and there have been suggestions that it was simply another of Newton's transcriptions (perhaps from Barrow or Starkey). But there are several aspects of the document that are clearly Newtonian. First, there is the most obvious Newton hallmark: meticulous instructions, precise measurements and recorded time periods. Newton's attention to detail can be seen in his laboratory notes, where he records mixing granules of components on a 'looking-glass that none of them might be lost', and dividing individual crystals 'with the point of a knife'.[21] Second, there are comparisons between this and the more expansive 'Praxis' – the apotheosis of his alchemical contribution, written some eighteen years later (see Chapter 10).

Although the 'Clavis' was a thorough and carefully contrived work of its type – a rare example of the empirical meeting the alchemical – it was only a stepping-stone for Newton, a summary of his preliminary forays into the alchemical world. From this he was to go on to build a body of profoundly important work which was to lead him both to the edge of acceptable inquiry and into scientific realms previously undreamed of.

The work was made easier by the acquisition of a genuine laboratory. In 1673 Newton and Wickins had moved to rooms at E-4 Great

Court, close to the Great Gate of Trinity. Attached to the rooms was a wooden 'shed' clearly visible in drawings made at the time. The rooms seem to have been chosen deliberately because of this annexe, which was quickly turned into a laboratory and filled with the furnaces and shelves of chemicals which had once cluttered up Newton's living-quarters. But still the work was often physically dangerous, and years of effort could be easily lost through negligence. A fire one morning sometime during the winter of 1677–8, when Newton made a rare appearance at the college chapel, destroyed a collection of papers both scientific and alchemical and almost burned down the entire laboratory.

For some early chroniclers of Newton's life, this incident explained the lack of material success during the decades of Newton's labours with alchemy. Stukeley, hearing from one of his sources that the great scientist had delved to the very heart of *chemistry* and had written a *Principia Chemicum*, says that Newton was 'explaining the principles of that mysterious art upon experimental and mathematical proofs and he valued it much, but it was unluckily burnt in his laboratory which casually took fire. He would never undertake that work again, a loss much to be regretted.'[22]

Others have suggested that the destroyed papers were an early version of the *Opticks*. The antiquary and Newton devotee Abraham de la Pryme, who was an undergraduate at Cambridge during the 1690s, was probably the first to circulate this idea when he wrote:

but of all the books that he ever wrote there is one of colours & light established upon thousands of experiments which he had been 20 years of making, & which cost him many a hundred of pounds. This book which he valued so much & which was so much talked of had the ill luck to perish, & be utterly lost just when the learned author was almost at putting a conclusion to the same after this manner: In a winter's morning leaving it amongst his other papers, on his study table whilst he went to chapel, the candle which he had unfortunately left burning there too, catched hold by some means of other papers, & they fired the aforesaid book, utterly consumed it, & several other valuable writings which is most wonderful did no further mischief.[23]

The lost document, if there was one, could have been a treatise on optics, but this story might also have been an attempt to disguise the truth. Both Stukeley and Pryme had almost certainly heard unpalatable rumours while Newton was still alive. It is quite possible that, by manipulating the tale, they managed neatly to dismiss Newton's alchemical interests (Stukeley referring to them as 'chemistry'), as well as quashing any hint that he continued with his experiments: 'He would never undertake that work again.'

Stukeley certainly could not have been more wrong. If nothing else, Newton's description of his experiments in the 'Clavis' confirm the direction in which he was heading during the mid-1670s. In his unique way, he was correlating traditional wisdom with his own findings. In another early paper, he drew up a list of forty-seven axioms derived from his alchemical reading and experiments.[24] Next he analysed their obscure imagery based upon his own observations.

> Concerning Magnesia or the Green Lion. It is called Prometheus & the Chameleon. Also Androgyne, and virgin verdant earth in which the Sun has never cast its rays although he is its father and the moon its mother: Also common mercury, dew of heaven which makes the earth fertile, nitre of the wise ... It is the Saturnine stone.[25]

Always critical, he could easily sort out the good from the bad. On one occasion, in the margin of a manuscript, he wrote, 'I believe that this author is in no way adept.'[26]

But what did he really gain from these experiments? They were preliminary, that much is certain, but even in these earliest attempts he was gaining an instinctive awareness of possible forces at work in the universe for which conventional mechanical theory (still dominated by Descartes's notions of matter and spirit) had no explanation. He was seeing things in the crucible that could not be explained by Descartes or by an orthodox appraisal of simplistic atomism. What we can discern from these early observations and from his writings about the mechanical philosophy demonstrates that Newton was then beginning to realise that the universe was a far broader canvas than even he had dared to imagine.

*   *   *

Newton was a good Puritan, but, craving learning for the greater glory of his God, he had dived into the murky, heretical pool of alchemy believing that there he could rediscover the secrets of Nature. Religious zeal was one of his prime motivators, and, like almost all other thinkers and uneducated folk of his day, he lived and died believing wholeheartedly in divine guidance. Writing in his notebook in 1664, he had commented, 'He [God] being a spirit and penetrating all matter, can be no obstacle to the motion of matter; no more than if nothing were in its way.'[27]

If anything, his fundamentalism became more pronounced in later life. He never swayed from his assertion that God was responsible for maintaining planetary motion through the device of gravity. Although he formulated a theory to describe mechanical motion, including the movement of planets in their orbits, he could not explain *how* this worked or *why*; he believed that a basic tenet of intellectual life was that some things will always be unknowable. Although in many intellectual areas Newton was generations ahead of his time, his notion of God was as simplistic and as orthodox as the next person's. In the notebook, under 'Of God', he wrote, 'Were men and beasts etc made by fortuitous jumblings of the atoms there would be many parts useless in them, here a lump of flesh there a member too much. Some kinds of beasts might have had but one eye, some more than two.'[28]

Within the broad sweep of Christianity there has always been room for certain forms of unorthodox beliefs, some of which conflict with the Established Church over what, to the outsider, seem only matters of detail. However, Newton's own brand of Christianity was one of the more extreme and was quite unacceptable to orthodox Anglicans of the seventeenth century.

Newton was an Arian. The Arian doctrine had its origins in the teachings of a fourth-century Alexandrian priest named Arius, and held (in defiance of orthodox Trinitarianism) that Jesus and God are not of one substance but that Christ, although divine, was created by God as *the first creature*.

Sometime between 1672 and 1675, Newton set out twelve points of faith that formed the foundations of his Arianism. These included the argument that:

There is nowhere made mention of a human soul in our Saviour [the Bible] besides the word, by the meditation of which the

word should be incarnate. But the word itself was made flesh &
took upon him the form of a servant . . . It was the son of God
which he sent into the world & not a human soul that suffered
for us. If there had been such a human soul in our Saviour, it
would have been a thing of too great consequence to have been
wholly omitted by the Apostles.[29]

To the dispassionate observer such distinctions between Arian
doctrine and orthodox Christianity may seem unimportant, but to
Newton a commitment to Trinitarianism was tantamount to blasphemy, and he found himself torn by the clash between his personal
devotions and the legal requirements of the college. In signing for
his BA in 1665 he had had to attest to an acceptance of the Thirty-
Nine Articles of the Anglican Church. In 1667 he had signed again
as he received his fellowship, agreeing to embrace the 'true religion
of Christ', and he signed once more upon receiving his MA the
following year. Finally, in 1669, he had promised, as a requirement
of the Lucasian Professorship, to take holy orders at some point in
the near future. But he continued to balk at the prospect. In a battle
between his conscience and his career, he was willing to sacrifice all
he had worked for. In a letter to Oldenburg in 1675 he informed the
Secretary of the Royal Society that he would soon have to cease his
payments of subscription: 'For the time draws near that I am to part
with my fellowship, & as my incomes contract, I find it will be
convenient that I contract my expenses.'[30]

Newton was walking on eggshells. During the seventeenth century,
religious orthodoxy was an important social factor, influencing
careers and the fulfilment of ambitions. Toleration on a wider scale
had grown out of the Civil War, allowing Protestants and Catholics
to live together with a degree of harmony. But within the individual
denominations adherence to the sacred texts was paramount, and
radical thinking along the lines of Newton's beliefs would have been
perceived as dangerous, even divisive. In the hands of a man of his
wilfulness and intellectual rigour, it would have been viewed by his
rivals as a potential threat to the status quo.

In 1674 Newton's friend the Cambridge fellow Francis Aston
applied for exemption from taking holy orders – for his own, still
rather obscure, reasons. His request was refused, because agreeing
was seen by the Trinity authorities as setting a precedent which

would destroy the college's reputation as a repository for young clerical talent. At the same time, Newton's own situation was becoming untenable. Tongues were wagging, and any further reluctance to be ordained would have signalled to his academic colleagues that he was trying to conceal something. But help came once more in the form of his old allies Humphrey Babington and, most importantly, Isaac Barrow.

At Barrow's suggestion, Newton applied directly to Charles II for a special dispensation to allow him to continue as Lucasian Professor of Mathematics but without a fellowship (the specific aspect of his college career that required the taking of holy orders). There was no precedent for this and his chances of success were slim, but he had little choice.

The situation was complicated by the fact that he could tell no one the real reasons for the difficulty he faced – not even Barrow, who was the King's Chaplain and a devout Trinitarian. Barrow must therefore have defended Newton entirely because of his unique talent, without knowing the reasons for the younger man's difficulty. Fortunately for Newton, Barrow was in the right place at the right time: as Chaplain, he had the ear of the King. Although less able to influence Charles, Babington was of assistance because he was still a senior fellow and involved in the decision-making processes of the college on such matters. Together, he and Barrow were able to smooth a very rocky path.

In early March 1675 Newton visited London to apply for a special dispensation and, astonishingly, by the end of April he had it. His Majesty had granted the Lucasian Professor and all subsequent holders of the chair exemption from holy orders and was willing 'to give all just encouragement to learned men who are & shall be elected to the said professorship'.[31]

No one was more surprised than Newton himself. He had expected a fight, and, true to form, he had gone to extraordinary lengths to prepare for it. Since 1672 he had spent time away from his alchemical experiments and had investigated deeply buried, anachronistic aspects of theology, ancient wisdom and the canon of comparative religion, immersing himself in the netherworld of biblical roots. And once again, true to his nature, he allowed it to take him over. What had begun as a means of self-defence had developed into an obsession taking him deep into the origins of modern civilisation.

He followed his established method. He studied the documentary evidence, analysed it, and then produced his own deductions in a notebook divided into headings.[32] By asking himself the direct question What does the Bible actually say on the matter of the Holy Trinity? he pinpointed what he saw as two major flaws in the Trinitarian credo. Together these led him to believe that Trinitarianism was a deliberate, calculated lie, perpetrated through the ages by a series of self-interested pontiffs.

The first fault he highlighted was that Trinitarianism cannot be supported completely by scriptural analysis. Newton filled his notebook with quote after quote extracted from the Bible supporting his case. From II Corinthians 4:4, for example, he listed the reference to Christ as 'the image of God'. He then added the first chapter of Paul's epistle to the Hebrews, in which it is said, 'Thou [Christ] has loved righteousness & hated iniquity, therefore God, even *thy* God, has anointed thee with the oil of gladness above thy fellows.' In the margin to this passage in his notebook, Newton wrote, 'Therefore the Father is God of the Son.'[33]

According to Newton, the second problem with Trinitarianism was that it was counter-logical. He shared the view of his contemporaries that certain phenomena described in the Bible could never be understood, or might be explained only with divine knowledge. Among these were the ability to turn water into wine, the gift of healing by touch, virgin birth and resurrections. These, according to Newton, did not defy logic; but three equalling one and one equalling three definitely did.

As he saw it, Trinitarianism was an institutionalised contrivance, and to prove it he reached further and further back, to the roots of the doctrine, to reveal the origins of the deceit.

Eventually he found it. The blasphemy, he concluded, had begun with the Council of Nicea convened in 325 by the Roman emperor Constantine. Constantine's objective had been to clarify various ambiguities within the Christian doctrine as it blossomed into a global religion. During the council, a dispute had arisen between the Alexandrian priest Arius, who believed that God and Christ were separate entities, and the Bishop of Alexandria, Athanasius (later beatified by the Church), who argued for the notion of Homoousion – the doctrine that God and Jesus are of the same substance. The council agreed with Athanasius, and the doctrine of Homoousion was generally adopted. But that was not the end of the matter: the feud continued for several decades, with the rival camps of Arius and Athanasius fighting over the conceptual framework of Christianity. Because of the self-interested scheming and the temporal ambitions of corrupt ministers, the Arian view gradually became outlawed and the official line of the Council of Nicea was adopted by the Roman Catholic Church.

The unravelling of this thread of corruption reinforced further Newton's hatred for Catholicism. A reading of The Revelation of St John the Divine – the Book of Revelation – led him to agree with radical Puritanism's identification of the Devil with the Catholic Church and the 'Day of Reckoning' with the eventual establishment of the 'True Church', after the destruction of the 'Evil One', Rome. Little wonder, then, that Newton felt an almost physical disgust with the notion of taking holy orders. The Anglican Church had retained so much of Catholicism that he found offensive, including the most despised concept of all, the belief in a Holy Trinity – Father, Son and Holy Ghost, one and indivisible, three in one, one in three: a notion which was in itself anathema to any self-respecting mathematician.

Newton's researches appeared in *Observations upon the Prophecies of Daniel*, published posthumously in 1733, but large sections of his writings have been lost. He is known to have prepared a text called the 'History of the Church', but it was never published and disappeared sometime during the nineteenth century. All that can be traced is a bundle of about 800 pages of rough notes and drafts sold in the Sotheby auction of 1936 to a Gabriel Wells as Lot 249. About half of this lot is now in the Yahuda Manuscript Collection of the Jewish National and University Library in Jerusalem, referenced as Yahuda MS 15. Other sections (in some cases mere scraps) are to be found in libraries scattered around Europe and secreted away in private collections. From what can be pieced together from these, 'History of the Church' embraced the twin obsessions of Newton's religion notebooks and other writings – his view of the incarnation of Christ and the creation of what he refers to as the 'false image' or 'the Whore of Babylon', the Roman Catholic Church.

But why was Newton so fanatically opposed to the concept of the Trinity? Such obsessiveness could not be based only upon a distaste for the illogical. Would it be unreasonable to suggest that he wanted to identify himself with Christ? After all, was he not the only child of a dead father, born on Christmas Day and (so he believed) possessed of unparalleled ability and unique talents? And could it be that identification with his long-dead father – a man he had never known, but whom he knew was uneducated – was so repugnant to him that he could not contemplate the notion of a Trinity, a concept involving not just attachment to 'the Father' but the sharing of identity? Newton, the man who had adopted the pseudonym 'One Holy God', was a perpetual loner. He could not even show love towards his mother; he felt only disgust at her betrayal. The possibility of unity or amalgamation could only engender further feelings of abhorrence in this most individualistic and private of men.

Isaac Newton's view of God and the universe was not pantheistic – he did not regard the concepts of 'God' and 'universe' as identical. Nor was he a Gnostic in the one true sense of the term – he didn't believe that the spirit or soul can be redeemed or liberated from the material world by the acquisition of 'spiritual' knowledge. But, although unconventional by any standards of the time, his belief was consistent. He was enamoured of the Puritan concept of the twinned worlds of God's works and God's word, and he saw the Creator's presence – if not his 'being' – in all things. Life was a riddle to be understood, a code to be cracked, as a duty to the divine. As God was part of all Nature and all learning, God was also history, the present and the future. Consequently, Newton believed the unravelling of the *true* past to be his duty in the same way that it was his destiny to understand the workings of light, of gravity, of alchemy and mathematics. He called this 'the duty of the first moment', and he believed in what can only be described as an 'alchemical history'.[34]

An understanding of alchemical history not only revealed the meaning and purpose of existence, but, through analysis of the 'true past' (as revealed through study of the Bible), it could lead to the decoding of biblical prophecy, and elucidate the wisdom of the ancients.

Newton shared the view of many intellectuals of the period that the most ancient civilisation was also the most knowledgeable, the most pure, the most advanced. 'So then the first religion', he wrote, 'was the most rational of all others till the nations corrupted it. For there is no way without revelation to come to the knowledge of a deity but by the frame of nature.'[35] According to his own religious faith, this ancient civilisation was that of Israel, the land of the Old Testament, the world of Moses and Ezekiel. The rest – the Romans, the Greeks, the Assyrians – were mere copyists, followers. The chronology of the history of humankind had therefore to be reappraised and rewritten.*

Although this may seem madness to us, it was of the greatest importance to Newton, for the simple reason that he needed to justify his whole world-view, to reveal the falsehood that was Trinitarianism, because in so doing he could confirm his subconscious belief that he was superhuman.

The way into the chronology was of course the Bible, and in particular the Book of Revelation, about which Newton wrote in his notebook, 'There is no book in all the scriptures so much recommended & guarded by providence as this.'[37] And, as always, the only way he could analyse the text properly was to establish rules. He created fifteen of them, and the key to the system was the same as his approach to science – to reduce everything to its simplest form. An example is the ninth rule:

It is the perfection of all God's works that they are done with the greatest simplicity ... And therefore as they that would understand the frame of the world must endeavour to reduce their knowledge to all possible simplicity, so it must be in seeking to understand these visions.[38]

His conclusions were nothing if not diverse. He reasoned that because God's work and God's word came from the same Creator, then Nature and Scripture were also one and the same. Scripture was a communicable manifestation or interpretation of Nature, and as such could be viewed as a blueprint for life – a key to all meaning.

He also reached conclusions about the theological detail. He

---

* Newton wrote a long and clumsy book on the subject called *The Chronology of Ancient Kingdoms Amended*, which was published posthumously in 1728 by his nephew-in-law John Conduitt. As the title suggests, its contents reappraise the time-scales and the interdependence of the great civilisations of history; however, one scholar has said, 'Newton's *Chronology* is such an incredibly difficult work to decipher, that it will probably have to be passed one day through a computer.'[36]

believed in the literal truth of the Creation story – that the world was made in seven days by a divine hand – but he qualified this with an ingenious addendum. Nowhere in the Scriptures, he reasoned, does it say that all seven days were of equal length: because during the first two days there was no Earth and therefore no twenty-four-hour day based upon planetary rotation, the length of a 'day' could be anything the Lord wished.

He went against the intellectual trend of the time by dismissing the fashionable simplistic notion of devils and demons and held the rather modern view that evil spirits are manifestations of a disordered mind and that the Devil is an illusory figure created by human imagination. Yet he also followed what would now seem totally irrational theories in order to develop his fifteen rules. In this he was greatly influenced by the ideas of the writer and mystic Joseph Mede.

Mede was a Puritan proselytiser who developed a technique for dating the events of the Bible and interpreting the prophecies. He wrote a book called *A Key to the Apocalypse*, published in 1627, which proposed a method by which 'days' in the Old Testament could be interpreted as 'years', so enabling him to chronicle the various proclamations of the prophets and, most importantly, date the 'end of the world'.

Blending Mede with his scientific understanding and powers of critical analysis, Newton arrived at some ingenious interpretations of biblical events. A prophecy concerning the Saracen monarchy serves as an example.

The Old Testament refers in detail to the torment inflicted by a wrathful God in the form of a plague of locusts. For reasons that remain unclear, Newton decided that the locust could be taken as being symbolic of the Saracen monarchy. Knowing that the life-span of the locust was five months, two life-spans equalled ten months, or 300 days, and so he deduced in his amended chronology of the past that the Saracen Empire had lasted 300 years – a figure that fitted neatly with the historical fact that the Saracens had acted as aggressors towards the Roman Empire between 637 and 936, for 300 years (give or take a year).

Using similar reasoning, and juxtapositioning chronologies derived from their iconoclastic interpretation of the Scriptures, Joseph Mede (and others, including Henry More) had concluded that the end of the world would come sometime during the second half of the seventeenth century. This was arrived at by the following argument.

According to their Puritan ideals, the Antichrist was the Roman Catholic Church. In the Book of Revelation, the reign of the Antichrist was prophesied to last 1,260 days (mentioned as forty-two months, 1,260 days or three and a half years). This was interpreted as 1,260 years. The date at which the Roman Catholic Church took control of the Christian movement they agreed to place during the reign of the Roman Emperor Theodosius (about AD 400). Consequently, the end of the world was expected sometime during the final decades of the seventeenth century, and certainly no later than AD 1700.

Newton strongly disagreed with much of this scheme. For him, the start of the evil reign of the Roman Catholic Church was to be dated not from the Church's earliest origins but from its peak – a date that varied according to different versions of his scheme. In one set of notes he placed this date around AD 800 and used Mede's figure of 1,260 years (taken from Revelation) to place the end of the world sometime during the twenty-first century. In another version he dated the height of the Roman Catholic Church precisely to the year AD 609 and took a period of 1,290 years from Daniel to place the Jews' return to reclaim Jerusalem during 1899. He then took a further prophecy from Daniel – that the return to Jerusalem would be followed precisely forty-nine years later by the second coming of Christ, placing this singular event sometime during 1948.

In still further refinements and adaptations he used a variety of starting dates and alternative numbers of days/years taken from various prophets and arrived at different dates for the end of the world.*

Initially, these prophetic interpretations appear strangely at odds with Newton's normally strict rationalism, but he had given them very careful thought. Following a line of intellectual development dating back to Paracelsus, Newton did not see this form of prophecy in the way we perceive it today.[39] To him it was not a fairground gimmick or party trick but a demonstration that humankind was governed by Providence.

Intellectuals had long tried to 'prove' the existence of an all-

* Some of his reworking of this theme is recorded in his notes in Keynes MS 5 in the library of King's College, Cambridge.

pervading Creator by reason alone. Descartes and the philosopher and alchemist John Dury had many arguments about how 'scepticism' (perceived at the time as an insidious plague leading to atheism and hell) could be fought. Descartes chose mathematics; Dury, like Newton, relied upon biblical prophecy.[40]

In spite of his obvious conviction and dedication, Newton's surviving writings on the subject are a rambling muddle (like his shambolic *Observations upon the Prophecies of Daniel*). Yet this work and other seemingly illogical pursuits occupied him until the day he died. Even during the last weeks of his life he was tinkering with his ideas of prophecy and biblical interpretation, constantly reappraising the date at which the Day of Judgement would come, seeing it as an event that was preordained but at the same time one whose date humans might deduce.

This demonstrates both the radical difference between the intellectual perspectives of the seventeenth and twentieth centuries and, perhaps more importantly, the dark well of Newton's own personal insecurity – his need to prove himself different from other men.

Just as science and alchemy offered shelter and solace from a world which, at the root of his being, he did not much care for, Newton also found a home within religion. He believed wholeheartedly in the notion that God had created the world and controlled events with a divine hand, but he did not like the outcome – principally because, to him, humankind had the habit of constantly tarnishing Creation. Aside from his desire to affirm his own unique nature, by retreating into the Scriptures he could identify a purer universe and could delight in predicting when the whole ugly edifice of the contemporary world would fall.

The clearest manifestation of this desire is the enormous effort he poured into his reconstruction of the plan of the Temple of Solomon, seeing this as a paradigm for the entire future of the world.

Newton, like many other thinkers of his era, believed the religious edifices of ancient civilisations were more than mere places of worship. The ancients had not expended such enormous effort simply to preserve their culture and their world-view: they had constructed temples and monuments as Earthly representations of the universe. (Examples of those still in existence may include stone circles found throughout Europe and the Great Pyramid at Giza.[41])

To Newton, the most profound ancient was King Solomon, whom

he called 'the greatest philosopher in the world'.[42] The temple that Solomon originally built around 1000 BC, on a site in Jerusalem already sacred to the Jews, was the most hallowed symbol of wisdom and faith long before Newton put his own personal interpretation upon it. Almost from the time of its construction until the Enlightenment, some 3,000 years later, it was as revered as the Pyramids or Stonehenge had been by the pagan faithful who built them.* Newton, however, asserted that Solomon had deliberately designed his temple with privileged eyes and the guidance of the Lord, not only as a paradigm for the universe but as a pattern for the future of the human race. So, taking descriptions of the temple from the Book of Ezekiel, in not one but three different languages, he set out to draw its floor plan, seeing this as a key to the wisdom of the ancients and a route to elucidating biblical prophecy. What he ended up with was an elaborate diagram (Plate 11) which he believed represented a plan for the future of the world.

Newton believed that the dimensions and geometry of the floor plan gave further clues to time-scales and pronouncements of the great biblical prophets (especially Ezekiel, St John the Divine and Daniel). Combining this floor plan with his interpretations of Scripture allowed him to produce an even more detailed outline for a revised chronology of both the past and the future, with new dates for such events as the end of the Catholic Church's domination on Earth, the second coming of Christ and the Day of Judgement.

As well as providing information for his prophetic interpretations, the configuration of Solomon's Temple also influenced Newton in developing his image of gravitation.

Newton described the heart of the ancient temple as 'a fire for offering sacrifices [that] burned perpetually in the middle of a sacred place',[44] visualising it as a fire around which the believers assembled. He called this arrangement a prytaneum, and described the way it represented the cosmos by saying, 'The whole heavens they reckoned to be the true & real temple of God & therefore that a prytaneum might deserve the name of his temple they framed it so as in the

---

* So potent has been the notion that God and his universe can be represented by geometry that the principle has survived into relatively modern times – it is visible in the construction of cathedrals and churches throughout the world.[43]

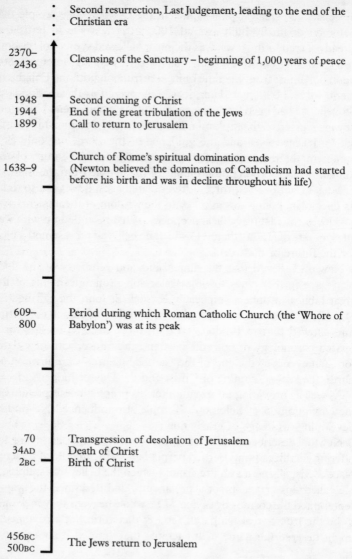

:
:
: Second resurrection, Last Judgement, leading to the end of the
  Christian era

2370–
2436   Cleansing of the Sanctuary – beginning of 1,000 years of peace

1948   Second coming of Christ
1944   End of the great tribulation of the Jews
1899   Call to return to Jerusalem

        Church of Rome's spiritual domination ends
1638–9 (Newton believed the domination of Catholicism had started
        before his birth and was in decline throughout his life)

609–   Period during which Roman Catholic Church (the 'Whore of
800   Babylon') was at its peak

70   Transgression of desolation of Jerusalem
34AD   Death of Christ
2BC   Birth of Christ

456BC
500BC   The Jews return to Jerusalem

*Figure 6. Newton's chronology in brief outline. (Although begun during
the 1670s, Newton continued his work of interpretation until the month of
his death in 1727.)*

fittest manner to represent the whole system of the heavens.'[45]

Although Newton was not alone in his belief that the temple was a microcosmic model, it has been suggested by at least one Newton scholar that the image of a fire at its centre with disciples arranged in a circle around the flames acted as another stimulus to Newton's concept of universal gravitation.* Key to this is the idea that, instead of simply seeing the rays of light as radiating *outwards* from the fire, Newton might instead have visualised them as a force *attracting* the disciples *towards* the centre. It is further implied that the similarity between this and the 'lines of force' suggested by the shape of the Star Regulus of Antimony could not have escaped him. In this scheme, the parallels between the solar system and the temple are apparent: the planets relate to the disciples, and the temple fire (sometimes called 'the fire at the heart of the world'[46]) is the model for the Sun.

During the mid-1670s, when Newton was first investigating ancient lore, his view of gravity was still a long way from the all-embracing concept expressed in the *Principia* a decade later. He had concluded that the receding force experienced by a planet in circular motion was described by the inverse square law, but he had yet to reach a detailed conclusion about the nature of an attractive force that countered this. His deep immersion in alchemical experiment and the ancient roots of theology must now have influenced his thinking towards a broader view of the universe, offering him possibilities beyond the realm of orthodox teaching and accepted philosophy. If the suggestive pattern of the Star Regulus and the arrangement of the prytaneum at the heart of Solomon's temple offered nothing in the way of quantitative empirical information for his elucidation of universal gravitation, they may nevertheless have provided signposts along the way to his grand conclusions.

These markers may have been subconscious and only manifested themselves later, as the ideas that led Newton to the principle of universal gravitation came together during the 1680s. There is no

---

* This suggestion is explored in great depth by the late Professor B. J. Dobbs in her highly academic treatise *The Janus Faces of Genius: The Role of Alchemy in Newton's Thought* (Cambridge: Cambridge University Press, 1991), but Newton scholars around the world are still divided as to the validity of the idea.

surviving record of an explicit reference to the Star Regulus or the Temple of Solomon to support the idea that they may have symbolised an attractive force, but, as we have already seen, the concept of gravitation did not come to him in a blinding flash of inspiration: it was arrived at by a process of gradual awakening over a period of two decades. And in one respect it is not surprising that there is no explicit record of how Newton's religious discoveries led directly to the formulation of the concept of universal gravitation. By the time these subconscious triggers had played their part (perhaps as late as the mid-1680s, when Newton was starting to write the *Principia*) he had already encapsulated the groundwork in strict mathematical language and could express it in orthodox terms. He then had no need, nor incentive, to explain this source of inspiration.

Newton perceived himself as the new Solomon and believed that it was his God-given duty to unlock the secrets of Nature, whether they were scientific, alchemical or theological. Such efforts were his reason for living, his mission, and he could not rest until he had fulfilled his dream. Even though he was submerging himself in the arcane world of alchemy and the Scriptures, he could still find stimulation in orthodox science, for this too was a manifestation of God's meaning and God's plan. It was also a crucial ingredient in the development of his most important ideas. But in its more open world he would, by the very nature of his exposure, meet unexpected opposition and some most unwelcome challenges.

CHAPTER 8

Feuds

He struggled to keep the parcel of himself from becoming unwrapped and scattered.

LEON EDEL ON HENRY THOREAU[1]

Newton's first lecture as Lucasian Professor took place at Trinity College in January 1670. He began with the subject of optics – an area of study with which his predecessor, Isaac Barrow, had concluded the previous year. For all his talents, Barrow had delivered a series of talks based upon a far simpler and largely incorrect set of theories; Newton's own discoveries, supported by rigorous mathematics, were in a totally different class.

The late invention of telescopes has so exercised most of the geometers, that they seem to have left nothing unattempted in optics, no room for further improvements . . . It may seem a vain endeavour and a useless labour, if I shall again undertake the handling of this science. But, since I observe the geometers hitherto mistaken in a particular property of light, that belongs to its refractions, tacitly finding their demonstrations on a certain physical hypothesis not well established: I judge it will not be unacceptable if I bring the principles of this science to a more strict examination, and subjoin, what to be true by manifold experience, to what my reverent predecessor has last delivered from this place.[2]

Despite the boldness of his manner and the conviction of a man who knows his own talents, the tiny band of students hardly took notice of him as he described the basis of theories that would form one of the twin pillars of his great contributions to the world of physics – material that would find its way into the *Opticks*.

Not a single student showed up for Newton's second lecture, and throughout almost every lecture for the next seventeen years (when he gave up all pretence of teaching and turned his position into a sinecure) Newton talked to an empty room, listening merely to his own voice bouncing back at him. According to his subsizar assistant Humphrey Newton (who arrived in 1685), during one term of each academic year, whenever he was in residence at the university, his master would dutifully set off across the Great Court and deliver his talk. If no one appeared, as was almost always the case, he would reduce the lecture from the intended thirty minutes to fifteen and return quickly to his rooms, where he would continue with what he considered his real work – attending his furnaces and his alchemical and scientific experiments. As Humphrey Newton describes it, 'So few went to hear him, & fewer that understood him, that oftimes he did in a manner, for want of hearers, read to the walls.'[3]

That the university students gave his lectures a wide berth meant nothing to Newton. Aside from the fact that he was only fulfilling the obligations of his tenure, the absenteeism of the student body was by no means limited to his lectures, nor even to those subjects considered especially demanding. When nobody appeared at one of Edmund Castell's first lectures after he was appointed the first Adams Professor of Arabic in 1666, he pinned to the door of the lecture theatre a note which read, 'Tomorrow the Professor of Arabic goes into the wilderness.'[4]

So Newton lectured to the walls – firstly on optics and some mathematics, and then on the principles later enshrined in the *Principia*. Following Barrow, Newton too was formally obliged to deliver notes covering a minimum of ten lectures per annum to the university library; but, like his predecessor, he rarely complied. Yet the quality of the work he did deliver more than compensated for the lack of quantity.

Soon after taking up the professorship, Newton reduced the times he lectured to one term a year, and as he began to take more and more time away from the university the number of his appearances in the lecture theatre dropped accordingly. He never enjoyed teaching and cared little for students. Like many men of his stature, he found it difficult to bring his intellect into line with young students or those of far lesser ability, even for a short period.

He is known to have only ever had three tutees in his charge throughout his entire academic career, spaced at irregular intervals from 1669 to 1696. The first of these was a fellow-commoner with the wonderful name of St Leger Scroope, who arrived in the spring of 1669, shortly before Newton's elevation to the Lucasian chair. The second was one George Markham, whom Newton taught for a brief spell during the summer of 1680. And finally there was a William Sacheverell, a transitory charge a few months after the publication of the *Principia* in 1687. None of them went on to achieve anything of intellectual merit, and little is now known of any of them. Surprisingly, perhaps, even less was recorded by them of their experiences as Newton's students. Considering that the Lucasian Professor was, by the end of the seventeenth century, the most famous intellectual in the world, it is odd that almost nothing was recalled by any of those who may have received instruction from him in their younger days. We can only conclude that the level of interest shown by the average Cambridge student towards the material that Newton taught was matched by his talents as a teacher.

Newton's overriding concern was for his own researches. His first scientific endeavour as Lucasian Professor was a return to optics – the subject of his first set of lectures. During the mid-1660s he had demonstrated that white light was composed of many colours which were split by a prism, giving red light at one end of the spectrum and blue at the other. During late 1669 and early 1670 (contemporaneously with his first forays into alchemy) he returned to these experiments to clarify his earlier ideas and to demonstrate a theory of colours beyond any reasonable doubt. Using little more than a few pieces of card and a couple of glass prisms, the first of this new set of experiments, later known as the *experimentum crucis*, was, belatedly, the first sign the scientific world had of his genius as an experimenter, for it was as beautiful in its simplicity as it was effective in encapsulating Newton's theory.

What Newton was attempting was to *prove* that white light was composed of the spectrum of colours and that the prism split 'pure' sunlight into these colours because the different components of the spectrum were 'bent' or refracted to different extents. He had observed this phenomenon back in 1664 by conducting two simple experiments using prisms, and had recorded his thoughts in his 'Quaestiones', the Philosophical Notebook. After passing the refracted light from one prism into a second, he was able to say in an essay entitled 'Of Colours' (written in 1666) that 'The purely red rays refracted by the second prism made no other colours but red & the purely blue ones no other colours but blue ones.'[5]

This demonstration of the way in which coloured light is refracted ran counter to the prevailing view which held that light is modified or altered in nature by passing through a 'dark' medium such as glass. Light is not modified but simply split into its component parts by refraction. But these simple procedures did not satisfy him: he wanted clearer, quantifiable evidence to support his embryonic theory.

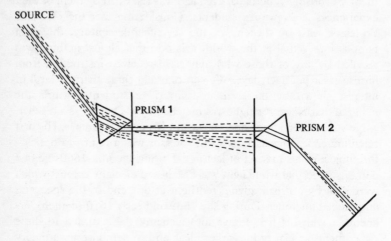

SOURCE

PRISM 1

PRISM 2

Figure 7. Newton's first experiment in the series, known as the experimentum crucis.

The first of the new experiments (the *experimentum crucis*) again involved passing sunlight through a prism. Parts of the spectrum produced were then made to pass through a tiny hole in a card placed close to it. Newton was thus able to let just one of the colours of the spectrum pass through the hole. This light was then sent through a second pierced card and into another prism, and the refracted light from that was allowed to fall on a white card behind it. Newton found that if he allowed only light from one end of the spectrum (the blue light) to pass through the system, it was refracted far more than if he only allowed red light to pass through. 'To bring forth my opinion more distinctly,' he wrote, 'in the first instance I find, that rays which are refracted more than others of the same incidence, exhibit purple and violet colours, while those exhibit red which are least refracted, and those blue, green, and yellow, which have intermediate refractions."[6]

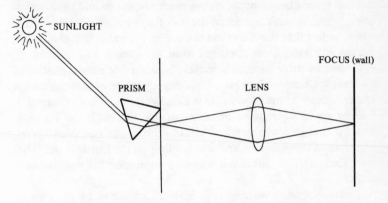

Figure 8. Passing light from a prism through a lens.

Still not content, he then set up another ingenious experiment. Using a prism, he produced a spectrum and passed this through a lens. The lens was moved so that it focused the beam on the far wall of Newton's room, where the light appeared as a white spot. The individual colours of the spectrum produced by the prism had been reconstituted as white light.

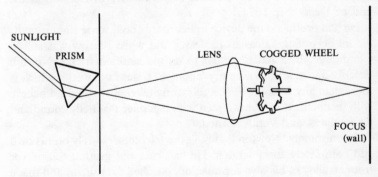

Figure 9. Producing single colours using a cogged wheel.

Finally, he mounted a specially constructed wheel between the lens and the wall at a position such that when he turned it the cogs intercepted different colours emerging from the lens before they reached the focus to produce white light. Turning the wheel slowly, a succession of different colours appeared at the focus, because the cog had taken out one of the components of what should have constituted white light. A further variation of this involved spinning the wheel so fast that the eye could not follow the individual changes in colour, when the focus appeared white once more.

In spite of these successes, initially Newton told almost no one of his findings. Barrow, however, was certainly aware of Newton's work and encouraged him. Now living in London (as the King's Chaplain), Barrow could have acted as an influential link with the scientific world if that role was required of him. But it was not. Newton felt unprepared for publicity, and was fending off the insistent publisher John Collins, who harboured hopes of publishing his mathematical work.

Today, scientists wishing to publish their work will write a paper and seek publication through a learned journal such as *Nature* or *Science*. In the late seventeenth century, the usual way to publish scientific work in England was through the Royal Society. Newton now had access to the Royal Society if he wanted it, through Isaac Barrow, but instead he chose silence. The first anyone outside Cambridge heard of him was at least two years after the *experimentum crucis*, when a reflecting telescope he had constructed was presented to the Royal Society in London at the end of December 1671.

He had produced the device in February 1669, some eight months before becoming Lucasian Professor and while Barrow was still in Cambridge. In a letter written to an unidentified friend, Newton described it as able to magnify 'about 40 times in diameter which is more than any 6 foot tube [i.e. refracting telescope] can do, I believe with distinctness . . . I have seen with it Jupiter distinctly round and his satellites, and Venus horned.'[7]

Astonishingly, Newton had built the telescope entirely on his own and completely from scratch. He had cast and ground the mirror from an alloy of his own formulation, polished the mirror, and made the tube and the mount and fittings. In old age, he discussed making the telescope with John Conduitt. 'I asked him where he had it made,' Conduitt recounts, 'he said he made it himself, & when I asked him where he got his tools he said he made them himself & laughing added if I had stayed for other people to make my tools & things for me, I would have never made anything of it . . .'[8]

It is difficult to overemphasise the great skill and application this required. For some years, craftsmen in London had been trying to produce a usable reflecting telescope and had failed utterly. In his book *Optica Promota*, published in 1663, the mathematician James Gregory had described a design for one and had later employed some of the best craftsmen in the country to produce a working version of the design, but in vain.

At Barrow's insistence, towards the end of 1671 Newton reluctantly agreed to let him show the telescope to a small group of associates at a meeting of the Royal Society. The instrument was little more than six inches long, but of exquisite design. As Newton had boasted, it was more powerful than a refracting telescope a dozen times its size, and it caused an immediate stir at its first showing. No less a figure than John Flamsteed (shortly to be appointed Astronomer Royal) was captivated by it; Christopher Wren, Robert Moray and Sir Paul Neile took it to Whitehall on behalf of the Royal Society to give Charles II a personal demonstration, and within a few days of its reception, in early January 1672, Newton received a gracious and flattering letter from the Secretary of the Royal Society, Henry Oldenburg, which reported that the telescope had been 'examined here by some of the most eminent in optical science and practice, and applauded by them'. Oldenburg then went on to explain that the Royal Society wished, on Newton's behalf, to 'secure this invention from the usurpation of foreigners'.[9]

To establish priority, they wrote to the Dutch scientist Christiaan Huygens, who was considered the leading optics expert of the time. Shortly after, the society's council ordered the construction of two larger models – one four feet in length, the other six.

Newton's device was a significant technical achievement, and of unsurpassed quality and power for the time, but, for all the overzealous claims of the Royal Society, he never pretended to be the first to design a reflecting telescope. That honour belonged to James Gregory, and the concept, if not a practical design, had been around for over a century.

One of the reasons why Newton had constructed the instrument was because he was disheartened by the poor quality of the commonly used refracting telescope invented by Hans Lippershey in

1608 and wanted to produce a more efficient instrument. Because they use lenses, the images produced by refracting telescopes are easily blemished by a phenomenon called *chromatic aberration*. Unless the lens is perfect, light rays of different wavelengths falling on it are not brought to exactly the same focus and a distorted image results. A reflecting telescope uses a mirror to focus the light, and if constructed properly it produces no chromatic aberration. Although no one but Newton was aware of it at the time, chromatic aberration could be explained by his own theory that white light was composed of all the colours of the spectrum, and it only later became clear to Newton's contemporaries that he had actually gone to the trouble of constructing the telescope principally to illustrate this.

Within days of receiving Oldenburg's first letter, Newton replied with more than a little false modesty, saying:

> At the reading of your letter I was surprised to see so much care taken about securing an invention to me of which I have hitherto had so little value. And therefore since the R. Society is pleased to think it worth patronising, I must acknowledge it deserves much more of them for that, than of me, who had not the communication of it been desired, might have let it [the reflecting telescope] still remained in private as it has already done some years.[10]

Within days, Oldenburg passed on more news – late the previous December, Newton had been nominated for election as a fellow by one of the founders of the Royal Society, the astronomer Seth Ward, and by 11 January 1672 he had been duly voted in. Newton was delighted, and in a letter arriving in London on 19 January he felt sufficiently emboldened to mention the optical theory linked to his creation of the telescope:

> I desire that in your next letter you would inform me for what time the Society continue their weekly meetings, because if they continue them for any time I am purposing them, to be considered of & examined, an account of a philosophical discovery which induced me to the making of the said telescope, & which I doubt not but will prove much more grateful then the communication of that instrument, being in my judgement the oddest

if not the most considerable detection which has hitherto been
made in the operations of Nature.[11]

Then on 6 February his ego had been pampered sufficiently for
him to send Oldenburg a description of his theory of light in a lengthy
letter which has since become known as the 'Theory of Light and
Colours'. Within a week a reply had arrived in Cambridge, and a
nervous Newton – unsure of the response he would receive for such
a revolutionary and complex theory – opened a letter filled with praise.
His paper, Oldenburg informed him, had been read to the Royal
Society on 8 February and had been exceptionally well received.

The Royal Society had begun life in the rooms of the Oxford don
John Wilkins in Wadham College, Oxford, in 1648 and had grown
quickly to include important natural philosophers of the day, including Robert Boyle, Henry Oldenburg and Seth Ward. Boyle dubbed
the group the 'Invisible College', and they gathered at irregular intervals to discuss the latest ideas emerging from Europe and from their
own academic contacts.

By 1659 the group had taken on a more formal structure and had
acquired a regular meeting-place in Gresham College in Bishopsgate,
London. The college was the former home of Sir Thomas Gresham,
who had stipulated in his will that the building should become a
place of learning. Three years later, in 1662, as the twenty-year-old
Newton had entered his second year as an undergraduate at Trinity,
the Royal Society received its official name and sanction – a royal
charter from Charles II. A second charter the following year bestowed
its arms and a motto: '*Nullius in Verba*' (from Horace, and loosely
translated as 'Take no one's word for it').

The fellows of the Royal Society during the seventeenth century
were almost all gentleman scholars and academics from Oxford and
Cambridge. This was an age racked by plague and constant petty
wars between European states, an era during which 90 per cent of
the population lived in what we would now consider abject poverty
while the lucky remainder dabbled in business affairs and intellectual
pursuits. As Molière and Dryden enjoyed their most successful years
and Spinoza delivered his *Tractatus Theologico-Politicus*, the
philosophically minded gentlemen of England were emulating the
early Continental societies such as the Pinelli Circle of Galileo's
Padua some eighty years earlier.

Once the institution moved to rooms in Gresham College, meetings became more formal and were carefully organised around the reading of papers followed by general discussions among the members. Later came the addition of regular demonstrations, and within a few years of its foundation the Royal Society had become the arbiter of all scientific life in England and anyone recognised as having any scientific value was invited to join.

The earliest remit of the Royal Society was, according to one of the founders, Joseph Glanvill, 'to defend against the attacks of the Aristotelian traditionalists, to enlarge knowledge by observation and experiment'.[12] This was a noble and far-sighted manifesto, and it has guided the society for over three centuries, but it was certainly not adhered to wholeheartedly in the early years.

The society was open-minded enough to entertain numerous crackpot ideas along with serious scientific endeavours: its members conducted crude vivisection experiments, and on one occasion they tried to make insects from cheese (following the fashionable belief in spontaneous generation). However, the work of some natural philosophers was considered to be outside the bounds of acceptability. John Dee, who had a great deal to offer the world of official scientific inquiry, was discredited posthumously by the Royal Society and his reputation was destroyed completely. Even today, almost four centuries after his death, Dee is seen primarily as a crazed alchemist and mystic, yet he had formulated a number of sensible scientific theories years before his time and should be seen more as a man led along a twin path of scientific and alchemical interest – not so very different to a man who went on to become the President of the Royal Society, Isaac Newton.

Religious tolerance also wore thin at the Royal Society. The great English philosopher Thomas Hobbes was never invited to be a fellow because of his professed materialist views, and Edmund Halley, a man instrumental in bringing Newton's *Principia* to the attention of the scientific community, was tolerated only so long as he kept his atheism to himself.

Following the Great Fire, in September 1666, the Royal Society was obliged to vacate its meeting-place temporarily to make way for the City of London authorities, who needed Gresham College as their headquarters for the planned rebuilding of London. Having connections in high places, as well as the patronage of the King himself, proved helpful and the learned gentlemen very quickly found a makeshift home. Henry Howard (later the Duke of Norfolk) offered them Arundel House almost before they were asked to vacate Gresham College and provided them with their first library, which marked the beginnings of the Royal Society's collection which today numbers some 200,000 volumes.

At the time of Newton's election, several now legendary names were members of the society. Christopher Wren was a fellow, and in 1681 took the mantle of President; Robert Boyle was a committed fellow who offered papers and experimental schemes on a regular basis; and a scientifically illiterate Samuel Pepys had been elected a fellow in 1665.*

From its earliest days as a formal organisation, one of the most important functions of the Royal Society had been publication of the *Correspondence* – a journal through which members could communicate ideas. By 1665 the *Correspondence* had evolved into the *Philosophical Transactions*, which became the forerunner of the scientific journal, and included papers, letters and minutes of meetings.

An indication of the revolutionary nature of Newton's 'Theory of Light and Colours' is that, before its publication in the February 1672 edition of the *Transactions*, not one of seventy-nine issues of the journal previously published had contained an experiment-based radical revision of an accepted scientific theory. The *Transactions* was full of articles based entirely upon casual observation or speculation. These ranged from medical reports such as 'An Observation Made upon the Motion of the Hearts of Two Animals after Their Being Cut Out' to 'New Observations of Spots in the Sun'. There were even papers concerning werewolves, two-headed calves, hermaphrodites and a singular item: 'An Account of a Foetus That Continued 46 Years in the Mother's Body', but nothing that produced a *verifiable* challenge to orthodoxy.

As Lucasian Professor at Cambridge, Newton had almost certainly seen copies of the *Transactions* before he was made a fellow, and he

---

* He became President in 1684, putting his imprimatur to the cover of the first edition of Newton's *Principia*, published by the Royal Society three years later.

must have been aware of the archaic *modus operandi* of contemporary natural philosophy. He should therefore have been prepared for the difficulties he was to face in trying to establish his own unique approach.

The writing and distribution of first the *Correspondence* and later the *Transactions* were originally the sole responsibility of the secretaries of the Royal Society (of which there were two at the time of Newton's election). The first edition of the *Transactions*, in March 1665, was masterminded by Henry Oldenburg, who also acted as Newton's point of contact with the society during his early years as a fellow. A German, who had lived in England for most of his adult life, Oldenburg had moved to Oxford in 1656. He was a lifelong supporter of Newton's ideas and, with great patience and diplomacy, acted as a buffer between the oversensitive Cambridge professor and the society's hot-headed and scheming Curator of Experiments, Robert Hooke. Indeed, the value of Oldenburg's patience and his frequent, measured intercession cannot be overestimated. Newton began to fall out with Hooke within just a few weeks of receiving the news that he had been invited to join the ranks of the scientific élite, and little more than twelve months later, before he had attended a single meeting, he was on the verge of resigning his fellowship.

Until his acceptance into the Royal Society, Newton had led a life of extreme intellectual seclusion in Cambridge. Not only was he cloistered from the rest of the world in a physical sense, he had almost no contact with anyone approaching his intellectual calibre. He had frequently aligned himself with older men who had been useful guides – in particular the triumvirate of Babington, More and Barrow – but otherwise his closest associates were Francis Aston and his room-mate, John Wickins. His arrival on the larger scientific scene brought him into immediate contact with a man with whom he clashed instinctively.

Born in 1635, Robert Hooke, the son of a clergyman, was a man who had experienced many childhood tribulations similar to those of his adversary, Newton. His father had committed suicide by hanging himself in 1648, when Robert was still in his mid-teens. As a boy, Robert had displayed an aptitude for drawing and painting, and after receiving a modest inheritance of £100 he had been packed off to London to be taught by the painter Sir Peter Lely. By good fortune, he came to the attention of Richard Busby, a master at Westminster

School, who realised that the boy's intellectual capabilities extended beyond his ability as an artist. Under Busby's tutelage Hooke received the best education available at the time and secured a place at Christ Church College, Oxford. There is no record of his obtaining a bachelor's degree, but he was nominated for an MA in 1663.

Hooke's university career followed a similar pattern to Newton's. He was obliged to make his way through his undergraduate studies as a servant, yet the combination of this with a fatherless adolescence nurtured a quite different personality. Newton's experiences as an undergraduate had encouraged his oversensitivity and introversion. A very different character by nature, Hooke's reaction to humiliation had been the very opposite. Whereas Newton shunned the company of others, Hooke was gregarious in the extreme.

After graduating, Hooke became Robert Boyle's paid assistant and worked in his Oxford laboratory. From there he became involved with the Invisible College and began to associate with the influential thinkers who later created the Royal Society. It was Boyle who later secured him the position of Curator of Experiments in London in 1662.

The Curator's job entailed the preparation and supervision of at least three or four demonstrations for each weekly meeting of the fellows, as well as extras suggested at short notice; but Hooke worked best under pressure and excelled in his position. Almost like a Renaissance figure, at one time or another he had produced theories of mechanics, gravity and optics. He also developed hypotheses in such disparate fields as geology, botany, cartography, anatomy, telescopes, microscopes and the workings of engines; he had constructed ingenious mechanical experiments and chemical apparatus, and had even proposed a collection of his designs for a flying machine.

Although he was not Newton's equal in terms of mathematical virtuosity or intensity of intellect, Hooke had a mind which in some ways worked in an equally brilliant, if entirely different, fashion. Whereas Newton could concentrate on a single problem for many decades, Hooke possessed a restless energy and flitted from one enthusiasm to another. He could never give his undivided attention to anything for long, and so to many he appeared something of a dilettante. His greatest work, *Micrographia*, was ostensibly a treatise about microscopy, but it also included a number of original theories concerning the nature of light. Published in 1665, it was a book that Newton knew well and secretly admired.

So, Hooke and Newton clashed on two levels – as personalities and as natural philosophers. Hooke loved the coffee-house, the gossip of his friends over a bottle of port, and the attentions of at least one mistress at a time – he recorded his sexual exploits and the quality of his orgasms in his diary. Newton lived a life of austerity and isolation within the walls of Trinity College, Cambridge. On a scientific level, Newton had only contempt for anyone who merely dipped into learning and did not drain it of blood or throttle it into intellectual submission.* For his part, Hooke saw Newton as a dried-up husk of a man – someone who was admittedly brilliant, but also obsessive, self-opinionated and with an excessively rosy self-image. They would inevitably have clashed professionally and personally, but their immense egos exaggerated their differences. Each subconsciously defended his own way of working, neither was able to allow the other credit, and they became and remained bitter enemies until Hooke's death in 1703.

The conflict began over Hooke's cursory treatment of Newton's 'Theory of Light and Colours'. It was the custom of the Royal Society to attempt to verify any serious scientific ideas presented to it, just as modern scientific journals require scrutiny and validation by a set of referees before a paper is published. As Curator of Experiments, it was Hooke's job to analyse Newton's theory and to write a report on it. Problems began when he decided to give the work only a cursory perusal (he later admitted spending just three hours analysing the complex paper). The reason for this only became clear later: Newton's theory was at odds with his own cherished ideas about the fundamental nature of light. Hooke's report began:

> I have perused the excellent discourse of Mr Newton . . . and I was not a little pleased with the niceness and curiosity of his observations. But although I wholly agree with him as to the

---

* In his *The Diaries of Robert Hooke: The Leonardo of London*, Richard Nichols makes the valuable point that, unlike Newton, Hooke could not devote himself entirely to research and was compelled to work hard at the Royal Society simply to make a living. In 1666 he was also one of three surveyors appointed by the City of London to initiate the rebuilding programme following the Great Fire.[13]

truth of those he has alleged, as having by many hundreds of trials found them so, yet as to his hypothesis of solving the phenomenon of colours thereby I confess I cannot yet see any undeniable argument to convince me of the certainty thereof. For all the experiments & observations I have hitherto made, nay and even those very experiments which he alleged, do seem to me to prove that light is nothing but a pulse or motion propagated through an homogeneous, uniform and transparent medium.[14]

Not surprisingly, this reaction infuriated Newton, who had never before opened himself up to his peers. His initial response, filtered as always through Oldenburg, was firm but cautious. 'I received your Feb 19th,' it began. 'And having considered Mr Hooke's observations on my discourse, am glad that so acute an objector has said nothing that can enervate any part of it . . . I doubt not but that upon severer examinations it will be found as certain a truth as I have asserted it.'[15]

Cautious the response may have been, but Hooke was instantly offended by Newton's superior tone. Perhaps because he knew he had not given enough thought to the Lucasian Professor's work and feared the younger man would, in his anger, stir up trouble for him, he immediately went on the attack.

A practised debater and possessed of considerable intellectual dexterity, Hooke immediately saw the weakness in Newton's paper and singled it out. Despite holding a different view concerning the nature of light, he could find nothing wrong with the professor's reasoning or observations. Instead he centred his attack upon what he saw as a contradiction in Newton's approach to his findings.

At the start of his paper Newton had made a point of stating his belief that hypotheses should never be mixed with verifiable facts. Yet, a little further in, it appeared he had done precisely that. Among the clearly described details of the *experimentum crucis* Newton had announced that 'it can no longer be disputed . . . whether light be a body'.[16] This was a clear pronouncement that light is corpuscular in nature rather than wavelike. Not only did Hooke take a very different stance on this matter, but he saw Newton's comment as a mere hypothesis and quite unproven.

Receiving a letter pointing out this apparent error of judgement,

Newton went into a brooding silence, angry with himself for making himself vulnerable. Meanwhile, Hooke played the advantage. When Newton's telescope had first been demonstrated at the Royal Society in January that year, he had grumbled that he had produced a similar device before Newton. John Collins, who had been at that meeting, later recalled:

> Mr Hooke moreover affirmed *coram multis* [in the presence of many] that in the year 1664 he made a little tube of about an inch long, to put in his fob, which performs more than any telescope of 50 foot long made after the common manner; but the Plague happening, which caused his absence, and the fire, which demanded his employments about the City, he neglected to persecute the same, being unwilling the glass grinders should know anything of secret.[17]

As Newton licked his wounds at Trinity College, Hooke now attempted to claim that he had produced a working reflecting telescope first.

Initially, Hooke's claim might be dismissed as egotistical fantasy, but he was after all the Curator of Experiments at the Royal Society and he possessed considerable practical skill. He was also a microscopy expert and, although he was slow to realise the revolutionary aspect of Newton's work and had little of the Lucasian Professor's mathematical or intuitive genius, he was certainly a very capable man. The claim that he had produced a powerful telescope small enough to fit into his watch is an exaggeration stated deliberately to impress, but he may have built a crude reflecting telescope a year or two before Newton.

On the other hand, Hooke had already gained something of a reputation for making unsubstantiated claims. On one occasion he had announced that he had developed over thirty different means of flight but could not go into details of them for fear his techniques would be stolen. This time he sought to prove his priority by trying to convince his friends at the Royal Society (respected figures such as Christopher Wren and the society's President, Lord Brouncker) to see things his way. But, unable to prove anything in writing or to produce any sort of working example as evidence, his efforts came to little and his colleagues saw his complaints as nothing more than his usual bluster. This served only to antagonise him further.

For a while, Newton ignored Hooke's claims. At the time, he was buoyed up by support from Huygens, who commanded far greater respect and influence within the scientific community, and was a man whom Newton admired greatly. Huygens had not only praised the telescope but was instantly (if temporarily) captivated by Newton's accompanying theory in the *Transactions*, declaring to the Royal Society that 'The theory of Mr Newton concerning light and colours appears highly ingenious to me.'[18]

But soon other critics began to comment upon Newton's theory. A Jesuit priest and professor of rhetoric in Paris named Ignance Gaston Pardies could not accept what he called Newton's 'very extraordinary hypothesis'.[19] He was a traditionalist who believed that white light was split into its individual colours because of the way it was incident on the prism (the way it entered the prism) and not because of the nature of the different types of light (different wavelengths) within the incident beam. He wrote to Newton via Oldenburg, arguing this point and politely asking for an explanation.

Newton's reaction, coming at a time when he was already extremely agitated by his scientific colleagues, was none too gentle. He started by dismissing Pardies as an amateur and then, when the professor persisted in contesting the issue through further letters, decided to give him a lecture on scientific method to shut him up:

In answer to this, it is to be observed that the doctrine which I explained concerning refraction and colours, consists only in certain properties of light, without regarding any hypothesis, by which those properties might be explained. For the best and safest method of philosophising seems to be, first to inquire diligently into the properties of things, and establishing those properties by experiments and then to proceed more slowly to hypotheses for explanations of them. For hypotheses should be subservient only in explaining the properties of things, but not answered in determining them; unless so far as they may furnish experiments. For if the possibility of hypotheses is to be the test of the truth and reality of things, I see not how certainty can be obtained in any science.[20]

After several more exchanges in which Newton explained his work in detail, Pardies finally accepted the theory. But Hooke, who was looking for any opportunity to escalate the feud and had been following the exchange between the two men in the *Transactions*, had taken exception to what he saw as Newton's superior and patronising attitude in his dealings with the Jesuit. In April 1672 he complained to Oldenburg, who then wrote to Newton outlining his objections.

In a letter written in May 1672, Oldenburg suggested that the Lucasian Professor should tone his letters down a little. The function of the Royal Society, he explained, was to act as an intermediary between parties who disagreed over philosophical issues. In particular, the *Transactions* was a forum for debate, and if contributors wished to have their material printed in it they should follow a code of practice and moderation. He then went on to stipulate that in future all correspondence would be printed without the names of objectors.

The letter caused such outrage in Cambridge that Newton could not bring himself to reply for two weeks. When he did, he rewarded Hooke's spite with a cartload of his own. 'I told you that it was indifferent to me whether they [his critics' letters] were printed with or without the author's names,' he declared angrily to Oldenburg. He went on to proclaim that he would now deliberately hold back material he was about to send him. 'Upon the receipt of your letter I deferred the sending of those things which I intended, and have determined to send you alone a part of what I prepared.'[21]

This was a reference to a new set of results from his optical experiments. He refused pointedly to allow Oldenburg to publish these, and he continued to withhold them until the *Opticks* appeared some three decades later. Consequently, his clash with Hooke deprived the scientific world of an important and complete set of theories until its creator felt he had won his battle with the scientific Establishment and his most voracious antagonist was safely buried.

Newton was quite clearly disorientated by this growing conflict. On the one hand, he believed that by acting as he did he could overcome the criticisms levelled against his work. Unable to deal with fault-finding in any form, from anyone, he concluded that he must withdraw, to put up the barricades and to defend his position with resolute silence. But, at the same time, he found he could not isolate himself for long. He had tasted recognition, even if it had been tainted by criticism. Part of him still needed to prove he was right.

In June 1672 he finally responded in writing to Hooke's criticisms of his paper and, with one brilliant blow, silenced his critics and demolished Hooke's claims. The letter was read before a meeting of the Royal Society (which Hooke was obliged to attend), and the minutes were later published in the *Transactions*.

'The first thing that offers itself is less agreeable to me, & I begin with it because it is so,' Newton began:

Mr Hooke thinks himself concerned to reprehend me for laying aside the thoughts of improving optics by *refractions*. But he knows well that it is not for one man to prescribe rules to the studies of another, especially not without understanding the grounds on which he proceeds. Had he obliged me by a private letter on this occasion, I would have acquainted him with my successes in the trials that I have made of that kind . . .[22]

He then went on to counter Hooke's claims that he had included hypotheses in his work by admitting that he *had* argued for the corpuscular nature of light but that 'I do it without any absolute positiveness, as the word *perhaps* intimates, & make it at most but a very plausible consequence of the doctrine, and not a fundamental supposition.'[23]

To his opponent's growing discomfiture, Newton proceeded with page after page of answers to each and every one of Hooke's original complaints, dissecting and dispatching each one with alacrity and succinctness. When the meeting was over, Hooke was officially chastised and instructed to conduct a complete reappraisal of Newton's original paper. This was to involve the reconstruction of the experiments Newton had himself described in detail, to verify his theory, and these were to be presented before the Royal Society at the earliest opportunity.

For a while the battle was over, but, embittered and seething in silence, Hooke was not going to take the matter lying down. This was merely the first loss of blood in what was to be a bitter contest lasting the rest of his life.

As Hooke licked his wounds, Newton still could not find peace. In a series of letters (channelled through the Royal Society and written between the autumn of 1672 and early 1673), Christiaan Huygens, who had shown such enthusiasm for his ideas a year earlier, now turned against him.

Although almost identical to Hooke's criticisms, Huygens's objections were based upon more considered judgement and were offered without any malicious motive. Newton's theory of colours, Huygens believed, was no theory at all, but an hypothesis.

To the modern mind, it seems odd that this same attack should be made without any form of experimental back-up from either dissenter. Huygens, like Hooke, could not accept what was then a completely novel approach – that a hypothesis is tested by experiment and dismissed only if the experiment shows that it is wrong. By 1673, Huygens had not conducted a single experiment in an attempt to prove or disprove Newton's theory. Instead he based his response upon *a priori* reasoning alone.

For some years the Royal Society had advocated *observation* to verify ideas. However, natural philosophers of the seventeenth century had still to realise that, if a series of experiments supported a hypothesis, then a law governing a phenomenon could be established by mathematical derivation from the experimental data. To the twentieth-century scientist, this is how all science is conducted – it is the modern scientific method – but Newton was the first to apply this method fully.* In spite of their claims to the contrary, natural philosophers of the 1670s were still thinking in ways reminiscent of Aristotle. More than any other thinker of the time, Newton (taking his lead from the innovative work of Galileo) was responsible for leading the way from the remnants of Scholasticism to the modern scientific age.

Today, it is taken as accepted practice that if a new hypothesis is proposed it can be tested by repeatable experiments. If these experiments support the hypothesis, then the hypothesis becomes a theory and a law describing the process may be created which can then be

---

* Francis Bacon is rightly considered the father of inductive scientific reasoning and Galileo had placed great importance upon both mathematical interpretation and experimental verification, but the combination of these methods reached full maturity with Newton.

1. *The traditional image of Newton watching the apple fall. For over three centuries this has been perceived as the moment of inspiration for the theory of gravity, but the story was almost certainly fabricated by Newton to disguise the truth.*

2. *The Woolsthorpe Manor House in which Newton was born and spent his early life.*

3. *The Royal Observatory at Greenwich during Flamsteed's time as Astronomer Royal.*

4. *Copernicus's heliocentric system.*   5. *Ptolemy's geocentric system.*

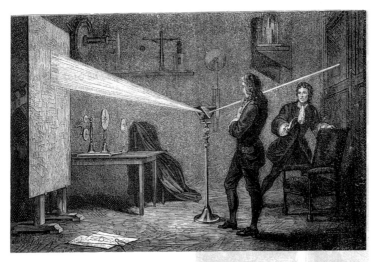

6. Newton conducting his earliest experiments on the nature of light in his rooms in Trinity. The other figure in this portrait is Newton's room-mate, the mysterious John Wickins.

7. A view of London produced during the early part of the 18th century, a city almost totally reconstructed after the devastation wrought by the Great Fire of 1666.

8. *Crane Court, the first permanent home of the Royal Society, purchased in 1710.*

9. *Opposite top: The Star Regulus of Mars.*

10. *Opposite bottom: The final page of Keynes MS.5 in which Newton prophesies the date for key future events described in the Bible, such as the resurrection of Christ and the end of the world.*

11. *Newton's floor-plan of King Solomon's temple, which he believed acted as a template for the chronology of an alternative world history.*

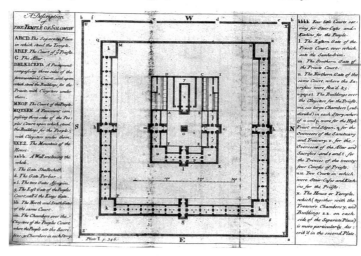

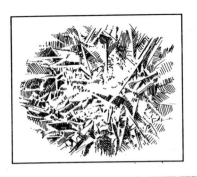

And at that time shall Michael stand up [that Michael who fought with the Dragon]
the great Prince who standeth for the children of the people [that Michael who anoints
the Dragon, that Prince of Israel whom Daniel calls the Prince, the King of Kings
& Lord of Lords who comes to ye battel of the great day will be an army on white horse
& a two edged word in his mouth] And [when the king of ye north goes forth wth great
fury to make away many] there shall be a time of trouble such as there never was
since there was a nation till that same time. And at that time
delivered [who are each ordeined from the nation
shall be delivered from that army of
be found written in the book. And many of
awake some [who have been marked or have not worshiped the Beast & his Image
ceived his mark] to everlasting life & some to everlasting shame & contempt. And they that
be teachers shall shine as the brightness of the firmament & they that turn many to righteous
ness as the stars for ever & ever. Thus Daniel brings down his Prophecy to the
was as the stars for ever & ever.

After this prophesy has this in continual order of time been brought down to ye first
resurrection the Gentile asked How long shall it [take the time of] the end of these
wonders, the Gentile being given for a time times & an half, & when he shall have
accomplished to scatter the holy people [that is finished] — And from the time
then back from their beginning at these things shall be finished
that the daily sacrifice shall be taken away, & the abomination of the desolation
be set up [sainte the relion of the captivity is the captivity
be 1290 days. Blessed is he that waiteth till 1335 days [when the people shall
be come and great tribulation with these shall rest & stand in his lot at
shall arise] but ye time thy way, Daniel for thou shalt rest & stand in thy lot at
the end of the days. Which end is to be imputed seven weeks of years after the
going forth of the commandment to cause ye return & up the end & by standing of the abomi
These numbers relating to ye time of the end & by standing of the abomi
nation in that time they seem to me to begin either with that time A.C. beg
or perhaps a little later, suppose for ye York Emperors in ye 8th century &
beginning of the ninth opposed the worship of ye images & saints & until that
opposition was over, which was not till the reign of the Empress Irene & second
Council of Nice A.C. 788 or rather not till the reign of the Empress Theodo
ra & her son A.BC. 841 & Council of Constantinople for setting up this worship
the year following.
The Little horn of Daniels fourth beast was to root up three of the
first kings wch stood in its way, & after it was grown up & established, the times &
laws were given into his hand for a time times & half a time. The three kings
were not rooted up before A.C. 774. At that time the Pope against his temp
was assisted by the grant of Charles the great & thereby became a king like of
rest of ye Rome. Afterwards A.C. his subjects were compelled to swear alle
gience to him. & A.C. he was declared above king judged by any power on
earth & the Pope any accusation he was only to purge himself upon oath, & at
the same time he took upon him to create Charles the great Emperor of the west
The Gentiles trode down ye Holy City 42 months & seem to have the
name of Gentiles from their worshipping the abomination of desolation all that
Beast [who is these Gentiles] is said to act 42 months & to bring
time. Whence the Beast [who is these Gentiles] is said to act 42 months & to bring
he was worshipped by all whose names are not written in the book of life.

12. *Above left: Nicholas Fatio de Duillier.*

13. *Above right: One of the many thousands of allegorical works of art depicting alchemical lore. Here we see the passing on of secret knowledge.*

14. *An artist's impression of an alchemist's laboratory of the 16th or early 17th century. Although Newton would have used similar equipment to some of that shown, he worked alone or with a single assistant.*

15. *Until the modernisation of the Royal Mint shortly before Newton's arrival, coin-making was a labour-intensive and very slow process.*

16. *Below left: A drawing of the reflecting telescope Newton presented to the Royal Society at the end of December 1671.*

17. *Below right: The title page of Newton's second masterpiece, the* Opticks, *first published in 1704.*

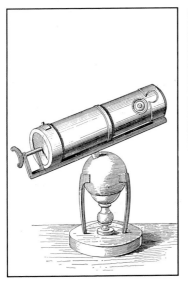

OPTICKS:

OR, A

TREATISE

OF THE

REFLEXIONS, REFRACTIONS,
INFLEXIONS and COLOURS

OF

LIGHT.

ALSO

Two TREATISES

OF THE

SPECIES and MAGNITUDE

OF

Curvilinear Figures.

LONDON,
Printed for Sam. Smith, and Benj. Walford,
Printers to the Royal Society, at the *Prince's Arms* in
St. *Paul's* Church-yard. MDCCIV.

18. *A fire in Newton's laboratory in 1677 is thought to have destroyed many irreplaceable documents including some of his alchemical writings.*

19. *Newton in his final years, painted by John Van der Bank.*

used to attempt to explain allied phenomena. If the experiments do not support the proposal then it will not be promoted from a mere hypothesis.★

Oldenburg dutifully passed on to Newton a very polite letter from Huygens, who, although he understood the work, genuinely wanted to receive a coherent explanation that he could accept. But again Newton took it very badly. A few days after receiving the letter, in early March 1673, Oldenburg opened a shocking reply from Cambridge. 'Sir I desire that you procure that I may be put out from being any longer a Fellow of the R. Society,' Newton began. 'For, though I honour that body, yet since I see I shall neither profit them, nor (by reason of this distance) can partake of the advantage of their assemblies, I desire to withdraw.'[24]

The idea that he should leave the Royal Society because it was difficult for him to reach London for meetings was nothing but a lame excuse, though it was true that he still had not attended a single meeting more than a year after acquiring his fellowship. But, if we ignore his familiar dash for the melodramatic gesture and the impetuous snapping of all ties, in this case it is possible to sympathise with him. Newton was confident that what he had discovered and had verified was of the utmost importance, yet no one would give him a fair hearing. For reasons that remain unclear, the experimental verification suggested by Oldenburg was still nowhere near completion nine months after Hooke had received his orders. (It was not presented to the Royal Society until April 1676.) But, worse still for Newton's fragile temperament, an esteemed scientist like Christiaan Huygens could not even visualise his methods. It must have seemed to him that nothing had changed: that even now, as he was accepted into the ranks of the élite, the whole world was still his enemy.

★ A modern example is the case of cold fusion. In 1989 two scientists, Professors Fleischmann and Ponns, announced that they had achieved the process of nuclear fusion at around room temperature. However, before this staggering claim could be accepted, teams of scientists around the world tried to repeat the experiments. When these attempts failed completely, the idea was considered to be almost certainly false and tests eventually pinpointed the fault with the original experiments. Although Professor Fleischmann is still researching cold fusion at a centre financed by a collection of Japanese corporations, the cold-fusion hypothesis has been widely discarded by orthodox science.

Thanks to the patience of Oldenburg, however, the problem was effectively smoothed over. He recognised Newton's bluster for what it was and called his bluff by offering to cancel his quarterly fees. This Newton ignored, and the issue was allowed to drop. But in a letter that he wrote to Collins two months later, on 20 May, his real reasons for offering to relinquish his fellowship are revealed. He confessed:

> I suppose there has been done me no unkindness, for I met with nothing in that kind besides my expectations. But I could wish I had met with no rudeness in some other things. And therefore I hope you will not think it strange if to prevent accidents of that nature for the future I decline that conversation which has occasioned what is past.[25]

Newton's wish to resign had been little more than a yell of frustration and anger. Shortly afterwards he managed to control his temper and frustration enough to write a polite and detailed letter to Huygens suggesting that, if he did not accept the validity of his 'Theory of Light and Colours', he should conduct the experiments himself. At which point the Dutch scientist withdrew from the discussion.

As Huygens glossed over Newton's theory, so too Newton and Oldenburg chose to forget the resignation threat. But, as he had announced he would, Newton continued to keep himself to himself in Cambridge and cut all ties with the scientific Establishment. Forgoing any further dramatic gestures, he merely wrapped himself in a blanket of isolation some eighty miles away from the discussions and controversy of the Royal Society.

And so it continued. During the first half of the 1670s Newton devoted increasing efforts to alchemical research (alongside his scientific endeavours) and simultaneously explored the arcane world of the Old Testament. This was the period during which he accumulated the theological information he thought he would need to defend his Arian faith against the Establishment and the King. This battle was never fought, however, and almost three years after cutting himself off from the Royal Society, in December 1675, with his professorship secure, he felt ready to send a new theory to Oldenburg.

The new work came in the form of two papers: a hypothesis and a set of observations. The first was 'An Hypothesis Explaining the

Properties of Light', part of which explained how reflection, refraction and diffusion of light were all caused by the corpuscles of light being made to change speed and direction by different media. Newton made it explicitly clear in this paper that colours were not produced by these effects, reiterating his point that white light was composed of many different colours and that it was only phenomena such as refraction, reflection and diffusion that came about as a result of this alteration to the particles of which light was constituted.

The second paper, 'Discourse of Observations', was a set of experiments in which Newton attempted to demonstrate and prove his hypothesis. This immediately sparked a fresh battle with Robert Hooke, who claimed that Newton had not only taken inspiration from his *Micrographia*, but had used it to demonstrate an opposing theory of the nature of light.*

Fond of intrigue and thriving on gossip, Hooke did not go through the official channels to rile Newton, but instead convened informal meetings with his friends in London coffee-houses, allowing news of these chats to leak out and work their way to Newton's lonely rooms on the edge of the Fens, chafing further the raw mood of animosity between them.

For his part, Newton had made no attempt to credit Hooke's concept as it had appeared in *Micrographia*, nor did he now recognise him as the source of his original inspiration concerning the experiments in 'Discourse of Observations'. He was never a man to give credit unless absolutely necessary, and, once crossed, he never forgave. The stage was now set for an almighty clash of egos. But again a public row was averted – this time by another long-standing rivalry.

It was no secret within the Royal Society that Oldenburg and Hooke were less than friendly. They had clashed earlier in 1675 over another charge of plagiarism, when Hooke claimed priority over Huygens for the invention of the spring-balance watch. To Hooke's lasting disgust, Oldenburg had taken Huygens's side in the

---

* Newton believed that light is made up of invisible particles; Hooke supported the view that it is composed of waves. This argument is still not completely resolved, but modern science suggests that light has a dual nature: that it is both wavelike and corpuscular, depending upon the experimental circumstances.

argument.* Furthermore, Hooke was still seething over his treatment at the society's meeting in June 1672, during which he had been embarrassed by Newton's attack. For these reasons, Hooke now decided to bypass Oldenburg and correspond directly with Newton.

Although unprecedented, and intended as a snub to the Secretary, this decision in fact suited Oldenburg, because it released him from any responsibility over the matter. He was also now able to warn Newton of Hooke's vitriolic mood and to pass on news of his scheming. Newton wrote thanking the Secretary for his 'candour in acquainting me with Mr Hooke's [most recent] insinuations'. He went on:

> It's but a reasonable piece of justice I should have an opportunity to vindicate myself for what may be undeservedly cast on me ... I desire Mr Hooke to show me therefore, I say not the sum of the hypothesis I wrote, which is his insinuation, but any part of it taken out of his *Micrographia*.[26]

And so, ironically, a public dispute was actually averted by Hooke's move to marginalise Oldenburg. Instead of expressing their fury openly in the *Transactions* or through letters addressed to a third party, Newton and Hooke were restricted by the custom of the time that personal letters between gentlemen should remain outwardly polite and mutually respectful. There began a correspondence in which the two rivals staked their claims and vented their frustrations by subtly sublimating their mutual disgust through polite, flattering, often obsequious language that always contained a subtext of mutual loathing and distrust. Why Hooke chose this route is unclear. Perhaps he believed that by removing Oldenburg from the argument he would gain more than if he feuded with Newton under the gaze of his contemporaries.

Hooke began the exchange towards the end of January 1676. 'I in no way approve of contention or feuding and proving in print,' he oozed, 'and shall be very unwillingly drawn to such a kind of war.'

---

* Hooke may actually have been the first to suggest it, but Huygens had made the first working model in 1674.

Then, disguising his patronising manner with sugar-coated linguistic finery, he insisted:

> I do justly value your excellent disquisitions and am extremely well pleased to see those notions promoted and improved which I long since began, but had not time to complete. That I judge you have gone further in that affair much than I did, and that as I judge you cannot meet with any subject with a fitter and more able person to inquire into it than yourself, who are every way accomplished to complete, rectify and reform what were the sentiments of my younger studies which I designed to have done somewhat at myself, if my other troublesome employments would have permitted, though I am sufficiently sensible it would have been with abilities much inferior to yours.[27]

Newton's reply, posted on 5 February, was, if anything, filled with even more false flattery and almost mocking overfriendliness. He began by agreeing with Hooke about the futility of public argument: 'There is nothing which I desire to avoid in matters of philosophy more than contention, nor any kind of contention more than one in print,' he declared. He continued, 'What's done before witnesses is seldom without some further concern than that for truth: but what passes between friends in private usually deserves the name of consultation rather than conquest, & so I hope will it prove between you & me.' But then he went on to write a sentence that has been quoted so often yet has been largely misunderstood for over three centuries:

> What Descartes did was a good step. You have added much several ways, & especially in taking the colours of thin plates into philosophical consideration. If I have seen further it is by standing on ye shoulders of Giants.[28]

In that last sentence Newton revealed the truly spiteful, uncompromising and razor-sharp viciousness of his character, for Hooke, once described as 'crooked' and 'pale-faced' and a man who 'is the most and promises the least of any man in the world I ever saw',[29] was so stooped and physically deformed that he had the appearance of a dwarf. The phrase 'standing of ye shoulders of Giants' was a perfectly double-edged comment, designed deliberately to mislead. On the surface, it appears a compliment – Hooke is called a giant – but Newton meant quite the reverse.*

If Hooke understood the subtext, there was no outward sign of it. In April the Royal Society finally witnessed his experiments to demonstrate the validity of Newton's 'Theory of Light and Colours', four years after the paper was first received.

⌈ If the arrival of the modern scientific age could be pinpointed to a particular moment and a particular place, it would be 27 April 1676 at the Royal Society, for it was on that day that the results obtained in a meticulous experiment – the *experimentum crucis* – were found to fit with the hypothesis, so transforming a hypothesis into a demonstrable theory. ⌋

For a time, this was the end of the Newton–Hooke débâcle, but it was not the end of Newton's ill feeling towards the scientific world, nor of the flow of argument and counter-argument between the Trinity professor and the noble gentlemen of the Royal Society.

Isaac Barrow died a year after the Royal Society demonstration, in May 1677, leaving Newton isolated. Later the same year, Henry Oldenburg, Newton's point of contact with the Royal Society and the supreme smoother and scientific diplomat, followed Barrow to the grave – a loss that left Newton even more exposed. To compound matters, some critics persisted in disagreeing with Newton's work even after the demonstration of the 'Theory of Light and Colours' in 1676. When Oldenburg was succeeded as Secretary by Robert Hooke, the hypersensitive Newton felt the atmosphere had soured beyond endurance. After this period of brief communications with the wider world of the scientific community, he now relapsed into total silence, cutting himself off completely from his peers and critics.

It was the beginning of a long period of isolation. Despite several attempts to draw him from self-imposed seclusion, Newton maintained a brooding silence. Within the isolation of his laboratory at Trinity, the theoretical ideas that were to coalesce in the *Principia* were coming together. Newton believed he could not develop his intellectual masterpiece under the gaze of an unsympathetic scientific

---

* By this time Newton had met Hooke on at least one occasion, having attended his first meeting of the Royal Society on 18 February 1675, when Hooke was also present.

community – one whose members were so far behind him they could not even grasp his methods. Following this first unhappy excursion into the world beyond Cambridge University, he was now content to explore the far shores of knowledge alone, until chance once more exposed him to conflict, and to glory.

CHAPTER 9

To the *Principia*

Chance favours only the prepared mind.
LOUIS PASTEUR[1]

The origins of the ***Principia Mathematica*** – probably the greatest single work of science ever written – can be traced to the coffee-houses of fashionable London. Although the mathematical, alchemical and religious ferment of Newton's imagination gave the book form, it was up to others to bring it into the world. And, if there was a single man who could be said to have played the role of facilitator, it was the astronomer Edmund Halley.

The story of how the world learned of Newton's shrouded thoughts begins with three men meeting in a City of London coffee-house in the depths of winter, during January 1684. One of them was Newton's most reviled enemy, the boastful, extrovert Robert Hooke; the other two were Sir Christopher Wren and Edmund Halley.

Over bowls of black coffee they fell into a conversation about gravity. Halley, who had for some time been interested in the nature of planetary motion, put a question to the other two. Could the action that keeps the planets in motion around the Sun, he wondered, decrease as an inverse square of the distance? Expecting a puzzled silence, Halley was surprised, and for a moment offended, when both Wren and Hooke burst out laughing. Finally, Hooke explained their odd reaction. 'Mr Hooke,' Halley wrote later, 'affirmed that upon that principle all the laws of celestial motions were to be demonstrated.'[2]

Wren concurred, and went on to profess that it was relatively easy to reach this conclusion as an hypothesis, but quite another to prove it. He had apparently spent some time on that very effort without the slightest success. In typical style Hooke then interrupted and insisted he had managed to prove it some years before but had revealed the secret to no one: 'he would conceal the solution for some time,' he said, 'that others trying and failing, might know how to value it, when he should make it public'.[3]

Now it was Halley's and Wren's turn to scoff, and, according to Halley, Wren declared that he would 'give Mr Hooke or me 2 months time to bring him a convincing demonstration thereof'. He further asserted that, if either of them managed it in that time, he would reward the winner with a book worth forty shillings.[4]

But it was not to be. The two months soon passed, and none of the three came forward with a proof of the inverse square law. Wren grew tired of the problem and weary of Hooke's bravura, later dismissing this with typical understatement: 'I do not yet find in that particular that he [Hooke] has been as good as his word.'[5]

Halley too grew impatient. As March slipped away and spring turned to summer, he decided to take the matter into his own hands. He knew that Newton was interested in gravity and he was aware of the Lucasian Professor's work in both optics and mechanics. He had also heard of the man's testiness and cherished monastic lifestyle. He considered writing to arrange a meeting (he had met Newton only once, two years earlier, as we know from a reference Newton made in his 'Waste Book' describing a meeting with Halley in 1682), but eventually decided against it. Instead, ignoring the blustering Hooke, who still claimed to know the secret but would not discuss it, and knowing that Wren was now too busy with other matters, he packed a bag and set off for Cambridge. It was a move that would put him into the history books.

Born into a wealthy family, Edmund Halley was a gentleman who knew well the art of discretion. Handsome and tall, with dark hair and a finely crafted nose, he possessed perfect manners, was a keen judge of character and, as his dealings with Newton were to prove, was a clever manipulator. He was also a respected astronomer who had spent two years during his early twenties gathering data for an accurate catalogue of stars from an observatory on the South Atlantic island of St Helena (where, a century and a half later, Napoleon Bonaparte would spend his final days).

Newton was forty-one at the time, shorter than Halley, and already almost totally grey – his shoulder-length hair was prematurely coloured, so he half-jokingly told Wickins, from experimenting with quicksilver so much that 'he took so soon the colour'.[6] Halley was quite aware of the animosity between Newton and Hooke and that the Lucasian Professor preferred not to discuss his theories, but by giving him no warning of his visit he had done exactly the right thing: it would have been unthinkable for Newton to turn away a visitor without first hearing the reason for his call. The Newton devotee Abraham Demoivre later wrote of the encounter:

> In 1684 Dr Halley came to visit him at Cambridge, after they had been some time together, the Dr asked him what he thought the curve would be that would be described by the planets supposing the force of attraction towards the Sun to be reciprocal to the square of their distance from it. Sir Isaac replied immediately that it would be an ellipsis. The doctor struck with joy & amazement and asked him how he knew it. Why said he, I have calculated it, whereupon Dr Halley asked him for his calculation. Without any further delay, Sir Isaac looked among his papers but could not find it, but he promised him to renew it, & send it.[7]

Newton did not search hard for the lost paper. Wary of his dealings with the outside world, and intimidated still by Hooke, he would have pretended to have mislaid the document so he could go through the work again. However, as Newton subsequently took Halley at least partially into his confidence, it is clear that the Lucasian Professor liked and trusted the astronomer more than many of the other scientists with whom he had had dealings. This is because Halley had asked him for mathematical verification of an hypothesis – the approach Newton himself applied to all scientific questions. By the time he returned to London, Halley had been assured that Newton would soon contact him on the matter and deliver a mathematical demonstration of the inverse square law of planetary motion.

For Halley this marked the beginning of a trial of patience. For three months he resisted the temptation to contact Newton, realising that any overeagerness would put him off. But eventually his subtlety paid dividends. He had judged Newton's temperament perfectly, and in return he was rewarded with a document that far exceeded his expectations – a nine-page treatise entitled *De Motu Corporum in Gyrum – On the Motion of Revolving Bodies* – hand-delivered by a mutual academic acquaintance, the mathematician Edward Paget. It was this paper which led, in little more than two years, to the appearance of the completed manuscript of the *Principia*.

To understand how Newton arrived at his grand synthesis and then refined and developed his ideas to the point where they were ready for publication as the *Principia*, we need to turn back the clock to the end of the previous decade, some six years before Halley's visit, to reconstruct the confluence of events both in Newton's troubled domestic life and in his multifarious intellectual interests that led to his scientific zenith.

By the mid-1670s Newton had long since established mathematically that an inverse square law governed the receding force experienced by a planet in orbit around the Sun (or the Moon about the Earth), and he may have begun to realise that an attractive force operated which held the planets in their orbits. Clues for this latter force would have come both from his alchemical discoveries and from his grasp of hermetic tradition as it extended into theology and ancient lore. What was lacking was any form of mathematical or observational evidence to support the notion of an attractive force – something that would give the concept substance beyond a mere idea. It was this evidence that was provided by a series of events during the 1680s.

To Newton, the final years of the 1670s must have seemed like the season of death. Barrow and Oldenburg had both died in 1677, nudging him further into isolation; he saw his few friends rarely, and, with the possible exception of Wickins, he never allowed anyone into his private thoughts and fears. But by 1679 even Wickins was slipping away from him – spending more and more time away from Cambridge. Then, towards the end of May of that year, Newton returned to Cambridge from a nine-day trip to London to learn that his mother was gravely ill.

Hannah had fallen sick after attending her son Benjamin, who had been suffering from a fever. The boy was recovering, but Hannah was not. Newton set off for Woolsthorpe immediately, not even stopping to sign the college's Exit and Redit book as he left.[8]

Arriving at the manor two days later, he took charge of his mother's treatment, concocting his own medicines from years of experience and research. 'He sat up whole nights with her,' John Conduitt reported many years later, 'gave her all her physic himself, dressed all her blisters with his own hands, & made use of that manual dexterity for which he was so remarkable.'[9]

His attentions did little good. Hannah died in late May or early June and was buried beside her first husband, Isaac, on the north side of the church at Colsterworth, a couple of miles from the manor. The parish register records the event with the entry 'Mrs Hannah Smith [widow]. Was buried in woollen June the 4th 1679.'[10] Her death was recorded officially in the name of Barnabas Smith, but at least Newton had the satisfaction of seeing his mother buried beside his natural father, rather than his despised stepfather.

It is impossible to guess what Newton may have thought as he sat beside his mother's bed late into the night. Notwithstanding his impermeable self-confidence, he may have known that she had little chance of survival, but disillusionment and anger with himself when his treatments failed to save his own mother must have further compounded the pain he felt at losing her again.

With Hannah's passing, Newton became a wealthy gentleman. The will was proved at Lincoln on 11 June 1679. As required by tradition, money was left for the poor of Colsterworth and Woolsthorpe – in this case £5: two and a half times the forty shillings that Isaac Newton senior had left at the time of his death thirty-seven years earlier, and a sure indication of how Hannah's fortune had increased in the intervening years. Each of the servants received forty shillings (£2) except for Hannah's oldest and most loyal domestic, one William Cottam, who received a generous £5. Newton's half-siblings received only small inheritances, as they had been given the major share of Barnabas Smith's fortune upon his death. The remainder – all Hannah's lands, goods and chattels – went to her first son. At the age of thirty-seven, Isaac Newton, gentleman and scholar, had become master of the family estate.

The villagers of Woolsthorpe must have viewed Newton with mixed feelings. There would have been many who remembered the arrogant boy who disappeared from their world at the age of nineteen and returned with almost indecent rarity. The younger servants and village folk would have viewed him as a distant, proud figure – distinguished and unapproachable.

He stayed several months in Woolsthorpe, arranging business affairs and ensuring that the manor was looked after, that the handover to new residents was conducted smoothly, and that bills were settled. Hannah had died in the late spring, but Isaac stayed to supervise the cultivation during the summer as well as the autumn harvest.

There were also other domestic problems to address. Hannah had let business matters slip during her final years, and the family was owed money from several bad debtors, including Eduard Storer – stepson of Clark the apothecary. Another debtor was a Mr Todd, who owed the estate £100. Newton pursued these bills assiduously. Perhaps because of his long association with the Clark family he did not harass Storer and the bill was paid without too much trouble, but it took threats of legal action to bring Mr Todd to heel. In 1679 Newton wrote to him:

> About your pretences of the moneys being ready long since & of a judgement which you would have me believe I had against you I do not think it material to expostulate. I shall only tell you in general that I understand your way & therefore sue you. And if you intend to be put to no further charges you must be quick in payment for I intend to lose no time. I desire you therefore to pay it to my sister Mary Pilkington at Market Overton as soon as you can & take her acquittance for your discharge.[11]

By the time Newton returned to Trinity it was late November and the flatlands of Cambridge were freezing cold and wind-swept; the room he had shared for so long with Wickins was empty and chill. Awaiting him was a pile of mail. In it lay a recent letter from the Royal Society. It was addressed in the hand of Robert Hooke, who was breaking their three-year silence.

At first Newton resisted the temptation to be drawn by Hooke and ignored this attempt to engage him in serious scientific correspondence. But, provoked by his colleagues at the Royal Society, the Secretary was obliged to persist and eventually succeeded in drawing Newton out of his shell and into the spotlight once more.

In the first letter, received in November 1679, Hooke cheerfully reported his latest ideas, most especially on the subject of planetary motion, and asked Newton's opinion of them. At the same time he asked whether the Lucasian Professor had any original findings he might want to share, promising to keep any such material to himself. 'I hope therefore that you will continue your former favours to the Society,' he said, 'by communicating what shall occur to you that is philosophical . . . Such correspondence shall be no otherwise further imparted or disposed of than you yourself shall prescribe,' he added

– concluding, 'I shall take it as a great favour if you shall be pleased to communicate by letter your objections against any hypothesis or opinion of mine.'[12]

Newton eventually responded and explained that he had no time 'to study or mind anything else but country affairs', claiming improbably that 'I am backward in engaging myself in these matters.'[13] But, as he later told Halley, to 'sweeten my answer, expecting to hear no further from him',[14] he offered Hooke a solution to a minor scientific puzzle: 'a fancy of my own,' he declared, 'about discovering the Earth's diurnal motion [the daily rotation of the Earth on its axis]'.[15]

The problem Newton had tackled was an old chestnut which had been discussed frequently in scientific circles for many years: if an object was dropped from a tall tower, philosophers wondered, would it be left behind by the Earth's rotation? If so, would it land to the west of the tower as the Earth spins beneath it?

Newton calculated that, if air resistance was ignored, the object would land slightly to the east of the tower, and he drew a little diagram showing a spiral path for the descent of the object, based upon his calculations, and gave a detailed description of the experiment that could be set up to prove it.

Without intending to, Hooke had opened up a can of worms, and in his keenly felt attempt to avoid contact, to 'hear no further' from Hooke, Newton had inadvertently opened himself up to attack. With this little 'sweetener' he had made a serious miscalculation, and Hooke pounced upon it. The Secretary realised that the falling object would land slightly to the east of the tower only if the tower stood precisely on the equator. In London it would fall more to the south than to the east. More importantly, Hooke showed that the path of descent would not be a spiral at all, but, under the influence of orbital motion, the object would follow an elliptical path to the ground.

Breaking his promise, Hooke immediately and gleefully read aloud Newton's work at the next meeting of the Royal Society. Writing to Newton on 9 December, he told the Lucasian Professor how it had gone, reporting that most of those gathered there had agreed with Newton – until, that is, he, Hooke, had exposed the error and described the corrected version to the assembled gentlemen. 'My theory of circular motion,' Hooke crowed, 'makes me suppose that it would be very different and nothing at all akin to a spiral but rather a kind of ellipsoid.'[16]

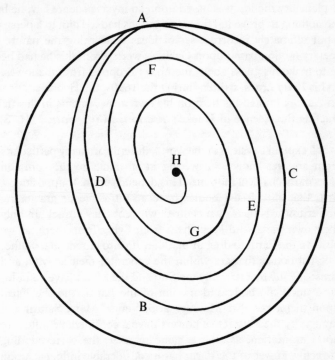

Figure 10. Hooke realised that an object falling to the Earth would follow an elliptical path of descent. The path ABC is the one the object would follow if it was falling through a vacuum. If it was falling through a resisting medium (such as air) it would follow the path ADEFGH.

By publicly parading this error, Hooke had not only broken Newton's trust but had deliberately tried to damage his reputation in the eyes of his scientific colleagues. Worse still, Hooke's calculation had actually been based upon nothing more than a lucky guess.

Whereas Newton had been preoccupied with family matters for the best part of the year and heavily involved with alchemy for much of the past decade, Hooke had become obsessed with the nature of planetary motion and had spent what was for him an inordinate amount of time thinking about it. However, unlike Newton, when Hooke became interested in a concept, he wasted little ink upon any form of detailed mathematical analysis. Concluding independently that planetary motion was based upon an inverse square law, he had done nothing to prove it. His investigations had led him to a persuasive but ultimately incorrect set of ideas concerning the nature of planetary motion based upon Galileo's work. From this he had been able to make his guess about the object dropped from the tower.

In his *Brief Lives*, written during the 1690s, the Hooke-supporter John Aubrey quoted his hero on his own achievements in this field, extracting the essence of Hooke's *Lectiones Cutlerianae* (1678):

I shall explain a system of the world, differing in many particulars from any yet known, answering in all things to the common rules of mechanical motions. This depends upon 3 suppositions; first, that all celestial bodies have an attractive or gravitating power towards their own centres, whereby they attract not only their own parts, and keep them from flying from them, as we observe the Earth to do, but that they do also attract all the other celestial bodies that are within the sphere of their activity, and consequently that not only the Sun and the Moon have an influence upon the body and motion of the Earth, and the Earth upon them, but that Mercury also, Venus, Mars, Saturn and Jupiter, by their attractive powers have a considerable influence upon its motion, as, in the same manner, the corresponding attractive power of the Earth has a considerable influence upon every one of their motions also. The second supposition is this, that all bodies, that are put into direct and simple motion will so continue to move forwards in a straight line, until they are by some other effectual powers deflected and bent into a motion describing a circle, ellipsis, or some other uncompounded curve line. The third supposition is, that these attractive powers are so much the more powerful in operating, by how much nearer the body wrought upon is to their own centres [i.e. their attraction is greater at close quarters].[17]

At first glance this sounds very much like Newton's own theory of gravity as it has been handed down to us, but a gaping chasm lay between the work of the two men. It is one thing to make guesses about the way the world might be, but quite another to *show* how the world actually is. By the time he was ready to publish his own findings, in 1687, Newton had achieved the latter – following years of practical research and the use of complex mathematics (which he had himself developed). He had made use of concepts and images derived from his alchemical studies and researches into ancient theology to construct a workable celestial mechanics, and to complete the process he had then devised repeatable experiments and extrapolated his concept of gravity to a revolutionary description of universal gravitation. Hooke did none of this: he was the medieval alchemist to Newton's empirical 'scientist'.

With surprising self-control, Newton responded coolly to Hooke's letter of correction and admitted his mistake. In a letter dated 13 December 1679, in which one can almost hear the writer clenching his teeth as he puts pen to paper, he said, 'I agree with you, that the body in our latitude will fall more to the south than the east if the height it falls from be anything great.' But he then went on to insist that the object would *not* fall via an elliptical path, and enclosed a brief explanation to support his claim. In conclusion he added that the matter was 'of no great moment', and invited Hooke to correct him if he was wrong.[18]

Unfortunately, he was. Employing his lucky guesses with the skill of the practised self-publicist and driven by a hatred born of humiliation, Hooke again turned the tables on the Lucasian Professor and demolished his argument before the next gathering of the Royal Society.

To reach his conclusion Hooke had employed two different mathematical relationships handed down from an earlier generation of natural philosophers. These formulae were based upon Galileo's theory of uniformly accelerated motion from rest and Kepler's law of velocities. Ironically, while he had arrived at a correct solution to this puzzle, his mathematics had been flawed and he had used both Galileo's and Kepler's formulae inappropriately. Captivated by his own vengeful eloquence and determined to persuade the scientific Establishment that the great Isaac Newton was erring badly, he did not realise his double error – but, luckily for him, nor did anyone else at the Royal Society.*

---

* Presumably Hooke's view was accepted over Newton's because none of his colleagues could find a flaw in his mathematics and because Hooke was on the scene and used his considerable powers of persuasion at Royal Society meetings.

Revelling in the discomfiture he knew he would cause, Hooke dispatched a third letter to Cambridge, detailing where Newton's explanation had gone wrong and highlighting the support and approval he had garnered for his own solution to the puzzle from their colleagues at the Royal Society. This time Newton was too stunned even to reply.

Sensing that he had perhaps gone too far, a week later Hooke sent a fourth letter, in which he attempted to mitigate the earlier response by praising the virtues of Newton's suggestions during the course of their short correspondence, saying that if the experiment was ever to be performed, the Royal Society would certainly follow his excellent instructions.

This letter was also ignored. And Newton not only failed to respond to Hooke, he refused to write to anyone for over a year. Yet this silence belied the fact that he was far from convinced by Hooke's approach. Rather, he was already plotting his revenge.

Although Hooke had been treacherous in publicly exposing Newton's mistakes and had been following a fallacious line of reasoning during their exchanges, he had inadvertently highlighted an important scientific point. He had reached the correct conclusion that objects falling under the force of gravity follow an elliptical path to the Earth. This meant that, if an accurate description of planetary dynamics was to be formulated, elliptical motion rather than circular motion had to be matched with the inverse square relationship. This is because the movement of planets orbiting the Sun is governed by the same force as objects falling to the Earth – gravity.

Newton had shelved this problem thirteen years earlier as an approximate calculation using circular motion. Although Hooke was adept enough to realise that any theory of gravity as applied to the planets had to be based upon elliptical motion, he had no clear understanding of how he could begin to demonstrate this mathematically. In January 1680 he wrote to Newton, 'I doubt not but that by your excellent method you will easily find out what that curve [the ellipse] must be, and its proprieties, and suggest a physical reason of this proportion [the inverse square law].'[19]

Hooke was clearly unable to master the mathematics of the problem, and, despite his flattery, he must have also doubted whether Newton could resolve the matter. But this time he was quite wrong. The *Principia* would become the answer to Hooke, and Newton began immediately to marshal the arguments that would eventually lead to his theory of universal gravitation.

Having derived the mathematics of the inverse square law for circular motion some thirteen years earlier, he now applied the calculus to show that the motion of a body in an elliptical path around an attracting body located at one focus of the ellipse entails an inverse square attraction. Demonstrating this first for the apsides of an ellipse (the points most distant from the centre), without letting on to anyone what he was doing, he went on to solve the mathematics for every point on the orbital path.*

Newton never forgave the Secretary of the Royal Society for his handling of their correspondence during the final weeks of 1679. Six and a half years later he detailed their exchanges to Halley, and thirty years on, in old age, and with his enemy long dead, he continued to harbour ill feelings towards Hooke, suggesting to Demoivre that his own original drawing of the spiral descent of the object from the tower had been a 'negligent stroke with his pen'.[21] But Newton knew he was indebted to Hooke for inspiring him to resolve the problem of matching the inverse square law with elliptical planetary motion, and in a letter to Halley written several years later he confessed, 'Hooke's correcting my spiral occasioned my finding the theorem by which I afterward examined the ellipses.'[22]

His fury had spurred him on, but there were other influences leading him to a theory of gravity. In November 1680 the first of what were originally thought to be several comets appeared in the skies over Europe. A visible comet is a rare enough astronomical event, so when a second was observed only a month after the first the fellows of the Royal Society and amateur astronomers around the country were understandably surprised.

Robert Hooke was up many a late night during November 1680, peering at the night sky above London. He first noted the appearance of the comet in his diary on the night of 12 November, and reported his nightly observations until the end of the month, when the comet seemed to disappear completely.[23]

---

* Two copies of this proof have survived to the present day: one in Newton's hand, the other copied later by the amanuensis of Newton's friend John Locke.[20] The former was almost certainly written in early 1680.

Edmund Halley had also become obsessed with the comet, and conducted observations from his house in London. Then, in December, he travelled to France, where he observed what he believed to be a second comet over Paris, writing to Hooke on Christmas Eve with a description of what he had seen.

Meanwhile, John Flamsteed (whom Charles II had appointed Astronomer Royal in 1675) also witnessed the event from the Royal Observatory in Greenwich, and was the only astronomer in Europe to realise that the two separate sightings were in fact observations of the same comet. On 15 December 1680 he wrote to a friend, the fellow of Jesus College, Cambridge, James Compton:

> This night I have seen the tail again which is 50 degrees in length. I believe scarce a larger has ever been seen. It is above a Moon broad [i.e. it appeared to be wider than the Moon as observed from the Earth], its motion decreases in swiftness, and I believe we shall see it longer than has been seen of late.[24]

Newton too had been captivated by the celestial spectacle during the final weeks of 1680, and, in typical fashion, he pushed himself to the limit of endurance to study the phenomenon. He stayed up all night every night making detailed observations in the 'Waste Book' (the notebook he had inherited from his stepfather), including the movements of the comet and the times of the observations. Not content with his own sightings, he traced every observation he could from astronomers throughout Europe, and even received a report from Arthur Storer (his likely adversary in the churchyard after school, and brother of the errant Eduard Storer), who was then living in Maryland, America. Storer's report came from the banks of the Patuxent river and described the comet as having a 'Form like a sword streaming from the horizon'.[25]

James Compton, who lived a short walk from Newton's rooms, continued to receive letters from Flamsteed during January and February 1681, and in one of them the Astronomer Royal made a favourable remark about Newton's work which prompted Compton to copy parts of the letters he had received and to pass them on to the Lucasian Professor.

Newton was grateful, but did not take seriously Flamsteed's suggestion that the two comets were the same object. He was probably also hearing reports of the astronomer's views concurrently from Edmund Halley, who had received detailed accounts of the Astronomer Royal's theory. The problem with Flamsteed's ideas was that they began with a great insight – that the two comets were the same object – but led on to what was for the time far-fetched speculation.

Although quite correct in suggesting to both Compton and Halley 'that having passed the Sun it [the comet] would appear after his setting in December',[26] Flamsteed then conjectured that, approaching the Sun, the comet was drawn into a 'vortex' by some strange and unspecified magnetic effect which then switched polarity and repelled it from the Sun. In a letter to Halley written in February 1681, Flamsteed speculated that the comet had come from 'some planet belonging to a vortex now ruined: For worlds may die as well as men'.[27]

Failing at first to notice the nugget of truth at the core of Flamsteed's argument, Newton viewed the astronomer's ideas of vortices as nonsense; nevertheless, through Compton, the pair began to correspond. Adopting an unnecessary air of condescension, Newton began with a response to Flamsteed's theory, in which he wrote:

The instance of a bullet shot out of a cannon & keeping the same side forward may be a tradition of the gunners, but I do not see how it can consist with the laws of motion, & therefore dare venture to say that upon a fair trial it will not succeed excepting sometimes by accident.[28]

But he had by no means lost interest in the comet, nor in the explanation for it, and even as he was dismissing Flamsteed's notions Newton was pumping him for observational information from the Royal Observatory. Although he had already calculated the inverse square relationship for planetary orbits, he, like all other astronomers of the time, had yet to realise that the motion of a comet is influenced by the Sun. Only by doing so could he then see that the force holding the planets in their orbits also determines the path and behaviour of comets.

Newton might never have resolved the matter if it had not been for fortuitous timing. In the early autumn of 1682 another comet, even brighter than the 1680 object, appeared in the sky. Once again astronomers all over Europe trained their telescopes on it, and each night Newton sat in the garden outside his rooms at Trinity College, wrapped in blankets, plotting the path of a celestial object for the third time in less than two years.

This time things were different. It was immediately realised by Flamsteed, Halley, Newton and others that the comet was retrograde: that is, it was moving away from the Sun. Although it was not appreciated by observers immediately, the comet had already travelled around the far side of the Sun and was now journeying towards the outer rim of the solar system and beyond.

Newton had not been idle. Between the sightings of 1680 and 1682 he had spent some time studying the phenomenon of comet motion. In the spring of 1681, no more than two months after the final disappearance of the 1680 comet, the Italian astronomer Giovanni Domenico Cassini had published his *Observations on the Comet* and made the revolutionary suggestion that the comet was in fact the same as the one observed by Tycho Brahe in 1577. But an even more influential work was a book by Robert Hooke called *Cometa*, in which Hooke attempted to address many of the questions that had arisen from data collected from a large number of sightings through the ages. In one passage he asked, 'What kind of motion was it carried with? whether in a straight line or bended line? and if bended, whether in a circular or other curve, an elliptical or other compounded line . . .' Later in the same volume he added that it might be difficult to explain the movement of the comet, 'but this is no more than is usually supposed in all planets'.[29]

The jigsaw pieces were falling into place, and the picture they would reveal was not just an explanation for the movement of comets: the behaviour of comets also provided a further clue to the concept of universal gravitation – if the movement of comets could be evaluated accurately, this could lead to a fuller understanding of planetary motion and the forces involved in celestial mechanics.

Persuaded by hints from Flamsteed, Cassini and Hooke, Newton had begun to calculate the movement of the recent comets using the methods he had developed during 1666 and had recently applied to the elliptical path of the planets as they orbit the Sun. A copy of these calculations survives today, and a close study of them shows that Newton spent some energy considering a variety of trajectories, eventually settling on an elliptical orbit for comets with a gravitational force acting between the centre of the comet and the centre of the

Sun. Naturally he came to the conclusion that the motion of the comets obeyed the same inverse square law as was demonstrated by the planets and the Moon.

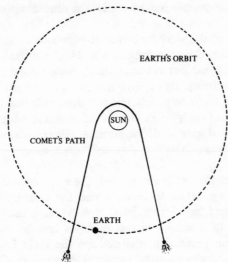

Figure 11. Newton's representation of the comet's elliptical path.

What role did alchemy play at this stage of Newton's reasoning? He never stopped experimenting, but interspersed periods observing comets and calculating their trajectories with a devotion to the furnace as intense as ever. And from this direction came the final clue.

During the 1680s the notion of attraction and repulsion was still seen as 'occult' – it implied that some kind of force was operating at a distance without the involvement of an observable medium or mechanism. Descartes's mechanical philosophy still prevailed, which described gravity as arising from the behaviour of clusters of matter and spinning vortices acting like whirlpools within the ether (which was visualised as an invisible, weightless, neutral medium pervading all space and facilitating all action).

By now Newton had rejected many aspects of Descartes's mechanical theory, including his notions of gravity. But, despite this, he had begun the 1670s with a traditional view of the ether as a corporeal medium which was intimately involved in the way in which some form of 'force' maintained planetary motion. He was sure that gravity was not a simple mechanical process – a process in which matter pushed upon other matter in the way of a whirlpool, as Descartes would have it – but he could not offer a viable alternative theory.

As a result of the work he conducted on comets, and of his discoveries prompted by the dispute with Hooke at the end of 1679, by the early 1680s Newton was gradually approaching an alternative to Descartes's theory. By the time he completed his masterwork, the *Principia*, he had completely rejected the traditional image of the ether in favour of gravity operating by 'attraction at a distance' (which perhaps required some ill-defined form of ethereal medium as facilitator – a medium via which this mysterious force, gravity, could operate).

But then, as the 1680s progressed, the role of the ether became less and less important to Newton and the concept of what the alchemists called 'active principles' took on far greater importance and led him to a radical reassessment of how gravity operated. This change in perception was perhaps the most important step in Newton's development of universal gravitation. One Newton scholar has gone so far as to say that Newton could not have visualised attraction at a distance had it not been for his alchemical work.[30]

The concept of active principles is an ancient one, rooted in hermetic tradition, though alchemists understood it in typically iconoclastic ways. Some believed that matter and spirit were interchangeable and imagined what they called a 'Universal Spirit'. Others held that matter and spirit were quite separate but that spirit could 'act upon matter'.

Henry More, Newton's mentor during his early career at Cambridge, wrote of a 'Spirit of Nature' which could act upon matter without interchanging with it, describing it as of 'great influence and activity in the *nascency* [origin] as I may so call it, & *coalescency* of things [the form of things]: And this not only in the production of plants, with all other *concretions* of an inferior nature . . . but also in respect of the *birth* of *animals*, whereunto it is preparatory and assistant'.[31] Here he attributes a single force, or active spirit, to what we know to be a diverse collection of phenomena, and he continues to the idea that life itself is created and sustained by this mysterious, singular power via spirit in some way acting upon matter.

The active principles of the alchemist were also a 'spirit' at work in Nature and not unlike More's concept. For centuries alchemists had observed how chemical reactions occurred and what facilitated them – noticing, for example, that some reactions required heat from a fire, while others did not. From these observations, made over more than two millennia, they came to believe there was a spirit at work in these operations. And this, they conceived, was how the philosophers' stone could perform its wonders – active principles guided by God and channelled through the alchemist allowed *spirit* to mutate matter.

By tracing the evolution of Newton's thoughts concerning gravity and ether through notes and papers written between 1672 and 1687, we can see how he came to perceive gravity as operating by action at a distance, made possible by a form of *active principle*. This was a mechanism which had no need for the corporeal ether of philosophical tradition.

First there is the paper 'Of Natures Obvious Laws & Processes in Vegetation', a 4,500-word document thought to have been written early in his alchemical career (possibly during 1672). In this Newton describes how metals and other substances are guided in their interactions by a subtle spirit (possibly analogous to More's Spirit of Nature) which he called 'vegetable action', or 'vegetation'. By this he meant that vegetable action was the mechanism that allowed changes to the materials in the crucible and among the furnace flames:

> There is therefore besides the sensible changes wrought in the textures of the grosser matter a more subtle secret & noble way of working in all vegetation which makes its products distinct from all others & the immediate seat of these operations is not the whole bulk of matter, but rather an exceeding subtle & unimaginably small portion of matter diffused through the mass which, if it were separated, there would remain but a dead & inactive earth.[32]

From here, the concept of spirit – the 'unimaginary small portion of matter diffused through the mass' – began to merge with Newton's changing vision of gravity. Modern physics tells us there are four fundamental forces in Nature – the strong and weak nuclear forces, electromagnetism and gravity. The alchemists saw the forces of Nature at work and put them down to a single divine or *spiritual* influence. Newton took this concept and concluded that gravity operated in a similar fashion – by an aspect of the Spirit of Nature (an active principle we perceive as 'attraction') which was guided by the divine hand of the Creator who maintained the ebb and flow of the universe.

From alchemy, Newton continued to find new stimulus, further nudges towards an all-embracing concept of gravity. From the alchemist Sendivogius, he noted, 'They call lead a magnet because mercury attracts the seed of Antimony as the magnet attracts the Chalybs [iron ore].' And 'our water' is drawn out of lead 'by the force of our Chalybs which is found in the belly of Ares [iron]'.[33] This prefigures the modern idea that most chemicals form chemical liaisons with others with varying ease. In some situations, one chemical can 'extract' or 'draw out' another from a solution or mixture because it can form stronger bonds with the solute (the material in the solution) than the solute can with the solvent (the dissolving medium). But to the alchemists this demonstrated the involvement of active principles.

Elsewhere, Newton recorded from his own experiments how certain metals or salts are 'drawn' or 'extracted',[34] that substances 'laid hold' of others when they reacted, and that when they failed to sublime or evaporate they were 'held down'.[35]

As early as 1675 Newton had begun to formalise his understanding of repulsion and attraction. In his paper 'An Hypothesis Explaining the Properties of Light' – the paper that had sparked the second phase of the Newton–Hooke dispute – he had coined the phrase a 'secret principle of unsociableness';[36] this principle, he believed, was what prevented certain substances mixing with others. In writing this paper, Newton was clearly influenced by the ever-shifting patterns of chemicals he observed in the crucible:

For nature is a perpetual circulatory worker, generating fluids out of solids, and solids out of fluids, fixed things out of volatile, & volatile out of fixed, subtle out of gross, & gross out of subtle, some things to ascend & make the upper terrestrial juices, rivers and the atmosphere; & by consequence others to descend for a requital to the former.[37]★

But even with this document Newton still held firmly to the traditional concept of ether, and in a letter to Boyle written three years later – early in 1679 – although the concept of active principles emerged again, the need for an ether remained. In describing how materials dissolve, he explained the phenomenon as due to a 'secret principle in nature by which liquors are sociable to some things & unsociable to others'. Elsewhere in the same letter he described 'an ethereal substance capable of contraction & dilatation, strongly elastic, & in a word much like air in all respects, but far more subtle'.[39]

But, though his alchemical practices directed him to these ideas, it was mathematics that finally led Newton to reject the concept of the ether as a necessary component in deriving a theory of gravitation.

Some sixty years earlier the German astronomer Johannes Kepler had devised his third law from the hypothesis that the velocity of a planet varies by inverse proportion to its distance from the Sun. This implies that planets travel fastest at perihelion (the nearest point to the sun in the elliptical orbit) and slowest at aphelion.† This was later supported by observation. Sometime in 1684 Newton decided to check the validity of Kepler's third law by using the calculus to calculate the velocity of planets at different points on the ellipse. What he found came as a complete shock: the mathematics tallied precisely with the observed facts.

This may seem a perfect result, but to Newton it meant that one of the most cherished notions of natural philosophy was completely false. If the calculation of the paths of the planets through the heavens (which was based upon mechanical movements through a vacuum) matched precisely the observed reality, it could lead to only one

★ This work has led some scholars to refer to Newton's ideas at the time as 'alchemical cosmology', almost akin to James Lovelock's notion *Gaia* – the principle that the Earth is a self-regulating entity which acts almost as a single, holistic organism.[38]

† As in Chapter 4, Kepler's third law may also be expressed as a relationship between a planet's mean distance from the sun and the time taken to complete one orbit.

conclusion – there was no such thing as a corporeal ether through which the planets moved, because if there was it would slow them down in their orbits.

This revelation was at once disquieting and liberating. For several years Newton had been struggling to accommodate the concept of a corporeal ether with his thoughts on the mechanism of gravity being an expression of an *active principle*, but without success. But even so, this new mathematical link still did not convince him fully: he needed to produce experimental evidence to prove or disprove the idea.

A famous thought experiment known to mechanical philosophers was the pendulum test, which was meant to prove the existence of the ether. The idea was that a pendulum slowed for two reasons – air resistance and the resistance caused by the ether through which it moved. The latter was demonstrated by observing the movement of a pendulum in a sealed container from which the air had been removed. It was seen that the pendulum still slowed over a period of time and eventually came to a halt. The orthodox conclusion was that the ether was able to penetrate all matter and to interact with it, slowing and finally stopping the pendulum. In order to verify that a corporeal ether did not exist, Newton conducted a series of further experiments using a pendulum.

First he carried out the experiment with an empty hollow pendulum in an air-free sealed cabinet and observed the rate of retardation of the pendulum swing. Next he filled the pendulum bob with varying amounts of different materials, to see if the retardation was affected. Taking into account the mass of the bob each time, he found that there was almost no difference in the rate of retardation. This contradicted the notion that the pendulum was moving through a corporeal ether: if it was, the ether should interact with the materials filling the bob.

These experiments were conducted as Newton was finishing a paper entitled 'De Aere et Aethere' – 'The Air and the Ether' – and as they verified the earlier mathematical discovery he began to realise that many of his previous statements upon the ether were now redundant. Early on in the text he had written:

And just as bodies of this Earth by breaking into small particles are converted into air, so these particles can be broken into lesser ones by some violent action and converted into yet more subtle air which, if it is subtle enough to penetrate the pores of the glass, crystal and other terrestrial bodies, we may call spirit of air, or the ether.[40]

A few pages on, the paper ends in mid-argument, leaving 'De Aere et Aethere' incomplete. Newton never returned to it, but later in 1684 he picked up where he had left off on the subject of the ether with the essay 'De Gravitatione et Aequipondio Fluidorum' – 'On Gravity and Equilibrium of Fluids' – and there his ideas were very different.

In this, the final document *en route* to **De Motu** (the paper Newton sent to Halley) and the earliest drafts of the *Principia*, he concluded that, if the ether existed at all, it was largely vacuum:

if the ether were a corporeal fluid entirely without vacuous pores, however subtle its parts are made by division, it would be as dense as any other fluid, and it would yield to the motion of bodies through it with no less sluggishness, indeed with much greater, if the projectile should be porous, because then the ether would enter its internal pores, and encounter and resist not only the whole of its external surface but also the surfaces of all the internal parts. Since the resistance of the ether is on the contrary so small when compared with the resistance of quicksilver as to be over ten or a hundred thousand times less, there is all the more reason for thinking that by far the largest part of the ethereal space is void, scattered between the ethereal particles.[41]

In other words, from his experiments it had become clear to Newton that if there was an ether then it presented very little resistance to bodies moving through it, which meant that it could not be corporeal or made of any known material, because, if it were, it would interfere with bodies moving through it.

With these results now in place, by the autumn of 1684 Newton could begin to rework the document he had promised Halley in August that year. The result was **De Motu**.

Halley's visit had acted as a catalyst for Newton's thoughts and had focused his attention upon the solving of a singular problem by bringing together disciplines and concepts he had been analysing for almost two decades. Fortunately, in Halley science had the perfect mediator between Newton and the outside world. Where others, such as Collins and, to an extent, Oldenburg and Barrow, had failed, Halley succeeded: his display of patience encouraged Newton to respond.

Receiving *De Motu*, Halley set off immediately for Cambridge to seek permission to announce Newton's findings to the Royal Society. He then travelled back to London, racing against the clock to make the meeting set for the following day, 10 December. Arriving late, his offering from the Lucasian Professor was the last on the agenda and many of the members had left for home.

Only a few of the gathering realised the importance of what they were hearing. But, despite the late hour, Halley's enthusiasm ensured the fellows' interest, and it was agreed that the society would publish the paper as soon as it could be arranged with the author. 'Mr Halley,' it was recorded in the minutes, 'gave an account that he had lately seen Mr Newton at Cambridge, who had showed him a curious treatise, *De Motu*; which, upon Mr Halley's desire, was, he said, promised to be sent to the Society to be entered upon their register.' The minutes then go on to report that '[The chair] desired Halley to put Mr Newton in mind of his promise for the securing of his invention to himself . . . until he could be at leisure to publish it.'[42]

But Newton was already beginning to consider a longer, more comprehensive, text. For most of 1685 he isolated himself from the rest of the world and communicated with no one, tearing himself away only twice – both times to visit Lincolnshire for about two weeks, first in April and again in June. Probably the only person he spent any time with during this period of intense work was his new assistant, Humphrey Newton.

John Wickins had finally departed in 1683, and Newton needed a replacement – someone to transcribe his notes and to attend the furnaces. Little is known of Humphrey Newton's life before he arrived in Cambridge. He was a Grantham boy like his master, and one of the brighter students at King's School; he was probably recruited for Newton by Dr Walker, Henry Stokes's successor at the school. Humphrey talked in old age of his tuition under Newton, so he may have been a sizar, but he was never registered as a student at Trinity. After serving five years with Newton, his recollections, recorded by both William Stukeley and John Conduitt some forty-five years later, give an intriguing picture of the minutiae of Newton's life during the writing of the ***Principia*** – a book that Humphrey Newton transcribed by hand.

According to Humphrey, Newton's regimen was severe. He was:

> So intent, so serious upon his studies that he ate very sparingly, he often forgot to eat at all . . . He very rarely went to bed, till 2 or 3 of the clock, sometimes not till 5 or 6, lying about 4 or 5 hours, especially at spring & fall of the leaf, at which times he used to imply about 6 weeks in his laboratory, the fire scarcely going out either night or day, he sitting up one night, as I did another until he had finished his chemical experiments, in the performance of which he was the most accurate, strict, exact. What his aim might be, I was not able to penetrate into . . . [43]

Not only was Newton isolated and working unceasingly: his entire correspondence was devoted exclusively to his work. The bulk of it was with the Astronomer Royal, John Flamsteed.

The exchange of letters lasted precisely one month, from December 1684 to January 1685. Newton was especially polite because he wanted something, and Flamsteed, who was perfectly affable, provided everything the professor asked for, including the location of stars and the position of planets in relation to one another, so that Newton could plot the path of the 1680 comet accurately.

By this point it is certain that Newton was visualising a more general theory than the simple application of the inverse square law might have implied. It is also clear from the Flamsteed–Newton exchange that Newton was working at the very limits of science. The Astronomer Royal, himself a very capable man, could not begin to imagine why his colleague needed the information he was requesting.

An example is an exchange of letters in January 1685 dealing with the movement of Saturn. Believing that Kepler's observational figure for the orbit of the planet was too small, Newton wrote to Flamsteed asking for the respective velocities of both Saturn and Jupiter as they came into conjunction (i.e. approached the same celestial longitude*):

---

* The ecliptic system used to designate positions of the planets in the solar system makes use of coordinates called the celestial longitude and celestial latitude.

This planet so oft as he is in conjunction with Jupiter ought (by reason of Jupiter's action upon him) to run beyond his orbit about one or two of the Sun's semidiameters or a little more & almost all the rest of his motion to run as much or more within it.[44]

Flamsteed wrote back in confusion:

I can not conceive of any impression made by the one planet . . . can disturb the motion of the other. It seems unlikely such small bodies as they are compared with the Sun, the largest and most vigorous magnet of our system, should have any influence upon each other at so great a distance.[45]

He then went on to remind Newton of experiments involving magnets which demonstrated that even the largest magnet could not disturb a small needle placed 100 yards away.

The intuitive and intellectual gap between the two philosophers is evident here, and Newton continued to give nothing away. He gave no reason for wanting specific pieces of information and offered no hint that the facts and figures Flamsteed volunteered confirmed his new ideas. Teasing, he responded to one letter with 'Your information about the error in Kepler's tables . . . has eased me of several scruples. I was apt to suspect there might be some cause or other unknown to me.'[46]

With no intention of sharing his thoughts until he was ready, he worked on frantically. As Humphrey Newton told Conduitt:

I never saw him take any recreation or pastime, either in riding out to take the air, walking, bowling, or any other exercise whatever. Thinking all hours lost that were not spent in his studies, to which he kept so close that he seldom left his chamber . . . So intent, so serious [was he] . . . that he ate very sparingly, nay, sometimes he forgot to eat at all, so that going into his chamber, I have found his mess untouched. When I have reminded him, he would reply: Have I! Then making to the table, would eat a bit or two standing, for I cannot say, I ever saw him sit at table by himself.[47]

And when he did break away from the furnace and the telescope for a while it was only through necessity, his mind still on other things, as Humphrey recounted:

At some seldom times when he designed to dine in the Hall he would turn left hand [instead of going ahead across the Great Court], and go out into the street, where making a stop, when he found his mistake, would hastily turn back & then sometimes instead of going into the Hall, would return to his chamber again . . . When he sometimes took a turn or two [in his garden], he made a sudden stand, turned himself about, ran up the stairs like another Archimedes, with an 'eureka!' fell to write on his desk standing, without giving himself the leisure to draw a chair to sit down on.[48]

Only one thing could and did – temporarily – shake Newton out of his single-minded obsession. At the beginning of February 1685 King Charles II died after a short but violent illness and England was gripped by fears that his brother, the unpopular and headstrong James II, would attempt to convert the country into a Catholic state.

Newton was horrified at the prospect. He had not agreed with many aspects of Charles II's policies, but had had much to thank the dead king for. In contrast, James II was to Newton little short of the Devil incarnate.

During the three years between Halley's visit to Newton in August 1684 and the publication of the *Principia* in July 1687 the country was almost plunged back into civil war. As Newton laboured over gravitation and celestial mechanics, James II was attempting to unleash the power of Rome – a force he imagined would transform the country for the better.

By the 1680s Puritanism had been reduced to little more than a 'peasant religion' and the attempt by the Protestant Duke of Monmouth to seize power from James was quashed with little effort in 1685. But even members of the Catholic nobility in Britain did not want to see a revolution. Anyone doubting the need to keep religious extremism at bay had only to cast their eyes across the English Channel to the horrors wrought by the French Catholic monarchy.

The turmoil of the time affected Newton more than almost any other political event in his life, and it is certain that the fear of impending doom drove him to even greater efforts. By the spring of 1686 a comprehensive text was already taking shape. It was to consist of three separate but linked books and an introduction. Newton's famous three laws of motion were stated in the introduction, and Books I and II dealt with forces and motions, a major part of Book I being based on *De Motu*, in which Newton explained concepts such as centripetal force and mechanical resistance. Book III described the application of many of the theoretical concepts postulated in Books I and II, and included Newton's theory of gravity.

Newton's great accomplishment – the concept of universal gravitation – was the icing on the cake. In Book III he unified the Earth-bound mechanics of Galileo (which could not be related to any form of celestial mechanics) with Kepler's laws (which had previously been thought to have no relationship to everyday Earthly existence). This achievement is comparable to solving the problem currently facing physicists – the unification of relativity and quantum mechanics. The level of sophistication of present-day mathematics and physics might be exponentially greater, but the problem has occupied the entire careers of hundreds of scientists around the world for over fifty years. Newton's unification of Galilean and Keplerian mechanics was achieved alone and, if we date its origins to 1665, in a little over twenty years.

Present-day scientists are keen to explain their work to the lay person, if for no other reason than to popularise their subject in an effort to increase funding. Newton took the very opposite stance: he deliberately wanted to make sure only the élite could read it. He wrote the *Principia* in classical Latin and suppressed publication in English until the final year of his life. Furthermore, the *Principia* was composed in the form of propositions which followed one from the other, so that the previous proposition had to be fully understood before tackling the next – it was not a book to dip into.

It is easy to see why Newton did this. He was tired of underqualified individuals questioning his great pronouncements and did not want a repeat of his earlier troubles:

To prevent the disputes which might arise upon such accounts [by those who] could not easily discern the strength of the consequences, nor lay aside the prejudices to which they had been many years accustomed . . . I chose to reduce the substances of this book into the form of propositions (in the mathematical way) which should be read by those only who had first made themselves masters of the principles established in the preceding books.[49]

He was being polite through necessity: what he really meant was summed up in a boastful admission, many years later, when, as a world-famous pontiff of science, he told a friend that he had made the *Principia* as unreadable as possible deliberately, 'to avoid being bated by little smatterers in mathematics'.[50] And, when asked by his friend Richard Bentley what should be read in preparation for the *Principia*, he replied:

> After Euclid's *Elements* [presumably all fourteen volumes], the elements of the conic sections are to be understood. And for this end you may read either the first part of the *Elementa Curvarum* of John De Witt, or Da la Hire's late treatise of the conic sections of Dr Barrow's epitome of Apollonius.
>
> For algebra read first Bartholin's introduction & then peruse such problems as you will find scattered up & down in the *Commentaries on Carte's Geometry* & other algebraic writings of Francis Schooten . . . For astronomy read first the short account of the Copernican system in the end of Gassendus's Astronomy & then so much of Mercator's Astronomy as concerns the same system & the new discoveries made in the heavens by telescopes in the Appendix.
>
> These are sufficient for understanding my book: but if you can procure Hugenius's *Horologium oscillatorium*, the perusal of that will make you much more ready.[51]

As Newton worked on, Halley was preparing the Royal Society for publication. As the first two books neared completion, Halley found himself falling upon hard times and was forced to resign his fellowship of the Royal Society and to take the vacant position of Clerk to the Society. This represented a serious drop in status, but it was also perfect timing for the advancement of the *Principia*, because it put Halley in the best possible position to influence the society and to fight off the objections of critics.

Even so, he faced an uphill struggle to acquire the finances to

publish Newton's work. Acceptance of a book for publication and
the costs involved was a decision for the council of the Royal Society,
but that spring the country was in the throes of political chaos, the
President of the Royal Society, Samuel Pepys, was attending the new
king, and both vice-presidents were on holiday. To make things
worse, the society had been financially embarrassed by its recent
publication of a worthy but commercially unsuccessful book entitled
De Historia Priscium – The History of Fishes – by the naturalist
Francis Willoughby and did not want to repeat this failure. As the
weeks passed, Halley knew he must act and that Newton would be
expecting a quick decision. At a meeting of the Royal Society held
on 19 May (chaired by one of the vice-presidents, Joseph Williamson,
newly returned from holiday), Halley pushed hard for a decision to
be made despite the absence of the President. He succeeded, but
there were other problems looming large.

As Halley had been trying to organise approval for publication of
Newton's work, Robert Hooke was once more attempting to cause
trouble for the Lucasian Professor. Coffee-house banter aside, he
was incensed that a small section of Newton's new treatise read at a
society meeting on 21 April 1686 made no mention of his own work
on gravity conducted some twenty years earlier. When, a week later,
Halley delivered a gushing appraisal of Newton's work in progress
and suggested that the Royal Society should publish the entire book,
Hooke's anger boiled over. In a letter written to Newton on 22 May,
describing the society's grateful receipt of the new theories and its
intention to publish, Halley had to risk all by reporting Hooke's
feelings:

> There is one thing more that I ought to inform you of, viz, that
> Mr Hooke has some pretensions upon the invention of the rule
> of the decrease of gravity, being reciprocally as the squares of
> the distances from the centre. He says you had the notion from
> him, though he accepts the demonstration of curves generated
> thereby to be wholly your own; how much of this is so, you
> know best, as likewise what you have to do in this matter, only
> Mr Hooke seems to expect you should make some mention of
> him, in the Preface which, it is possible, you may see reason to
> prefix. I must beg your pardon that it is I that sent you this
> account, but I thought it my duty to let you know it, so that you

may act accordingly; being in myself fully satisfied, that nothing but the greatest candour imaginable is to be expected from a person, who of all men has the least need to borrow reputation.[52]

Newton did not shoot the messenger; instead, he went through the entire manuscript deleting the name of Robert Hooke, and a few days later he wrote to Halley explaining that he considered that no fault lay with him but that he would not now mention Hooke's name at all.

But then, after sending the letter, Newton began to stew. It was one of the most stressful times of his life. On top of the strain of completing the *Principia*, he was anxious about the country's future as well as his own prospects. The attacks of a man he saw as an inferior and a dishonourable coward pushed him into a state of suppressed rage. On 20 June, only four weeks after Halley had finally secured the council's approval of the plan to publish the book, Halley's relief was smothered by an unexpected burst of anger from Cambridge. 'Mr Hooke could not from his letters [of 1679] which were about projectiles & the regions descending hence to the centre conclude me ignorant of the theory of the heavens,' Newton seethed:

That what he told me of the duplicate proportion [the inverse square law] was erroneous, namely that it reached down from hence to the centre of the Earth. That it is not candid to require me now to confess myself in print then ignorant of the duplicate proportion in the heavens for no other reason but because he had told me in the case of projectiles & so upon mistaken grounds accused me of that ignorance.[53]

Newton then went on to relate the entire painful sequence of events leading to Hooke's misjudgement and his own triumphant realisation. One can almost feel the anger radiating from the page:

. . . in my answer to his first letter I refused his correspondence, told him I had laid philosophy aside, sent him only the experiment of projectiles (rather shortly hinted then carefully described) in compliment to sweeten my answer, expected to hear no further from him, could scarce persuade myself to answer his second letter, did not answer his third, was upon other things, thought no further of philosophical matters, then his letters put me upon it, & therefore may be allowed not to have my thoughts of that kind about me so well at that time.[54]

Halley was naturally stunned, but it was Newton's conclusion that must have come as the most shattering blow. After mentioning that the work would be in three sections, Newton snapped:

the third [the all-important Book III] I now design to suppress. Philosophy is such an impertinently litigious lady that a man had as good be engaged in law suits as to have to do with her.[55]

It took over a week for Halley to respond, and when he did it was with a note of regret over the Lucasian Professor's offended sensibilities. Showing diplomatic tact reminiscent of Oldenburg, he wrote:

I am heartily sorry that in this matter, wherein all mankind ought to acknowledge their obligations to you, you should meet with anything that should give you disquiet or that any disgust should make you think of desisting in your pretensions to a lady, whose favours you have so much reason to boast of . . . It is not she but your rivals envying your happiness that endeavour to disturb your quiet enjoyment, which when you consider, I hope you will see cause to alter your former resolution of suppressing your third book.[56]

In conclusion, he went on to report that he had consulted Christopher Wren, who heartily agreed with Newton and dismissed Hooke's outrageous claims, and that the general gossip around the society and in the coffee-houses was that Hooke was making a fool of himself.

I found that they were all of the opinion, that nothing thereof appearing in print, nor on the books of the Society, you ought to be considered the inventor; and if in truth he knew it before you, he ought not to blame any but himself . . .[57]

How much Halley's words really soothed is unclear, but, as with Newton's threats to resign from the Royal Society some decade and a half earlier, the matter was never mentioned again. By the beginning of April 1687 Newton had completed the third book, and on 4 April it was in Halley's hands. '[The] world will pride itself to have a subject capable of penetrating so far into the abtrusest secrets of Nature, and exalting human reason to so sublime a pitch by this utmost effort of the mind,' Halley wrote in response.[58]

At last, the labour of twenty years and the amalgamation of so many diverse disciplines had reached fruition in a 550-page treatise that had taken just eighteen months to compose.

With the *Principia*, Newton not only unified the disparate theories of Galileo and Kepler into a single, coherent, mathematically and experimentally supported whole: he also opened the door to the Industrial Revolution. Along with solutions to age-old puzzles such as how the tides are produced and how comets travel through the heavens, Newton addressed more exotic ideas – for example he explained the Earth's 'wobbling' or precessing as it revolves as being due to the varying strength of gravity at different points on the globe. The *Principia* laid the cornerstone for the understanding of dynamics and mechanics which would, within a space of a century, generate a real and lasting change to human civilisation. Without being understood, the forces of Nature cannot be harnessed; but this, in essence, is what the Industrial Revolution achieved – it dragged humanity from the darkness, from the whim of Nature, to the beginnings of technology and the yoking of universal forces.

And this was the harvest of dedication, unsurpassed insight, peerless technical powers and a willingness to explore exotica such as alchemy. Newton saw the power of attraction and repulsion at the bottom of the alchemist's crucible as well as in the movements of heavenly bodies and was able to make the imaginative leap that linked the two, establishing that all matter attracts other matter. For Newton and humankind, the *Principia* was the fabled elixir, the alchemist's gold, the philosophers' stone.

Chapter 10

Breakdown

Whatever withdraws us from the power of our senses; whatever
makes the past, the distant, or the future predominate over the
present, advances us in the dignity of thinking beings.

SAMUEL JOHNSON[1]

For Isaac Newton, the decade between completing the *Principia* and leaving Cambridge to reside in London in 1696 was the most fraught and changeable of his entire life. Many of the changes came indirectly from the delivery of his masterwork, but they also sprang from a shift in the way he perceived himself.

Newton had lived in only two places in his entire life – his first nineteen years had been spent in or near Grantham, the following quarter-century at Trinity College, Cambridge. Even by the standards of the time, his circle of acquaintances was limited, and he had shown no inclination to become involved in college administration or any form of official position within the university. Towards the end of the 1680s all of this changed. During the following ten years he was to hold temporary offices both within Cambridge University and on a governmental level and he became an internationally famous figure, befriending some of the most important intellectual and Establishment figures of the time. On a personal level, he engaged in what was almost certainly the most emotionally charged relationship of his life, stepped to the brink of insanity, and changed his career radically – forsaking science, the mainstay of his existence, for the life of a civil servant.

The first step on this road was the publication of the *Principia*. Although no more than a few hundred copies of the first edition were sold within a decade of its publication, the *Principia* has since gone through over 100 editions and has been translated into almost every language on Earth.

At the time it was published very few people realised the revolutionary nature of the book, in much the same way that neither Darwin's *The Origin of Species* nor Einstein's special and general theories of relativity were fully understood when they first appeared. Reactions were varied and intemperate. One student, upon seeing Newton walking through Cambridge, was overheard to say to his friend, 'There goes the man that wrote a book that neither he nor anybody else understands.'[2] At the other extreme, upon reading the *Principia* the French natural philosopher the Marquis de l'Hôpital apparently 'cried out with admiration: "Good God, what a fund of knowledge there is in that book." He then asked the Dr [with whom he was discussing the book] every particular about Sir I, even to the colour of his hair, and said: does he eat & drink & sleep? Is he like other men?'[3]

The book was reviewed extensively. Halley was first, with an anonymous rave review in the Royal Society's own *Transactions* on the eve of publication, and during the following year learned publications across the Continent followed his lead. These reviews eventually established Newton's international reputation, but outside the élite circle of professional mathematicians and the fellows of the Royal Society the full depth of his achievement was realised only slowly. The mathematician David Gregory, who was then Professor of Mathematics at Edinburgh, was moved to write to Newton to tell him, 'Having seen and read your book I think myself obliged to give you my most hearty thanks for having been at pains to teach the world that which I never expected any man should have known.'[4] Meanwhile, a young teacher of mathematics who would several years later become a chronicler of Newton anecdotes and disciple of the ageing master, Abraham Demoivre, was so obsessed with the *Principia* that he tore pages from it so he could read it while travelling from one pupil's house to another.

There were, of course, those who did not welcome the book with open arms. Hooke continued to criticise and deride Newton's achievements in private while acknowledging the importance of the book in public. His bitterness over Newton's obvious success is evident in his diary entries of the time. In February 1689 he had attended a dinner at the home of his friend Dr Richard Busby and delighted in reporting that another guest, a Dr Hickman, had described Newton as 'the veryest knave in all the house'. A few months later Hooke reported, 'Royal Society met: Hoskins, Henshaw, Hill, Hall . . . New-ton and Mr Hamden came in, I went out. Returned not till 7.'[5]

The previous August a bad review of the *Principia* had appeared in the journal of the Académie Royale des Sciences. Although the identity of its author has never been established beyond doubt, it was quite possibly from the pen of Hooke. It bears his stamp – an odd blend of exuberant praise combined with overt but learned criticism – and most telling is the fact that the author of the review based his attack upon the very charge that Hooke had pressed so vehemently years before: that Newton's work was purely hypotheti-cal. At the time, there could have been few others in Europe who possessed both a searing hatred for Newton and the intellectual cal-ibre to have composed such a review.

But the *Principia* did far more than gain Newton an international reputation: it revolutionised not only the way the universe was per-ceived, but also the way in which science operated. By taking physics beyond the arena of ecclesiastical and philosophical debate, Newton unintentionally created a new intellectual realm. It did not gain the name 'science' for another century and a half, but the set of disci-plines which made the Industrial Revolution possible was one based not upon faith and guesses but upon hard mathematical facts and verifiable evidence. Newton's *Principia* was the first book to docu-ment the universe in this way, and it acted as a template for the practical application of mechanical theories. Because it dealt with mathematically supported principles, it bridged the gap between phil-osophy and mechanical engineering. Men of future generations, such as Watt and Brunel, could apply Newton's principles, and for almost all everyday situations Newton's laws and principles stated in the *Principia* are still used today.

It is, of course, impossible to say what would have happened to Newton's career if it had rested solely upon this book. In spite of its revolutionary nature and international impact, Newton might have simply continued living and working in seclusion as he had done for twenty-five years. Although the *Principia* was the summation of two decades of work and is seen as the pinnacle of his intellectual achievement, he did not perceive it in that way himself – at least not until the early 1690s. Around the time of publication, there was, he believed, still much to achieve, and he saw the first edition of the *Principia* as a report on work in progress. There are clear indications that as Newton was completing it he was seriously pursuing a form of what we would now call a unified theory.

In a 'Conclusio' intended for the first edition but held back by Newton and never sent to Halley, he points the way towards a theory of attraction which applied to the microcosm as well as the macrocosm:

> Hitherto I have explained the system of this visible world, as far as concerns the greater motions which can easily be detected. Whatever reasoning holds for greater motions, should hold for lesser ones as well. The former depend upon the greater attractive forces of larger bodies, and I suspect that the latter depend upon the lesser forces, as yet unobserved, of insensible particles. For, from the forces of gravity, of magnetism and of electricity it is manifest that there are various kinds of natural forces, and that there may be still more kinds is not to be rashly denied. It is very well known that greater bodies act mutually upon each other by those forces, and I do not clearly see why lesser ones should not act on one another by similar forces.[6]

Two versions of the 'Conclusio' have been found among Newton's papers; neither of them was ever included in the *Principia*, nor were they published in any form during the author's lifetime. This is because his thoughts on the nature of subatomic forces could not, at the time, be verified either by experiment or mathematically and he did not want to draw any criticism which might reflect badly upon the rest of the treatise.

Today, scientists regularly conduct experiments using particle accelerators at CERN and other high-energy research centres around the world, but the construction of such devices has been possible only in the latter half of this century and the search for techniques to develop still more powerful machines to test the behaviour of ever more exotic particles is ongoing. Newton had no way of confirming his suspicions, and the mathematical framework that supports modern research – quantum mechanics – lay more than two centuries in the future.

All Newton had to offer in the arena of particle research was hypotheses, and to deliver a hypothesis dressed up as a theory was anathema to him.* Furthermore, the basis upon which his ideas of subatomic forces operated was too obviously derived from alchemy and the hermetic tradition – he could not risk exposing his sources.

I am convinced that Newton did not become an alchemist because he believed it would provide an understanding of the macrocosm or, specifically, a theory of gravity. He did not realise that his alchemical discoveries would bear fruit in physics, although after the event he may perhaps have understood where one of the essential inspirational threads leading to the *Principia* had originated. But, by the time he had formulated a clear and mathematically and experimentally verifiable 'system of this visible world', as he called it, he must have started to conceive the idea that the principles operating in the macrocosm might be mirrored in the microcosm of the crucible.

These thoughts would have occurred to him as he was completing the first edition of the *Principia*, between 1685 and 1687, but the memory of his treatment by certain members of the scientific community during the 1670s still pained him. He would not have wanted to go too far with a published document, even if he could have verified his ideas. And, as he prepared a text that he did feel able to show the world, external forces were conspiring to draw him out of his scholastic shell. Beyond the ivory towers of Cambridge, England was on the eve of a political change that would alter Newton's life.

Spurred on by the anti-Protestant actions of Louis XIV in France, within two years of taking the throne James II had begun to implement a political agenda based upon his own allegiance to Rome. Realising that real influence lay in the universities and the education of his senior servants, James began a programme of infiltrating the great seats of learning. He changed the rules of the universities to allow Catholics on to degree courses and removed barriers to their holding high office.

James's first target had been Oxford, where he had forced through the appointment of a Roman Catholic, one Obadiah Walker, as Master of University College. When he then tried to install the crypto-Catholic Bishop of Oxford, Samuel Parker, as President of

---

* It was not until the second edition of the *Principia*, edited by Roger Cotes and published in 1713, that Newton made his famous pronouncement in the Preface: '*hypotheses no fingo*' – 'I do not invent hypotheses.'

the fellows, they protested so strongly that the appointment was possible only after he had dismissed twenty-five of them. By February 1687, just as Newton was putting the finishing touches to Book III of the *Principia*, Cambridge was drawn into the fray.

To gain a foothold within the strongly Protestant Cambridge Establishment, James wanted to admit a Benedictine monk, Father Alban Francis, to the degree of Master of Arts at Magdalene College. The letter informing the university of the King's wishes arrived on the desk of the Vice-Chancellor of Cambridge, the Master of Magdalene, Dr John Peachell, on 9 February 1687. With brutal directness, it declared that Father Alban Francis was to be sworn in as a student 'without obliging him to perform the exercises requisite thereunto . . . and without administering unto him any oath or oaths whatsoever'.[7]

Peachell was a weak man, an alcoholic and an ineffectual Master, who found himself caught between the demands of the King and the resistance of his fellows. Using his influential contacts at court, he tried to arrange a private meeting with James, but failed. He then organised a petition to the court, but that was met with a second letter demanding that Father Francis be admitted. The King's response concluded 'disobey at your own peril'.[8]

Every way Peachell turned, his path was blocked. He tried through his friend the Duke of Albemarle, the Chancellor of the University, to convene the university senate to arrange another petition, but the King had recently had another Catholic, Joshua Basset, appointed as the Master of Sidney Sussex College, Cambridge, and Basset was one of the six-man committee whose unanimous agreement was needed for a senate petition. Peachell was left with only one choice: Basset could stop a petition, but he could not prevent the senate from debating the matter in an effort to find an alternative route through the mess.

Meanwhile, news of the affair had filtered through to Newton. On 19 February he wrote to an unidentified correspondent, 'Those that counselled His Majesty to disoblige the university cannot be true friends . . . The Vice-Chancellor cannot by law admit one to that degree unless he take the Oaths of Supremacy & allegiance which are co-joined by 3 or 4 statutes.'[9]

Seemingly unaware of his hypocrisy in making such a comment, before long Newton was directly involved in the affair. Peachell convened a senate meeting on 11 March, and for perhaps the first time in his university career Newton contributed to the proceedings. We have no record of what he said, but by the end of the session he had become one of the Vice-Chancellor's two advisers.

Yet the senate could still provide no answers to the problem the university was facing. In desperation, Peachell arranged for a letter to be handed to the King by Albemarle himself, outlining the position of the university – that it could not allow Father Francis admission. This move escalated the dispute into a crisis.

King James, a man known for his impatience and impetuousness, immediately convened a newly reconstituted Commission for Ecclesiastical Causes – a body appointed by the King to investigate religious disputes. This was merely a front through which he could apply his Catholic agenda if menaces failed, and he commanded Vice-Chancellor Peachell and his delegation of fellows to appear before it.

As the news reached Cambridge, the senate was again convened at short notice. In near-panic, on 11 April 1687 the fellows reviewed their limited options. We know from the Exit and Redit book that Newton was away from Cambridge at the time – surviving correspondence shows that a week after delivering the *Principia* he was dealing with far more mundane ongoing problems involving bad debtors and tenants[10] – but his strong feelings and determination to be involved must have been clearly stated at the first senate meeting, because in his absence he was appointed as one of the eight representatives of the university to attend the commission. (Another was Humphrey Babington.) Hearing the decision when he returned to Cambridge a week later, on 18 April, he at once threw himself into preparing for the hearing, set for 21 April.

Newton's passionate anti-Catholic feelings and what he saw as the insidious attempts of King James to foist his religion on the English people are apparent in the way he volunteered himself; but it was an ironic move. Newton had himself refused to take holy orders and had survived in his position only because of a special dispensation from Charles II, yet he was arguing against the admission of anyone who had not taken an oath of allegiance to the Anglican faith. One of the strangest aspects of the entire affair is that at no point did anyone working against the university identify this discrepancy and make an example of the Lucasian Professor. It is possible that the King's commission overlooked his double standards in its rush to force the university to bow to its will. If so, it was extremely fortunate for Newton. Such a revelation not only would have brought the university's protest to a shuddering halt but would have damaged his career irreparably.

James's seven-man commission was headed by the infamous Judge Jeffreys, known throughout the nation for his ruthless handling of the 'Bloody Assizes' following Monmouth's doomed rebellion after the death of Charles. Newton and Jeffreys may well have met on a previous occasion. Although he had not completed his degree before leaving for London and the Inner Temple, the Lord Chief Justice had been admitted to Trinity College, Cambridge, as a pensioner in March 1662 and may have had his empty dinner-plate taken away to the kitchens by the nineteen-year-old sizar Isaac Newton.

The first meeting of the commission, on 21 April, was held before a large audience, most of whom were sycophantic supporters of the King. According to the minutes, Newton did not say a word at this or any subsequent meeting (there were four in all between 21 April and 12 May), but he was certainly active behind the scenes, writing speeches for Peachell. He was also the most radical of all of them in his opposition to the King.

In front of the commission, Peachell, ill with worry and alcohol-related complaints, crumbled under Jeffreys' questioning and could hardly string together two coherent sentences. The rest of the university delegation were powerless to help: any interjection they made was met by Jeffreys' icy gaze or crude suggestions that they were squabbling over who would take the Vice-Chancellor's mantle when Peachell was dismissed.

And dismissed he was. At the third meeting, on 7 May, Jeffreys finally lost his patience with Peachell, and in front of the gathering he stripped him of his office, his home and his income before calling the proceedings to a conclusion with a final order that the commission would reconvene two weeks later to pass judgement upon the remainder of the university delegation.

In one sense, Peachell's dismissal was a relief. He had been unable to stand up to Jeffreys and could not even convey the feelings of the fellows satisfactorily. With him gone, his former colleagues could at least answer Jeffreys' questions coherently and state their case in writing. Together they drafted endless versions of their official answers ready for the final meeting. Five of these have survived among Newton's papers, which suggests that he was prominent in their composition. In all the drafts, the university delegates attempted to convey the impression that a complex mesh of statutes and legislation tied their hands: that they could not admit a Catholic even if they wanted to. In one surviving draft, Newton went further with a stiffly worded addendum:

> They [the senate] were influenced also by their religion established and supported by the laws they are commanded to infringe. Men of the Roman faith have been put into Masterships of Colleges. The entrance into Fellowships is as open. And if foreigners be once incorporated it will be as open to them as others. A mixture of Papist & Protestants in the same University can neither subsist happily nor long together.[11]

Not surprisingly, this failed to make the final draft presented to the last session of the Commission for Ecclesiastical Causes, on 12 May 1687, and it probably would have had no influence upon the outcome if it had been – except to jeopardise the careers of the representatives still further.

After hearing the university's final thoughts, Jeffreys summed up. He swept aside all claims that the university could not allow a Catholic to enter for the MA on legal grounds, and he commanded, under the authority of King James II, that Father Alban Francis be allowed to continue at Cambridge University and to complete his degree unhindered. He then passed judgement on the representatives who had dared question the authority of their monarch. Newton's career and that of the others hung on a very delicate thread.

After the brushing aside of Peachell and the claims of the university, the King, through Jeffreys, was surprisingly lenient. To the relief of the senate representatives, there were no more dismissals – just a humiliating ticking off and, in conclusion, a warning not to interfere again:

> Gentlemen, your best course will be a ready obedience to his majesty's command for the future, and by giving a good example to others, to make amends for the ill example that has been given you. Therefore I say to you what the Scripture says, and rather

because most of you are divines; Go your way, and sin no more, lest a worse thing come unto you.[12]

The representatives could do nothing more. But during the following eighteen months – a period during which the *Principia* was published and Newton's name began a sharp ascent to fame – the lives of his enemies James II, King of England and Scotland, and Lord Jeffreys began a steep decline. By the end of 1688 James was an exile in France, and in April 1689, less than two years after reaching the zenith of his power, Jeffreys died at the age of forty-one in the Tower of London. Shifting political fortunes meant that Alban Francis never received his degree but left Cambridge before William of Orange was welcomed as King by the English people and Parliament in December 1688. The religious legislation of Cambridge University remained unchanged until 1858, when dissenters were finally admitted, and by 1871 all religious tests were abolished.

For Newton, the affair had lasting significance. Not only had he acquired a taste for official responsibility, he had also been noticed by the university authorities, who saw in him a hitherto hidden ability for marshalling arguments and taking a forceful lead in representing strongly held beliefs.

This episode had a profound effect upon Newton's self-image, imbuing him with a new-found sense of status beyond the scientific community. By 1688 the *Principia* was published and receiving the scrutiny of leading intellectuals throughout Europe, and as international recognition greeted his work he was becoming aware of the potential to operate within a wider social and political milieu. He knew that if he was ever to emerge from his sanctuary and take an active role in any form of politics, this was the perfect time. The country was ready for radical political change.

William of Orange landed at the head of a 600-strong fleet at Torbay in early November 1688. James escaped from London a few weeks later and was allowed to slip away by sea, making the French coast unmolested on Christmas Day. In December a special parliament known as the Convention Parliament was summoned to deal with the logistics of transferring power from the Stuart line to the house of Orange. In response to both his behind-the-scenes leadership in facing up to Jeffreys and his sudden heightened profile within the academic community, Newton was offered the position of representing Cambridge University at Westminster.

Newton was an MP for almost exactly one year, but he seems to have contributed nothing personally to the proceedings of the new parliament. He reported back to the Vice-Chancellor faithfully and in minute detail, but according to one anecdote, which may or may not be true, he said only one thing during the year-long proceedings. Feeling a draught down his back, he asked a nearby usher to close an open window.

But, despite this apparent passivity, appearing on a broader stage brought him influential contacts – something he had treasured since his first experiences as a sizar in 1661. Only two days after he had been elected to Parliament he dined with William of Orange himself, who had arrived in England little more than a month earlier.

With William's ascension to power, the entire political and social structure of the country had changed. Although the Tories were in power for several years at the beginning of this new era and Newton was an extreme Whig, he was still able to forge useful relationships with those more experienced in political matters as well as with foreign intellectuals now in favour at the new court.

The most important of his new contacts was Charles Mordaunt, the newly created Earl of Monmouth, who was a great supporter of the new regime and personally close to William. Others were Sir Francis and Lady Masham, a couple who had strong links with both politicians and intellectuals, and whose estate at Oates in Essex became a popular meeting-point for many of the most important people of the day. The network of political friends was completed by Charles Montagu, who had entered Cambridge as a fellow-commoner in 1678. He and Newton had been acquainted at the university, and by the time of the Convention Parliament they had become friends, just as Montagu's star was rising.

After completing his degree Montagu had become a fellow of Trinity and had soon gained a reputation as a literary wit after publishing a swingeing but subtle tract parodying a piece by the Poet Laureate, John Dryden, called *The Hind and the Panther* – a work that had sycophantically supported James's Catholicism. Montagu's satire – *The Country Mouse and the City Mouse* – had quickly become a best-seller, elevating him into the ranks of what could be described as the Young Turks of the seventeenth century.

This was one side of Newton's changing lifestyle, but as he entered a more cosmopolitan world and spent a year in the capital as a political representative for the university his circle of intellectual contacts also widened. Chief among these, and a man whom we can now see as the crucial link in Newton's social development, was the English philosopher John Locke.

The two men first met early in 1689 and discovered an immediate affinity for each other's ideas. Ten years Newton's senior, Locke had studied at Oxford and although he had not gained a medical degree he had acquired a medical licence and joined the household of the first Earl of Shaftesbury. Following a charge of high treason, Shaftesbury had fled to Holland in 1682, and Locke subsequently became his political adviser and lived on the Continent in self-imposed exile until the Glorious Revolution of 1688. Returning shortly after William's landing at Torbay, he had been given the responsibility of escorting the future Queen of England, Mary, on her journey to London.

Locke had become familiar with the *Principia* while still in exile and saw it as a work in tune with his own philosophical ideas. The immediate depth of mutual understanding between the two men is evident in Locke's homage to Newton in the Preface of his own masterpiece, *An Essay Concerning Human Understanding* (1690), in which he calls him 'the incomparable Mr Newton'.[13]

Locke had no mathematical training and found the demonstrations and many of the arguments in the *Principia* totally impenetrable. He asked Huygens for his opinion of the work and whether the formulae could be trusted. The Dutch mathematician's support was so positive that, according to one account, 'Being told he might depend upon their certainty, Locke took them for granted, and carefully examined the reasonings and *corollaries* drawn from them, became master of all the physics, and was fully convinced of the great discoveries contained in the book.'[14]

As this account suggests, Locke dedicated himself to understanding the book, but, because of his lack of mathematics, he studied only the text and ignored the mathematical demonstrations. However, this did not stop him reviewing the *Principia*. His was the first review the book received on the Continent, and it appeared, unsigned, in the March 1688 issue of the *Bibliothèque universelle*. Although flattering, the review was obviously written by someone who had not grasped the book's central theme – that gravity is a force experienced by all matter and acts at a distance. When he arrived in London, Locke sought out Newton in order to further his understanding, and the author obligingly drafted him a simplified proof.

It is easy to see why the two men had an immediate affinity for one another. Locke's philosophy (which became the basis of British empiricism) was based upon the idea that all our knowledge derives from experience. 'Let us suppose the mind to be, as we say, white paper, void of all characters, without any ideas,' he wrote; 'how comes it to be furnished?'[15] His answer was that knowledge comes from two sources of experience: sensation, which provides ideas about the external world, and reflection, which provides knowledge of the internal workings of the mind. But, he concluded, because we do not truly perceive an object, but only the *idea* of it, the true nature of anything can only be ascertained by mathematics.

Locke was also an influential political thinker, and if anything his political legacy has proved more lasting than his philosophy. As a Whig, and considered by many the father of liberalism in both Britain and America, Locke would have shared many of Newton's views. Almost half a century earlier, Thomas Hobbes had created a political philosophy centred around materialism – a doctrine denying the independent existence of spirit, a notion abhorred by Newton. Locke held the opposite view to Hobbes, believing that the first duty of any government was to serve the people. This became the cornerstone of modern democracy, and was a creed dear to Newton's own heart.

As well as being someone whose philosophical, religious★ and political views coincided to some degree with Newton's, Locke was also an amiable, open-minded man, who possessed both the circum-spection of Edmund Halley and an intellectual intensity equal to that of Newton himself. Until Locke's death in 1704, the two men

---

★ By the very fact that he had supported Shaftesbury – a keen denouncer of 'popish plots' – and had almost lost his own life in the process, Locke's anti-Catholicism was well known. Nevertheless, he was also a great supporter of religious freedom, and in his *A Letter Concerning Toleration* (1689) he expressed the view that, within limits, no one should dictate the religious beliefs of another – a doctrine given official status in the ground-breaking Toleration Act of 1689, which granted religious freedom to all Christians with the notable exception of Catholics and Arians.

corresponded on a wide range of subjects, and their relationship was certainly the most fulfilling that Newton ever maintained – no one else ever came so close to Newton on so many levels simultaneously. Barrow, More and Babington had acted as father-figures and career-enhancers. Montagu was a useful and witty young supporter, but did not share many of Newton's most cherished beliefs. Boyle, Wren, Halley and Oldenburg were scientific colleagues rather than close friends. Locke, however, acted as a bridge between the scientific world and the political; he not only acted as an intellectual companion for Newton, but also helped to mould his future. Together they were later seen as the twin pillars upon which the Age of Reason was built, but they were also intimates, and Newton opened up to the older man, revealing to him his own extreme religious views and his alchemical practices.

As the nucleus of a social web involving Charles Montagu, the Mashams, the Earl of Monmouth and William of Orange himself, Locke often acted on Newton's behalf, unfailingly supporting him in his later attempts to find permanent public office. Newton trusted Locke, and towards the end of 1690 he even allowed himself to be persuaded to take steps to publish a tract on his unorthodox views – nothing less than an Arian manifesto: a project from which he withdrew only at the last moment.*

Any discussion of Newton's adult emotional life (such that it was) must also take account of two men with whom he may have shared a sexual relationship. The first of these was the elusive John Wickins, with whom Newton lived for almost two decades. There is no hard evidence of their relationship being sexual in nature, only speculation surrounding the intensity of their bond as indicated by the absolute and clinical manner of its breaking. Far clearer is the emotional aspect of Newton's relationship with a young Swiss mathematician named Nicholas Fatio de Duillier.

---

* Newton drew back after going some way towards allowing it to be published in Holland through Locke's contact at the *Bibliothèque universelle*, the founder, Jean Le Clerc. After Le Clerc was refused permission to use the treatise, it was locked away in the Remonstrant's Library in Amsterdam and was published under Newton's name only some fifty years later (over a decade after the scientist's death), under the title *Origins of Gentile Theology*.

Fatio de Duillier was born into a wealthy family who mollycoddled and spoilt him for most of his early life. Although he became quite well known for a while among the intelligentsia of both mainland Europe and England, his reputation did not last and his appeal to Newton seems ambiguous. The seventh of twelve siblings and twenty-two years Newton's junior, Fatio enjoyed a childhood that could not have differed more from Newton's own; yet before the younger man's thirtieth birthday he had forged a close personal relationship with the Cambridge professor.

Fatio's upbringing was shaped by an ideological clash between his father, a wealthy Swiss landowner who wanted him to study divinity, and his doting mother, Catherine Barbaud, an intellectual who insisted that her favourite son should study in Paris before serving at a Protestant court in Germany. In old age, Fatio wrote of his youth:

> My father designed that I should study divinity; and accordingly, having been instructed, both at home and at Geneva, in the Latin and Greek tongues, I spent two or three years in the study of philosophy, mathematics, and astronomy; and began to learn the Hebrew tongue and go to the lessons of the divinity professors.[16]

By the time Fatio was eighteen, in 1682, his mother's wishes had prevailed and he found himself in Paris, living off a generous allowance. Already he was showing an early flair for self-promotion and outrageous self-assurance by corresponding with the great scientific figures of the time, including Huygens, Leibniz and the head of the royal observatory in Paris, Domenico Cassini.

It is clear he had some talent for mathematics and science, but quite how much is difficult to judge. He impressed a series of eminent philosophers with his youthful precociousness, and before de Duillier travelled to England in 1687 the historian Gilbert Burnet described him as 'one of the greatest men of his age, who seems born to carry learning some sizes beyond what it has attained'.[17] But later, especially within London society, many saw him as a buffoon, and he soon acquired the nickname 'the ape of Newton'. It is certainly possible that Fatio was a prodigy who burned bright for a short time as a youth, but, equally, his skills at self-promotion may have been his greatest attribute. Whatever his intellectual abilities, it is clear he had a great talent for impressing powerful men – often much older than he – and of keeping them enamoured long after impartial observers had tired of him.

Fatio arrived in England through just this talent. In 1685, as England lay in fear of revolution, William of Orange felt threatened by France. During a chance encounter with a Piedmontese count at his father's estate, Fatio heard of a plot to abduct the Dutch monarch and deliver him into French hands. Informing William at once, he gained great favour at The Hague and was even offered, as a reward, a professorship at the university. Instead, fearing reprisals from the thwarted French, in 1687, as Newton was awaiting publication of the *Principia*, de Duillier asked for and was granted permission to travel to England, where he would stay, he claimed, ''til the Prince of Orange was in full power of these kingdoms'.[18]

Arriving in London with a letter of introduction from the French natural philosopher Henri Justel, within two weeks, towards the end of June 1687, de Duillier was elected a fellow of the Royal Society. By the summer of 1689, a few months after Prince William had become the constitutional monarch of England, he was given the honour of escorting Holland's most celebrated intellectual, Christiaan Huygens, during his visit to London. And it was as a result of this appointment that he first encountered Newton, at the Royal Society on 12 June 1689, when Newton and Huygens also met for the first time.

The relationship between Newton and Fatio blossomed almost immediately, and the two men saw each other frequently during the summer of 1689. The Swiss mathematician's role as Huygens's guide brought him into contact with many of England's most distinguished intellectuals, and Newton attended several of the same functions as Huygens. Early that summer, Newton had heard that the position of Provost of King's College had fallen vacant. Huygens, who had been staying in Cambridge, travelled to London with him in support of his petition for the post, and Fatio was certainly with the party.

The earliest surviving letter from Newton to Fatio is dated October 1689 and, like later letters between them, is laced with innuendo and encoded subtext. The letter has been censored by an unknown hand at some time since its composition in 1689 – in three places words have been cut out of the letter. Yet, despite these cuts, it illustrates a remarkable degree of intimacy considering that the two men had known one another for only a few months. It begins:

> I am extremely glad that you . . . [first cut] friend & thank you most heartily for your kindness to me in designing to bring me acquainted with him. I intend to be in London the next week & should be very glad to be in the same lodging with you. I will bring my books and your letters with me.[19]

Newton then goes on to offer criticism of his friend Robert Boyle – one of the greatest thinkers of the time – something he was not to repeat to anyone other than Locke, and then only after Boyle's death in 1691:

> Mr Boyle has, at various times, offered to communicate & correspond with me in these matters but I have declined it because of his . . . [second cut] & conversing with all sorts of people & being in my opinion too open & too desirous of fame. Pray let me know by a line or two whether you can have lodgings for both of us in the same house at present or whether you would have me take some other lodgings for a time 'til . . . [third cut].[20]

Although it is almost impossible to unravel the words of a hidden censor so long after the event, Newton's reference to Boyle's attempts to 'communicate & correspond with me in these matters' which appears just before the second cut and the comment 'being in my opinion too open & too desirous of fame' suggest that the letter touched upon alchemy. We know from his letters to Oldenburg that Newton disapproved of Boyle's attitude to proselytising his alchemical ideas.

Newton also appears to be unusually preoccupied with the matter of where he and Fatio will be staying, suggesting twice in the same short letter that he would like to stay in the same lodgings as his friend. The third cut (and perhaps also the first) could have been personal comments that the censor (possibly Newton himself in retrospect) considered too embarrassing to leave untouched.

By November, letters were flowing between the two men and Fatio was singing Newton's praises in his correspondence with philosophers on the Continent. He had arrived in England a dedicated

Cartesian, but was soon drawn into Newton's mechanical universe. Newton was 'the most honest [*honnête*] man I know,' he declared to his friend Jean-Robert Chouet, a philosophy professor in Geneva, 'and the ablest mathematician who has ever lived'. He went on to describe the Cartesian system he had now discarded as 'an empty imagination', before concluding by waxing lyrical about how he would like to erect a monument to the great Newton and how he wished he could take up residence with the professor.[21]

Fatio knew he could gain by association in his relationship with Newton. He had shown an early flair for such opportunism, and later tried to make a career out of knowing the right people at the right time. Across a divide of three centuries it is difficult to assess how genuine were the feelings he expressed in his letters and how much was a mixture of bragging and delusion. For the first six months of their relationship he attempted to show the world just how intimate was his understanding of the *Principia* and to project the image of being Newton's favourite, an apprentice to the master (hence his sobriquet 'the ape of Newton'). But the relationship was not based upon shared scientific interests alone. By 1689 Newton was keen to find a position outside Cambridge, and Fatio gave the clear impression that he could be useful in trying to accomplish this. He even tried to take advantage of his familiarity with William and others at court. 'I did see Mr Lock above a week ago,' he wrote on 24 February 1690, 'and I desired that he should speak earnestly of you to My Lord Monmouth. He promised me he would do it . . .'[22]

He concluded with an unmistakable sign of their growing intimacy. Newton was due to visit him in London the following week, and Fatio was expecting the arrival of a copy of Huygens's book, *Traité de la lumière*, at the same time. 'It being written in French,' he added, 'you may perhaps choose rather to read it here with me.'[23] This may seem innocent enough, but who else could have said such a thing to Newton? It implies that Fatio was aware of Newton's less than perfect French and that they had perhaps joked about it on a previous occasion.

In early June 1690 Fatio began a fifteen-month trip around Europe, partly to promote his own scientific theories in the major centres of learning but also to visit his family in Switzerland. Any correspondence between him and Newton from this period has disappeared, but in October, only a few months after Fatio's departure, Newton wrote to Locke asking if he had heard word from him. Again, this would not seem unusual for anyone other than Newton, but he almost never enquired after an acquaintance. Significantly, when Fatio did return, in September 1691, Newton rushed to London to meet him almost as soon as his ship had docked.

The meeting between the two men in mid-September was so private that even their mutual friends in London were unaware that Newton had been in town. The Exit and Redit books at Trinity record the professor's departure on 12 September and his return on the 19th, but the mathematician David Gregory, whom Newton frequently visited in London, wrote to Newton in October to inform him that their friend Fatio de Duillier had been in the country 'some days'.[24]

After this reunion their relationship stepped up a gear. Although all Newton–Fatio correspondence between September 1691 and November 1692 has been lost, letters between Fatio and Huygens show that Newton and his young devotee met frequently during the year following the latter's return to England. During one visit to Cambridge, Fatio was shown some of the theories Newton had been working on during his absence; he reported back to Huygens that he 'was frozen stiff when I saw what Mr Newton has accomplished'.[25]

Then, on 17 November 1692, shortly after Fatio had returned to the capital following a visit to Cambridge, Newton received a letter that threw him into an uncharacteristic panic. 'I have sir, almost no hopes of seeing you again,' Fatio began:

With coming from Cambridge I got a grievous cold, which is fallen upon my lungs. Yesterday I had a sudden sense as might probably have been caused upon my midriff/diaphragm by the breaking of an ulcer or vomica, in the undermost part of the left lobe of my lungs. For about the place of my midriff/diaphragm I felt a momentary sense of something bigger than my fist moving and acting powerfully. That sense was distinct in all that region, but not troublesome to me, although my surprise caused my body to bend forwards, as I was sitting by the fire. What I felt next was only a gentle and easy sense of a natural heart in that region. My pulse was good this morning; It is now (at 6 pm) feverish and has been so most of the day. I thank God my soul is extremely quiet, in which you have had the chief hand. My head is something out of order, and I suspect will grow worse and worse. The imperial powders, of which I have taken today four of the weakest sort, and one of the best sort, have proved quiet insignificant. Which confirms to me that I have guessed right about the breaking of the vomica, or else my disease must be an ague. Were I in lesser fever I should tell you sir many things. If I am to depart this life I could wish my eldest brother, a man of an extraordinary integrity, could succeed me in your friendship. As yet I have no doctor perhaps with a paracenthesis they may save my life, which I am yet certain is in any danger. Mr Cunningham, sir will acquaint you with what shall befall me.[26]

This quite extraordinary letter could be explained by a genuine fever which left Fatio confused and rambling; however, the minute attention to detail is not only characteristic of much of his other surviving writing, but also demonstrates an acute hypochondria not dissimilar to Newton's own. Even the relatively wealthy of the seventeenth century could be forgiven an obsession with their own health, but Fatio is laying it on a little too thick here, and it raises the question, Why? Was he seeking attention by exaggeration? We know that he was a great self-publicist; is this another example of his attempts at manipulation? What is to be made of his suggestion that his brother could stand in for him should he die? Is it simply a further, rather childish, dramatic twist?

Another significant comment is Fatio's assertion 'I thank God my soul is extremely quiet, in which you have had the chief hand.' Does this mean that Fatio was troubled in spirit before meeting Newton: that their relationship had somehow begun to heal a wound? Or is it yet another flattering gesture?

It is quite possible that Fatio was manipulating Newton entirely for his own ends because it must have been obvious that the older man was infatuated with him. Yet, at the same time, Fatio would have felt intellectually insecure in Newton's presence. To compensate, he smothered any sense of inferiority with a show of bravado – even asserting at one point in 1690 that he would supervise the second edition of the *Principia*. At times this show went to his head. In a letter to Huygens, he declared blithely that his edition of the *Principia* would be far longer than the original because he would want to add so much. 'Nevertheless,' he boasted, 'it would be possible to read and to understand that folio in much less time than it takes to read or to understand Mr Newton's quarto.'[27]

Fatio did not die immediately but lived a further sixty-one years. His letter had taken three days to reach Cambridge, during which time the worst had already passed. Newton's anxiety for Fatio's health is evident from his replying immediately by special carrier:

> I . . . last night received your letter with which how much I was affected I cannot express. Pray procure the advice & assistance of physicians before it be too late & if you want any money I will supply you. I rely upon the character you give of your elder brother & if I find that my acquaintance may be to his advantage I intend he shall have it, & hope that you may still live to bring it about, but for fear of the worst let me know how I may send a letter &, if need be, a parcel to him . . .

He signed off in the warmest terms he had ever used to anyone openly: 'Your most affectionate and faithful friend to serve you. Is. Newton.'[28]

This is another puzzling letter. Although Newton's concern is declared in the opening line and the repeated hopes that Fatio will recover, there is an odd emphasis on the importance of Fatio's older brother. Even allowing for the seventeenth-century mind being more readily attuned to the inevitability of sickness and death than we are, Newton's insistence that he would contact Fatio's brother could really mean only one of two things. Either, unsure of how serious the illness could be, he is humouring a sick man to make him feel better, or else 'but for fear of the worst let me know how I may send a letter &, if need be, a parcel to him' could imply that he and Fatio had some materials or documents that needed to be preserved in the event of Fatio's death. This theory is supported by a letter crossing Newton's which arrived the following day. After announcing that the worst of the symptoms had passed, Fatio gives a lengthy description of his brother's character, saying, 'My brother, sir is without exception the most discreet and reserved man I know . . . He leaves everybody the liberty of their thoughts, and never has made any bad use of any man's confidence.'[29]

Newton and Fatio had discussed alchemy and religion at length during their correspondence. Fatio wrote to Newton with page after page of text dissecting religious ideas and alchemical theories, and in their earliest surviving correspondence Newton probably commented on Boyle's involvement with alchemy. As Newton's closest personal companion in adult life, Fatio may have been told of the professor's secret obsessions, and perhaps they conducted experiments together in Cambridge, and bought books and studied arcane texts together in London, just as they discussed their reading of more conventional tracts such as Huygens's *Traité de la lumière*. As a young, open-minded mathematician and natural philosopher, Fatio would probably have been happy to experiment, seeing his initiation into alchemical practices at the hand of the master as another bond between them. On Newton's part, it illustrates further his infatuation. There could have been little to gain from Fatio that he had not already discovered from his own lifetime of study and experiment: this association merely presented a further risk.

Throughout the remainder of 1693 and into the new year Newton and Fatio continued their emotionally charged correspondence. Fatio's illness flared up from time to time, and there is no evidence that they met up other than during a few visits Newton made to London, but their letters were full of the same caring sentiments: each remembered the other in his prayers.

Then, early in 1693, Newton suggested that Fatio should consider living in Cambridge. In a letter dated 24 January, he proposed that his friend's continuing poor health might be associated with living in the capital: 'I fear the London air conduces to your indisposition,' he wrote, '& therefore wish you would remove hither so soon as the weather will give you leave to take a journey. For I believe that this air will agree with you better.'[30]

Fatio did not respond immediately, but a few weeks later he informed Newton that he had received word that his mother, Catherine, had died and he was considering returning to Switzerland permanently. This threw Newton into a panic, and he responded by reiterating that Fatio should seriously consider moving to the university and living with him.

Again Fatio's reply was guarded, but he admitted that, if his inheritance allowed it, he would rather stay in England: 'chiefly at Cambridge, and if you wish I should go there and have for that some other reasons than what barely relates to my health and to the saving of charges [expenses] I am ready to do so; but I could wish in that case you would be plain in your next letter'.[31]

And, in a letter sent to his brother in Switzerland in February, Fatio hinted at an emotional dilemma which may have centred around Newton's offer: 'My pain comes chiefly from a cause that I cannot explain here . . . the reasons I should not marry will probably last as long as my life.'[32]

Convincing himself that Fatio was all but destitute, Newton then sent him £14 on the pretence that it was to cover some minor expenses the younger man had incurred when they had worked together briefly in Cambridge the previous autumn. Interspersed with talk of the prophecies (about which Fatio seems to have become overenthusiastic, with Newton suggesting he 'indulged too much in fancy in some things'[33]), they again touched upon the practicalities of Fatio moving to Cambridge. 'The chamber next to me is disposed of,' Newton reported in a letter dated 14 March 1693; 'but that which I was contriving was, that since your want of health would not give you leave to undertake your design for a subsistence at London, to make you such an allowance as might make your subsistence here easy to you.'[34]

Newton had to wait a month for a reply. Then on 11 April 1693, Fatio wrote, 'I could wish sir to live all my life, or the greatest part of it, with you, if it was possible, and shall always be glad to any such methods to bring that to pass as shall not be chargeable to you and a burden to your estate or family.'[35]

There then followed a brief exchange of letters, continuing into May, in which Fatio revealed that his inheritance had come to little. His latest plan, he announced, was to become a physician and to produce an elixir of his own design. The letter included calculations of start-up and running costs for the scheme, and how the product could succeed and make him a fortune. Mysteriously, all talk of his moving to Cambridge had dried up.

The last significant contact between the two men occurred a few weeks later, in late May and early June of that year, when Newton made two separate trips to London to visit Fatio.

Although Newton spoke to Fatio and corresponded with him over minor matters in subsequent years, June 1693 marks the abrupt end of their intense relationship and the beginning of one of Newton's blackest periods. Whatever was said during those two final meetings, the clinical nature of their later dealings shows that that second London meeting must have been the occasion for a final break.

So, what conclusions can be drawn about this relationship? And what brought it to such a sudden end?

For a time, at least, Newton was enamoured of Fatio. Just as other intellectuals of the period had believed the young Swiss to be far more able than he probably was, Newton too had become convinced of Fatio's abilities. He had also become infatuated with him, and in some respects it is easy to see why.

Until this time, Newton had known very few people of his own age or younger. The only younger men or contemporaries with whom he had been on friendly terms were Wickins, with whom he may have had a physical relationship, Halley, a man Newton admired for his willingness to move with new ideas and to see the value of innovative methods, and Charles Montagu. But Newton was an excessively competitive, oversensitive and introverted man. What was it about Fatio that seemed to dispel these barriers almost immediately?

If his portrait painted as a young man (hanging now in the Geneva University library) is anything to go by, it was not Fatio's looks that had won Newton over: he was a plain man with a large Roman nose and rather small eyes. Nor was Newton easily swayed by reputation alone, or by bearing or heritage, unless an individual was of very high social rank, which Fatio certainly was not. One can only assume that Fatio charmed the older man with a blend of intelligence, flattery and imagination which interacted with a nascent and largely suppressed sexual interest on Newton's part (something Fatio probably noticed and nurtured for his own ends). Good timing was also on Fatio's side. If they had met ten years earlier, this potent blend might not have worked, but, in the first flush of international recognition following the *Principia*, Newton, the man who had been at the centre of political life, the associate of the great and the good, and now a renowned intellectual, could have felt little threat from Fatio. And, despite the young man's constant bombast (such as his claim to have produced a theory to *explain* gravitation) and his occasionally patronising comments, Newton's interest and desire for Fatio survived for almost four years. Given the speed with which others less enamoured lost interest in the young man, this was far longer than one might have expected.

To judge Fatio's intentions and emotions accurately is almost impossible, because so little is known about him. Some have painted him as a temperamental genius, but a truer image may be that of a manic-depressive. The moods expressed in his letters swung from extreme excitement and overenthusiasm to fits of pessimism and angst. He pursued one ill-conceived plan after another, leaped from Cartesian philosophy to Newtonianism within a period of weeks, and was setting himself up as Newton's heir apparent only months after first meeting him. The advantages for him in this new-found relationship were obvious, and he did everything he could to nurture it. He knew Newton better than anyone ever had – partly because, at this point in his life, the older man at last wanted to be known. Fatio may not have been a genuine scientific genius but he was astute and clever, cunning and happy to oblige. He then trod a very fine line between seduction and overkill.

Newton wanted to believe in Fatio. He had never before needed anyone with whom to share his ideas, but as he emerged into public view certain aspects of his character were beginning to change; the success of the *Principia* was beginning slowly to metamorphose him. By supporting Cambridge University against James II he had pinned his anti-Catholic colours to the mast, and he had taken a year off from his alchemical and mathematical pursuits to sit in Parliament – moves which would have been inconceivable to him a decade earlier. And then came a young man, too inexperienced to be a danger, who believed in him, who wanted to be his *scientific son*, and towards whom he may have felt a certain sexual attraction or infatuation. Fatio could have spelt disaster for his reputation, and he almost did.

Newton had walked head-on into danger before, but had pulled back at the last moment. A few years earlier he had accepted Locke's invitation to publish an Arian tract, only to realise the possible consequences shortly before publication. It would seem reasonable to assume that by the summer of 1693 Fatio de Duillier had also become a danger and Newton had once again to scramble from a trap of his own making.

Fatio was becoming careless. He had begun to send his friend alchemical secrets by post, and he may have spoken of such things in the company of others – something for which Newton had criticised Boyle. And of course Newton had enemies all too ready to attempt to destroy him. Hooke disliked and ridiculed Fatio after hearing his cranky attempts to explain the cause of gravitation; at one point he labelled him the 'Perpetual Motion man'.[36] Even friends of Newton grew to see Fatio as more clown than 'genius', and, according to David Gregory, Newton himself had made at least a show of ridiculing some of his young friend's ideas: 'Mr Newton and Mr Halley laugh at Mr Fatio's manner of explaining gravity,' he commented in a memorandum of 1691.[37]

Whatever prompted it, during the early summer of 1693 the scales fell from Newton's eyes and he returned immediately to his paranoid, defensive self.

The plan to cohabit had little chance of ever becoming reality. Newton might have cited Fatio's health as the reason for the proposal, but his enemies would certainly have questioned the nature of their relationship if the young man had moved into rooms at Trinity. Newton had skated on thin ice for much of his career; he must have given thanks each and every day that enemies such as Hooke had not discovered his obsession with alchemy or his heretical religious views. As he began to realise the folly of his behaviour towards Fatio, he could not have failed to see the extra risk he had been taking.

Severing the relationship solved the practical difficulties but brought with it a fresh set of problems. Within weeks, his caged emotions overflowed into temporary insanity.

The first sign that something was wrong with his mental state came in a letter written on 13 September 1693 and addressed to Samuel Pepys. It said:

Sir,
Some time after Mr Millington [John Millington, a colleague of Newton's at Magdalene College, Cambridge] had delivered your message, he pressed me to see you the next time I went to London. I was averse; but upon his pressing consented, before I considered what I did, for I am extremely troubled at the embroilment I am in, and have neither ate nor slept well this twelve month, nor have my former consistency of mind. I never designed to get anything by your interest, nor by King James's favour, but am now sensible that I must withdraw from your acquaintance, and see neither you nor the rest of my friends any more, if I may but leave them quietly. I beg your pardon for

saying I would see you again, and rest your most humble and most obedient servant,

Is. Newton.[38]

Pepys was stunned, and at first thought he was the object of some peculiar misunderstanding. No longer active politically, he lived in seclusion and had not been in touch with Newton for some considerable time. He had certainly not been involved with the professor's attempts to make social or political contacts in the capital. Upon a second reading he must have noticed the signs of paranoia and the confused thinking (after all, King James had been in exile for almost five years) and reached the conclusion that Newton was ill. He quickly contacted his nephew John Jackson, who was then at Cambridge, and asked him to talk to Newton. He then followed this up with a letter to Millington himself, asking him to interview the Lucasian Professor as soon as he could. Newton was suffering from 'a discomposure in head, or mind, or both', he told Millington – continuing, 'I owe too great an esteem for Mr Newton . . . to be able to let any doubt in me of this kind concerning him lie a moment uncleared.'[39]

Millington complied readily and called on Newton a short time later, only to find the Lucasian Professor perfectly lucid. He, rather than Millington, broached the subject of the letter to Pepys, declaring that it had been 'a very odd letter', and asking Millington to pass on his sincere apologies. It had been due to a 'distemper that much seized his head and kept him awake for about five nights together', he confessed.[40]

But, while the distemper had lasted, Newton had sent another, far more disturbing, message to his friend John Locke. Written in a shaky uneven hand, it read:

Sir

Being of opinion that you have endeavoured to embroil me with women & by other means I was so much affected with it as that when one told me you were sickly & would not live I answered that it was better if you were dead. I desire you to forgive me this uncharitableness. For I am now satisfied that what you have done is just & beg your pardon for having hard thoughts of you for it & for representing that you stuck at the root of mortality in a principle you laid down in your book of ideas & designed to pursue in another book & that I took you for a Hobbist. I beg your pardon also for saying or thinking that there was a design to sell me an office, or to embroil me. I am your most humble & most unfortunate servant.

Is. Newton.[41]

It was as though all the fiends that Newton had ever harboured were let loose at once. Here is a match for the confessions notebook of 1662. The old demons are there – Hobbes and atheism, sexual repression, and his confused feelings for those whom he considered friends.

In one respect, Newton was lucky to have chosen two gentlemanly friends to attack: Locke was even more circumspect than Pepys, and quickly wrote a searching and worried letter to his disturbed friend.

Give me leave to assure you that I am more ready to forgive you than you can be to desire it, and I do it so freely and fully that I wish nothing more than the opportunities to convince you that I truly love & esteem you & that I have still the same good-will for you as if nothing of this had happened.[42]

This was posted on 5 October and received by a perfectly lucid Newton, now doubly embarrassed by what he had done. In his reply to Locke, he wrote:

The last winter by sleeping often by the fire I got an ill habit of sleeping & a distemper which this summer has been epidemical put me further out of order, so that when I wrote to you I had not slept an hour a night for a fortnight & for 5 nights together not a wink.[43]

But how much can be genuinely attributed a 'distemper'? Crucial to any analysis of Newton's mental state during the summer of 1693 is the timing of this breakdown, coming as it did at a point in his life when several disparate emotional crises had reached a peak.

First, the academic acclaim he received for the *Principia* must have led him to wonder whether he had reached the pinnacle of his career, a work he would be incapable of bettering. Secondly, although he was still pursuing his efforts in alchemy, they were not yielding the hoped-for unified theory. During the spring of 1693 he wrote what was to be the summation of these efforts, a treatise entitled 'Praxis'. But, rather than this being his *Principia Chemicum*, 'Praxis' is a harrowing complement to the deluded letters of September 1693. Now discernible as Newton's last-ditch attempt at a unification produced from disparate alchemical threads, 'Praxis' is little more than a blend of naked delirium and false conviction – the work of a man on the edge of madness. At one point he relates that:

Artefius tells us that his fire dissolves & gives life to stones, & [word deleted] Pontanus that their [illegible words deleted] fire is not transmuted with their matter because it is not of that matter, but turns it with all its faeces into the elixir. Which deserves well to be considered. For this is the best explication of their saying that the stone is made of one only thing.[44]

Earlier in the same treatise, Newton declares that he has achieved 'multiplication', the dream of the alchemist:

Thus you may multiply each stone 4 times & no more for they will then become oils shining in the dark and fit for magical uses. You may ferment it with gold by keeping them in fusion for a day, & then project upon metals. This is the multiplication in quality. You may multiply it in quantity by the mercuries of which you made it at first, amalgaming the stone with the mercury of 3 or more eagles and adding their weight of the water, & if you design it for metals you may melt every time 3 parts of gold with one of the stone. Every multiplication will increase its virtue ten times &, if you use the mercury of the 2d and 3d rotation without the spirit, perhaps a thousand times. Thus you may multiply to infinity.[45]

After five short chapters the work was totally abandoned.

The third factor – and perhaps most important of all – was his relationship with Nicholas Fatio de Duillier. It is difficult to see it as coincidental that Newton's breakdown followed so close on the ending of their relationship. Indeed, it would be fair to assume that the split served to tip Newton's already delicate emotional make-up into a form of temporary psychosis.

Finally there is the question of career pressure. Newton had been Lucasian Professor for almost a quarter of a century, but during the years leading to the crisis of 1693 he had sampled life outside Trinity College. The frustration he must have felt at having tasted the heady wine of political life and a broader agenda only then to return to the complete solitude of Cambridge must have been extremely unsettling.

Combined, these factors go a long way to explaining Newton's short-lived mental breakdown. Many researchers have tried to establish complex reasons for the unfortunate pair of letters of September 1693 and what was later dubbed Newton's 'Black Year'. These range from the idea that he suffered a form of poisoning, from the exotic chemicals he used in the laboratory, to simple overwork and self-neglect.

Superficially, these seem reasonable propositions. Newton was certainly exposed to a virulent cocktail of substances whose biochemical effects were quite unknown at the time. But a close scrutiny of the symptoms shows that this explanation falls some way short of a definitive cause.

Supported by tests conducted in the late 1970s on a sample of Newton's hair, two researchers, P. E. Spargo and C. A. Pounds, proposed the theory that unusually high levels of lead and mercury were responsible for Newton's breakdown.[46] We would, of course, expect such high concentrations of laboratory chemicals in Newton's hair – after a quarter of a century of alchemical experiments, any other result would have been surprising – but there is no proof that these chemicals were the cause of his illness. Although we have only scant details of his symptoms from letters to Locke, Newton displayed none of those usually associated with lead or mercury poisoning, save a mental imbalance which proved quite temporary. He did not suffer jaundice, flaking of the skin, premature ageing, weight-loss, or darkening of the nails. Nor did he develop kidney disease, slip into a coma, or become paralysed. He did not even suffer ulceration of the gums resulting in the loss of his teeth. When he died, aged eighty-four, Newton had a full set of teeth bar one.

As for stress and overwork, Newton did overwork during this period, but for him that was nothing new. Perhaps, for some reason, he could deal less well with stress at this time, but it is quite unsatisfactory as a sole reason for his mental breakdown.

Despite the circumspection of Pepys and Locke, news of Newton's illness slipped out and became greatly exaggerated. Tales soon began to circulate that the Lucasian Professor had lost his mental faculties. One German philosopher wrote to the mathematician John Wallis to say he had heard that Newton was 'so disturbed in mind . . . as to be reduced to very ill circumstances'.[47] And there were even reports (mainly from Europe) that Newton was dead. John Flamsteed wrote to him as late as February 1695 to say, 'The day after I received your last Mr Hanway brought me news from London that you were dead, but I showed him your letter which proved the contrary.'[48]

In spite of these rumours, the crisis soon passed. If we date its origins to late May 1693 and mark the end of the trauma as the latter part of September of that year, then Newton suffered for some three or four months. And, as he re-emerged, one of the problems that had so disturbed him and may have nudged him into the abyss had been resolved.

The question of where to go in his career had been troubling Newton since his sabbatical as a Member of Parliament. It had also occupied the thoughts of his friends. Millington, upon seeing Newton during the 'Black Year', commented in a letter to Pepys that 'it was a sign of how much it [intellectual life] is looked after when such a person as Mr Newton lies so neglected by those in power'.[49]

Between his leaving London in 1690 and the 'Black Year', 1693, there had been several attempts to find Newton a suitable position. Fatio de Duillier had tried unsuccessfully, but Newton's admirer and friend John Locke had offered the best hope and had made strenuous efforts to find him a civil position.

The problem had lain not with Newton but simply in inauspicious timing. From the end of the Convention Parliament in January 1690 until 1694 the Tories had remained in power, but with the election of 1694 everything changed: the Whigs were swept to government, and Newton's friend Charles Montagu was appointed Chancellor of the Exchequer.

By the end of 1695 rumours began to circulate in London that Newton was to be appointed Master of the Royal Mint, and from the capital the news spread quickly through the scientific community. From Oxford, John Wallis wrote to Halley on 26 November, 'We are told here . . . he is Master of the Mint: which, if so, I do congratulate him.'[50]

Unknown to all but Newton's closest friends, in September he had met with Montagu, who had indeed offered him a position at the Mint. But as the rumours reached Cambridge they were still vigorously denied. Almost up to the day when the official request for his services arrived, Newton was insisting to colleagues that nothing had been decided. Hearing from Halley as late as 14 March 1696 that rumours were still circulating, Newton appealed to his friend:

> And if the rumour of preferment for me in the Mint should hereafter upon the death of Mr Hoar [the Comptroller of the Mint] or any other occasion be revived, I pray that you would endeavour to obviate it by acquainting your friends that I neither put in for any place in the Mint nor would meddle with Mr Hoar's place were it offered me.[51]

Five days later the long-awaited letter arrived. It read:

> Sir
>
> I am very glad that at last I can give you a good proof of my friendship,* and the esteem the King has of your merits. Mr Overton the Warden of the Mint is to be made one of the Commissioners of the Customs and the King has promised me to make Mr Newton Warden of the Mint, the office is the most proper for you it is the chief officer in the Mint, it is worth five or six hundred pounds per annum, and has not too much business to require more attendance than you can spare. I desire you will come up as soon as you can, and I will take care of your warrant in the meantime . . . let me see you as soon as you come to town, that I may carry you to kiss the King's hand, I

---

* This was perhaps a subtle reference to the fact that a few years earlier, in January 1692, Newton had, in one of his typical, sudden fits of pique, written to Locke declaring that he had given up on Montagu's help in finding public office. In a letter dated 26 January 1692, he had declared, 'Being fully convinced that Mr Montagu upon an old grudge which I thought had worn out, is false to me, I have done with him & intend to sit still unless my Lord Montagu be still my friend.'[52]

believe you may have a lodging near me. I am sir your humble
servant.

Chas. Montagu[53]

It could not have come as a surprise, and Newton was fully prepared for a radical and lasting change in his life. He had been slipping
again into a mire of depression and was probably aware that his best
scientific work lay behind him. He knew that the development of a
unified explanation for all the forces of Nature was now, for him,
an impossible dream. He needed to banish the ghosts of the past,
and the stagnant environment of Cambridge.

Four days after Montagu's letter had arrived, Newton was in
London to accept the appointment, and four weeks later, on 20 April,
he signed the Exit and Redit books at Trinity for the last time.
Leaving Cambridge he took the coach at the Rose public house as
he had done on so many occasions during the past three decades.
But this time was different: now he was leaving behind the world of
academia, of science, of alchemy – leaving for good the tiny city that
had been his home for almost thirty-five years. He was stepping
instead into an unknown future, as a servant of the Crown.

Chapter 11

Metamorphosis

Forward, forward let us range,
Let the great world spin for ever down the ringing grooves of change.
ALFRED, LORD TENNYSON[1]

N ewton's decision to leave academe and remove to London marked a complete metamorphosis in his career, and the city in which he was to remain for the rest of his life had gone through no less a transformation contemporaneously. The Great Fire of 1666 had consumed an estimated 13,000 houses and 87 churches, including the beautiful old St Paul's. Gone was the city of Shakespeare, Chaucer, Raleigh, Elizabeth I and Marlowe, and the material loss has been put at £10 million during a time when the city's annual income was £12,000.

But the task of rebuilding had begun almost immediately. By the time William and Mary were proclaimed King and Queen a little over two decades later, some 8,000 houses and more than three dozen churches had been built to replace the old. As early as 1671, John Evelyn could report in his diary:

> Returning home, I went on shore to see Custom House, now newly built since the dreadful conflagration. Rebuilt Guildhall used for the Lord Mayor's Feast in November. The Royal Exchange and Blackwell Hall completed; halls for the Vintners, Drapers, Coopers, Parish Clerks and Skinners; repairs to the river stairs at Queenhithe, Trig Lane and the Old Swan. Work begins on four new churches, on the Fleet Canal, and, in Fish Street Hill, on the Monument to the late dreadful fire.[2]

The new houses and streets were of a totally different character to the old. Until the Great Fire, the city had been a mass of wooden cottages huddled around the old St Paul's, a hotchpotch of ram-shackle, tinder-box habitation, some of which dated back to the reign of Henry VIII. The new grand design had begun to produce a city of wide avenues lined with flat-fronted houses, improved sanitation and solid foundations.

By the last decade of the seventeenth century, within thirty years of the most devastating fire England had ever known, London was the largest city in Europe, with a population of 750,000 (around one tenth of that of the British Isles), and would soon outstrip Amsterdam as a commercial centre. Towards the end of William's reign, the War of the League of Augsburg – known as 'William's War', and involving a Grand Alliance of England and Holland – had shifted power away from Holland and towards England.* Partly as a result of this, London had become a boom city, growing fat on imported wealth, banking and European commerce, its administrators and businessmen invigorated by a sense that the country had entered a new, more prosperous, era.

It was also a city of brutal contrasts. The entire metropolis was sustained by an underclass of labourers and servants most of whom lived in the liberties to the east of the city. The liberties were no-go areas for the authorities and lay largely beyond the jurisdiction of the Lord Mayor of London. The Liberty of the Fleet, a ghetto built around the infamous Fleet Prison, was a lawless hellhole, the last area to be touched by the great reconstruction scheme, the home of countless whores and thieves who thrived on the rich pickings to be had a few miles away in the wealthier districts. By night the lawless worked 'up west'; by day they slept in windowless hovels lining streets awash with human waste and flood waters from the underground river Fleet. A couple of miles from the liberties stood the elegant homes of wealthy bankers and merchants, the open parks and the fields of Kensington and Knightsbridge, the rarefied atmosphere of the coffee-houses, where the idle rich spent long lunches in genteel chatter. Lloyd's was conceived at the coffee-house of the same name

* This trend was sharpened by the second war against Louis XIV, the War of the Spanish Succession (1701–13), which resulted in the Treaty of Utrecht – overseen by the Duke of Marlborough, who was accepted as diplomat-in-chief for the whole of Europe.

(in the City), Mus's was the venue for frequent astronomical discussions between Hooke and Flamsteed, while authors and wits of the day, including Dryden and Pepys, favoured Will's in Bow Street.

The London of the late seventeenth century was a polluted, congested city, with a single bridge – London Bridge – spanning the river. The Thames itself was a choked, grey ribbon crowded with ships, the water murky with bilge. In winter, rolling brown fogs and thick clouds of dirty smoke from 100,000 inefficient chimneys hung low over the entire area, sinking the homes of the illustrious into the same mire as the hapless masses of the East End.

Newton spent a week arranging his affairs in the capital before travelling to his new workplace in the Tower of London on 2 May 1696. The journey, in the dim light of dawn, would have taken him through the heart of the new city, along Threadneedle Street and past the elaborate façade of the new Royal Exchange completed only twenty-seven years earlier at the astronomical cost of £70,000.

The coach would then have turned towards the river and the ancient, crenellated outline of the Tower, dark against the river and the wastelands of the South Bank. Crossing the moat, still flooded in Newton's day, he would have passed along the echoing cobbled lanes between the outer and the inner walls of the fortress to where the rooms of the Mint faced the quadrilateral street between the two great stone walls.

The Tower of London had been the home of the Mint since 1300, and the indentured position of Master was established by Edward I soon after William de Turnemire had been made Master Moneyer of England in 1279. Recent holders of the position had treated the job as little more than a sinecure, receiving a handsome salary and a royalty on every ounce of gold and silver passing through the presses. By the time Newton came to the position of Warden, this formerly important and highly regarded job as second in command to the Master, who was supposed to supervise every detail of the running of the Mint, was seen as nothing more than a title. The day-to-day responsibility of running the place rested in the hands of a comptroller.

If Newton's arrival in London suited his purposes, the Royal Mint needed him in equal measure. The difference was that Newton knew he needed a fresh challenge, but the Mint authorities did not realise how much they would soon be relying upon his services.

Even Newton's friend Charles Montagu, who was a man of vision, had no idea just how much Newton would contribute to the job, or how essential he would prove to the Chancellor's plans to reconstruct the entire financial infrastructure of the country. It was only in retrospect that Montagu was to admit that Newton was indispensable to the success of recoinage – the cornerstone of his economic revolution.[3]

Recoinage was needed desperately. Many of the coins in circulation had been minted in Elizabethan times, and it was not unusual to find coins from the time of Edward VI, who had reigned some 150 years earlier. But old, worn-out coins were a minor difficulty: a far greater menace came from clippers and counterfeiters.

The techniques for producing coins in Britain had hardly changed since the Middle Ages; there was absolutely no form of quality control at the Mint, and coins could vary in weight substantially. The images stamped upon most of them were crude and easily copied, and the use of milled edges or more elaborate patterns had been introduced on only a relatively small scale during the 1660s. By using cheaper metals or blends, counterfeiting coins was simple and lucrative, and clipping, the practice of cutting bullion from the edges and reshaping the debased coin, needed nothing more than a pair of cutters and sufficient nerve. In the backstreet hovels of the liberties there was no shortage of desperate people willing to risk hanging for a few pennies.

Shortly before Newton arrived at the Mint, the Treasury had been forced to realise that mass recoinage was essential if the country was to be saved from bankruptcy. The situation on the streets and in the banks of the nation had reached crisis point; business simply could not function in a civilised manner – the element of trust in the coinage had been eroded and commerce was breaking down. Shopkeepers could no longer rely upon the cash of the purchaser, and so prices became inflated to compensate for the risk of accepting counterfeit coins. Workers were finding their weekly pay made up of tin coins or money clipped beyond recognition, and riots were becoming almost a daily occurrence.

In 1662 the Treasury had attempted to halt the downward slide by introducing new coins with milled edges, employing the inventor of the milling machinery already used in France, Pierre Blondeau. But this move had served only to exacerbate the problem, because the authorities in London had not taken the precaution of withdrawing the old currency as they introduced the new. Unscrupulous moneyers and goldsmiths (of which there were many) had secreted the new coins out of circulation, melted them down and sold them as bullion in Holland and other parts of the Continent where the metal commanded a higher price than that set by the Treasury. Thirty years after the introduction of milled coins, the currency crisis was worse than ever. To compound the situation, since 1689 England had been engaged in a full-blown war on the Continent, allied with Holland against the armies of Louis XIV. There was not even a temporary respite in this until the unsatisfactory and short-lived Treaty of Ryswick in 1697, by which time the national reserves had been depleted to within an ace of bankruptcy, stretching the national debt (a device newly created by Montagu) almost beyond control.

The process of recoinage had been set in motion before Newton's appointment, but little thought had been given to the detailed practicalities of this mammoth task. The Comptroller, James Hoare, was a capable man but possessed of little imagination or vision. The incumbent Master, Thomas Neale, was a lazy gambler and heavy drinker who made only rare appearances at the Tower and was content to cream off a sizeable income for the least effort. Blinded by panic, the Treasury had initiated a vast and complex process without realising that with men such as Neale at the helm they were heading for even greater trouble.

The recoinage of England may have lacked the intellectual depth or universal significance of the *Principia*, but if it had failed it could have broken the English economy and provoked social upheaval comparable to that of the Civil War. The Mint, the City of London and consequently the King and Queen of England and Holland were immensely lucky to have found in Newton a man who could turn his hand to almost anything with equal skill and intellectual rigour – a man whom, a few years earlier, none would have guessed to be the perfect administrator.

On his first morning at the Mint he took an oath of allegiance:

You shall swear that you will not reveal or discover to any person or persons whatsoever the new invention of rounding the money & making the edges of them with letters or graining or either of them directly or indirectly, so help you God.[4]

With this formality out of the way, he began an inspection of his new domain, probing into every aspect of the coining process.

One can imagine the shock Newton's arrival caused the authorities at the Mint. Because he was the laziest of them all, for some time Thomas Neale remained almost entirely ignorant of Newton's attitude to the job. Others at the Mint experienced his energy and drive immediately. During his first few weeks Newton was there when the presses began working at 4 a.m. and returned to see in the night shift. For a short time he even occupied the rooms and garden put aside for him next to the Mint rooms – something no Warden had done for generations.

The lodgings were cramped and the garden was little more than a patch of lawn running up to the outer wall and perpetually in shade. The accommodation was also noisy. Two shifts ran head to tail with only a four-hour break (between midnight and 4 a.m.) six days a week. The air was damp and dank, and the three hundred workers and scores of horses used to power the presses raised an almost intolerable, permanent stink. According to the Mint records, just the cost of hauling away horse manure from the site during the recoinage process came to a staggering £700.[5] Not surprisingly, by the end of his first summer in the new job, Newton had sublet the lodgings and purchased a small house in Jermyn Street in the fashionable suburb of Westminster.

The first stage in the recoinage process was to recall old coins as the new money was released. This was done largely as a simple exchange. After 1 January 1696 no clipped crowns or half-crowns were allowed in commercial transactions except for the paying of taxes and loans to the King. Clipped shillings were allowed to pass until 13 February, and sixpences until 2 March. By 2 April all clipped coin was to be declared unusable.

The plan almost spelled disaster, because new coins were slow to circulate and the population did not fully understand what was happening. On several occasions the Treasury was obliged to push back the deadlines to avoid serious civil unrest. To make matters worse, unscrupulous moneyers were fooling the poor and the uneducated into giving up their coins at well under face value and then cashing them in for the full amount. Trade began to break down further, and for a while during the spring and summer of 1696 a system of bartering reminiscent of the Dark Ages replaced conven-

tional commerce throughout the nation. Although his tone was mocking, John Evelyn's recounting of events in his diary was almost
understated: 'The Parliament's wondrous intent on ways to reform
the coin,' he wrote, 'made much confusion among the people.'[6]

During the period from January to the first deadline of April, a
mere £$\frac{1}{3}$ million of old coins were reclaimed, but, as the presses at
the Mint worked flat out and the new coinage began to circulate, old
money then flooded in. According to the contemporary observer
Narcissus Luttrell, by 24 June £4,706,003 had been taken in by the
Treasury.[7]

The Mint was now the engine-room of recovery – everything
depended upon the efficiency with which it could take in old coins
and pump out new. The ancient method of hand-struck coin-making
had been a labour-intensive process, slow and inefficient; the new
presses worked on the French principle introduced by Blondeau,
which had been in operation on a very small scale since Charles II's
time. In ten mills powered by a total of fifty horses, huge rolling-bars
produced sheets of metal of precise thickness. Blanks were cut from
these and taken to presses operated by workmen pushing a rotor
with heavy weights, or flies, at each end. A hapless junior coiner,
who might survive only a few days before losing at least one finger,
had to pop the blanks one at a time into a slot, and the press would
then crash down and imprint the image of the monarch on the metal.
The coins were then finished off by an edger, who produced the
unclippable milled edges.

Newton's first concern was to increase the efficiency of this process
by carefully watching each step and carrying out a time-and-motion
study on the system. He calculated where and how improvements
could be made, noting that, if the movement of the press and the
action of the coiner were coordinated properly, a single coiner could
flick out a coin and insert a new blank between fifty and fifty-five
times a minute. Writing in one of his many notebooks covering the
work at the Mint, he analysed the processes to the finest detail:

Two mills with 4 millers, 12 horses, two horse keepers, 3 cutters,
2 flatters, 8 sizers, one nealer, three blanchers, two markers, two
presses with fourteen labourers to pull at them can coin after
the rate of a thousand weight or *3000 lib* [pounds] of money
per diem.[8]

According to a clerk at the Mint, Hopton Haynes, who wrote an account of the recoinage, Newton's efforts were essential to the eventual success of the operation and his mathematical skills streamlined the Mint and increased efficiency enormously. Also, 'he could judge of the workmen's diligence . . .'⁹ One can imagine.

In order to spread the load of recoinage, Newton established country branch mints – smaller versions of the Royal Mint – at five sites around the country. These were built in Bristol, York, Exeter, Norwich and Chester during his first summer and autumn in the capital, but were eventually seen as a disappointment. They never reached the capacity expected of them – because, it could be argued, they were not run directly by the slave-driving Warden.

Incompetence, dishonesty and internal feuding also played their part in limiting the contribution of the five branch mints. The best of a poor bunch was Bristol, but even there, at the peak of national production of money (the summer of 1697), output barely met the minimum requirement – just under £77,000 of silver coin in June, compared to the London mint's June figure of £330,000 (and £360,000 the following month).

The other four mints were significantly worse, with output ranging from £15,000 to £25,000 per month throughout the summer of 1697. The Chester mint was the runt of the litter, as well as a headache for Newton's first publisher, Edmund Halley. Knowing that Halley was still in financial difficulties, Newton had finally rewarded him for his help with the *Principia* by making him Deputy Comptroller of the Chester mint, at a salary of £90 per year. Unfortunately, Halley soon became involved in a dispute with the Deputy Master, one Thomas Clarke, who was in league with two dishonest clerks to filter off a profit by inflating the value of handed-in coins. Initially grateful to Newton for the chance of a well-paid job out of London which he also believed would leave him time to pursue his scientific interests, Halley soon found himself working around the clock to make up for the inefficiency of his staff and trying in vain to bring his corrupt workers and immediate superior to justice. When the branch was eventually closed down, in 1698, he was more than happy to let this brief involvement in the world of commerce pass into bitter memory.

Not so his erstwhile scientific master. If Newton really was a slave-driver, he was no less harsh with himself than he was in his treatment of others. Not content with making a sinecure into a full-time job, within weeks of his arrival in the frantic climate pervading the Mint and the Treasury he had become indispensable. Yet, despite working sixteen-hour days at the height of production, he still found the energy and the drive to grab the power that had formerly belonged to the position – power he viewed as his by right.

The Master of the Mint, Thomas Neale, was no real hindrance in this: he was so complacent in his comfortable position – which was secure for life – that he paid little attention to Newton's gradual accumulation of responsibility. As Newton saw it, the post of Warden had once come with privileges and responsibilities that had been eroded by the indifference of his predecessors. He had always relished power and status and had fought and schemed for them through-out his life. Now that he had a modicum of power, he would do his utmost to build upon it. The wardenship of the Mint was a mere stepping-stone – the first stage in his plan to ascend the social ladder.

Although this preoccupation with social status and political power might seem at odds with the image of the man who had spent the past thirty years of his life in the isolated pursuit of purely academic goals, it should actually be viewed as a product of the same drives that fuelled his science. Newton had gone as far as he could with the work that had obsessed him for so long. Now, as he began a new life, he threw himself into it with the same vigour he had employed as an alchemist and natural philosopher. Newton might have been isolated for much of the three decades he spent in Cambridge, but he knew as well as the next person the importance of political man-œuvring and was quite capable of applying himself to achieving his new goals with the single-minded energy and dedication he had used formerly to reveal the secrets of Nature.

To this end, he moved on two fronts simultaneously. First, he researched ancient documents and charters recording the division of power at the Mint. He then pursued his right to an ever-widening set of responsibilities which had once been the preserve of the War-den. Here is the dark shadow of the obsessive Arian at work, prepar-ing to dispute the validity of the Trinity, scouring the Book of Revelation and unravelling prophecies from the Book of Daniel. Once again he was in search of validation for his claims, but this was no divine battle but a fight for privilege. God was not to be found in the Mint documents, but the elevation of Newton's own ego most certainly was.

As he found the facts, he passed them on to his bemused superiors at the Treasury, explaining details of lineage and responsibility, outlining the point at which a piece of the Warden's rightful power had been snatched away decades earlier. He studied every economics book he could lay his hands upon and picked the brains of the foremost financial thinkers of the day – Francis Brewster, William Lowndes, Jean Boizard and his friend John Locke. He especially treasured a collection of 180 official documents issued by the French government and bound together in a volume known as *Cour des monnoyes*. Judging by the description of their condition when Newton's library was catalogued, they appear to have been among the most well-used books in his collection (which at the time of his death contained thirty-one volumes on economics[10]).

As with his pursuit of the *prisca sapientia*, the philosophers' stone and the lies of Trinitarianism, as soon as he had accumulated the information he needed he began to write. In his limited spare time he filled page after page with the history of economics, the theory of commerce, the principles of the currency system of various countries; he charted the lineage of positions within the Mint and the privileges that had been lost. Reminiscent of his philosophical notebooks and his laboratory notes are pages of text with titles such as 'Observations Concerning the Mint', subheaded, 'Of the Assays', 'Of the Melting' and 'Of the Making the Moneys'. He employed several amanuenses working simultaneously to copy out his sketchy notes and make duplicates of everything he transcribed. Nothing was published, but copies of everything were kept. According to John Conduitt, box after box of papers – many of which contained duplicates of complex arguments for his superiors at the Treasury – were burned during the 1720s.

The authorities at the Treasury were understandably confused by much of Newton's propaganda but did not wholly disapprove of his quest for power. Their trust in Neale had almost certainly been shaken by the sweeping efficiency of the new Warden, for it was clear that the Master had had little to do with the recent success of the Mint. It is also quite likely Newton was encouraged by Montagu, who would have done his best to smooth the path for many of his friend's claims.

In 'The State of the Mint', his first report to the Treasury, at the end of 1696, Newton made his position clear, declaring:

And thus the Warden's authority which was designed to keep the three sorts of ministers in their duty to the King & his people, being baffled & rejected & thereby the government of the Mint being in a manner dissolved, those ministers act as they please for turning the Mint to their several advantages. Nor do I see any remedy more proper & more easy than by restoring the ancient constitution.[11]

The other means through which Newton knew he could accrue power to himself was that of direct confrontation with almost every senior official involved with the Mint. By ensuring that in any dispute he was clearly visible as the defender of the institution and the Treasury's ultimate aims, he could display his loyalty and value to the government.

First came clashes with greedy government contractors employed to supervise the melting and refining processes. Long years of study and practical work at a quite different furnace had enabled Newton to calculate that one troy pound of alloy for the production of coin could be produced at a cost of $7\frac{1}{2}d$. So, when the financiers Peter Floyer and Charles Shales submitted what they claimed to be a fair tender of $12\frac{1}{2}d$. per pound, Newton called their bluff and forced them down.

But his most protracted and bitter feuds at the Mint were with the Governor of the Tower, Lord Lucas. Trouble began when, during the worst months of panic over the recoinage, the works was obliged to expand into the quarters of the Tower garrison. Lucas immediately resented such an intrusion. Neale was happy to ignore the Governor's protests, but Newton revelled in conflict and crossed swords with Lucas at every turn.

When the Governor forcibly searched the house of one of the chief engravers at the Mint, an old man named John Roettiers, Newton hauled him over the coals for what he saw as a gross invasion of his authority and a risk to the security of the operation. '. . . if the military searches without us should be now allowed . . . we cannot undertake any longer the charge of the dies & puncheons & marking engines

& other coining tools & of the gold & silver which lies scattered about in all the rooms,' Newton raged.[12]

On another occasion, as the Mint operated at full capacity and exhaustion frayed even the coolest temper, one of Lucas's men was accused of attacking a Mint worker. Newton lodged an official complaint to the Treasury. When it took Newton's side, Lucas retaliated by blockading the Mint gates on the grounds of some supposed lapse of security. Food and drink were turned away, angry workers threatened to strike, and a shut-down of the Mint was averted only at the last moment.

Newton's documenting of every detail of the operation of the Mint and the clashes with his colleagues served the purpose of furthering his career, but other drives were also satiated during his early years at the Mint. He took it upon himself to seek out and prosecute the clippers and counterfeiters who had been partially responsible for precipitating the recoinage crisis in the first place.

Recoinage had broken the trade of the clippers – they could not deface coins with milled edges and elaborate engravings convincingly or cheaply – but the counterfeiters' art was relatively unaffected and Newton went after its practitioners with a revealing passion.

The bringing to justice of those who perpetrated crimes against the Crown through acts of fraud and counterfeiting was one of the official duties of the Warden, but, like many of the other responsibilities of the position, it had long been delegated to clerks. They received no funding from the Treasury and had to rely upon commission from reclaimed coins, which usually amounted to very little. Understandably, at first Newton was not keen to become involved. It was a dangerous, thankless job and, although the records of recoinage later showed that the total quantity of old coin recalled and melted down actually weighed only 54 per cent of its legal weight, the Treasury placed little importance on stopping the criminals.[13]

But it was not just the activities of the lawless and the loss of revenue that so disturbed Newton. Measures introduced to stop coiners and counterfeiters were already backfiring to such an extent that his Mint officers were becoming objects of ridicule. As an expediency measure, before recoinage had begun, Parliament had passed a law by which a reward of £40 was offered to any person informing on a clipper, and any clipper who informed on two of his colleagues was granted a full pardon. As Newton spelled out in a letter to the Treasury, the law had:

> now made courts of justice and juries so averse from believing witnesses & sheriffs so inclinable to impress bad juries that my agents & witnesses are discouraged and tired out by want of success & by the reproach of prosecuting and swearing for money. And this vilifying of my agents and witnesses is a reflection upon me which has gravelled me & must in time impair & perhaps wear out & ruin my credit.

In conclusion, he requested that the job of bringing prosecutions be passed over to the King's Attorney and the Solicitor-General, who, he stated curtly, 'are best able to go through it'.[14]

This time, the Treasury was not interested. Refusing his request, its only concession was to agree grudgingly to pay for an extra clerk. Furious, Newton threw himself into the attack with renewed enthusiasm. Much to the chagrin of the authorities, he unearthed a long-forgotten edict specifying the right of the Mint to confiscated goods currently held by the Sheriffs of London. In this way he obtained his funding, and the Treasury was obliged to reimburse the Sheriffs. Then, over a period of several months, he wheedled out extra funds from various departments within his domain, even occasionally supplementing expenses from his own pocket.

As well as providing him with the means to pursue his new passion, this internal battle had whetted his appetite. Having already cut the umbilical cord to academia to submerge himself in the world of high finance and economic theory, he now leaped into a new role – that of private investigator and prosecutor.

As the effort of recoinage began to relax in 1698, Newton expended increasing energy in his efforts to track down his new enemies, and he took personal responsibility for ferreting out suspects, using a network of agents in eleven counties. The man who had once despised his fellow students for their noise and drinking, gambling and womanising could now be found in some of the most sordid public houses in the capital, following leads in the heart of the liberties and arranging secret meetings with informants in gin-houses and brothels. One entry in the Mint records notes, 'paid £5 to Humphrey Hall to buy him a suit of clothes to qualify him for conversing with a gang of coiners of note'.[15] Another lists a claim for expenses of £120 spent 'in coach-hire and at taverns, prisons and other places in the prosecution of clippers and coiners'.[16] A third records Newton spending several pounds on becoming a justice of the peace in seven of the home counties, so that he could gain inside information on the activities of known criminals.[17]

And detection was only the start of the process. If his undercover sleuthing had exposed him to the dangers of law-enforcement, the interrogations that followed served only to confirm for him the depths of depravity and squalor which the back streets of London could harbour. Yet Newton tackled the task with enthusiasm, and was quite able to make the transition from genteel Cambridge University life to the horrors of the world beyond. Between June 1698 and Christmas 1699 he conducted some 200 cross-examinations of witnesses, informers and suspects, and in a single week in February 1699 he attended seven such sessions and had ten prisoners in Newgate Prison awaiting hanging. He treated the petty criminal and the grand larcenist with equal contempt, once commenting, 'Criminals, like dogs, always return to their vomit.'[18]

This was his first opportunity to wield real power – power over life itself. With a wave of his hand, he could have a man sent to the gallows, but he could also show mercy and offer life to those grovelling at his feet.

The detailed accounts of every interrogation recorded in the Depositions Book (1697–1704) now held in the Royal Mint library show that Newton sat through hour after hour of confessions and informants' reports. The majority of these reports were received within the Mint itself, but some were taken in Newgate Prison, in one-to-one interviews with criminals already facing charges, or else in taverns and billets during clandestine meetings with underworld double-dealers and bounty-hunters. In recounting the misadventures and the pathetic lives of hundreds of those involved, the Depositions Book makes for depressing reading.

On 25 July 1699 Elizabeth Sutton acted as an informant on an acquaintance called Elizabeth Pilkington and had her account transcribed by a clerk:

She said that about a fortnight before she was apprehended she became acquainted with one Elizabeth Pilkington who took a room in the Mint in Southwark and sometime after she had taken the said room she came to this informant and told her she must raise her bottom.[19]

It transpired that after a search of the unfortunate woman a few pennies had been found hidden in a napkin under her skirt.

An entry on 19 August 1699 records that Julian Tuffin accused Ann Pillsbury of trying to pass off a counterfeit sixpenny piece and her tiny daughter was subjected to a body search:

> This informant with others searched the said Ann Pillsbury and her said little girl at the said Warden's of the Mint and the informant found wrapped in a paper in the said girl 5d worth of farthings & 4 six pennies, two of which were counterfeit ones.[20]

At the other end of the scale there were the evil manipulators and schemers who for the most part escaped by turning King's evidence or evaded the law until greed overtook them. A Whitehall porter named John Gibbons made a handsome living by informing on small-time clippers and counterfeiters – usually single mothers from the poorest districts.

Most famous of all the cases was Newton's three-year battle with a charismatic rogue named William Chaloner – a figure cast in the romantic mould of the daring highwayman or a suitable model perhaps for Raffles, the fictional Victorian cat-burglar. Chaloner loved the thrill of the chase and pushed his daring to the limit, tugging the tail of authority whenever he could. But, within a year of arriving at the Mint, Newton's acute intellect had already begun to tackle the cases of most prolific criminals on record, and with such a methodical opponent, Chaloner's days were numbered. Newton had him sent to Newgate Prison in September 1697 for counterfeiting on a grand scale. But, to the Warden's astonishment and fury, through his contacts inside and outside the judicial system Chaloner somehow managed to secure an acquittal and was released. Eighteen months later an inmate at Newgate informed on Chaloner, and within a week Newton had the counterfeiter behind bars once more.

During the following two months, sitting in the dark, clammy recesses of Newgate, Chaloner wrote letter after letter to friends on the outside, telling them he would soon be free; but it was not to be. On the eve of his execution, in March 1699, his will broke and his final letter contained none of the self-assurance of its predecessors. Sent to his nemesis, Newton, who read it in his office in the Tower, Chaloner begged for clemency:

> Mrs Holloway swore false against me or I desire never to see the great God, and I desire the same if Abbot did not serve false against me so that I am murdered, Oh God Almighty knows I am murdered. Therefore I humbly beg your Worship will consider these reasons and that I am convicted without precedents and be pleased to speak to Mr Chancellor to save me from being murdered Oh! Dear Sir . . .[21]

Chaloner's pleas fell upon deaf ears. Newton had taken a total of thirty-four depositions detailing the man's every move, and every aspect of his counterfeiting operation had been documented by the Warden's ubiquitous agents – men whom even Newton's clerk, Hopton Haynes, described as 'very scandalously mercenary'.[22] These depositions were more than enough to have Chaloner convicted of the ultimate anti-Establishment crime of high treason, rather than the lesser charges of counterfeiting or forgery. On 22 March, the hapless prisoner was dragged on a sledge to Tyburn gallows (on the site of the present Marble Arch), where he was hung until almost unconscious and had his breathing torso cut open with a knife and his bowels removed, before meeting death as he was hacked into four quarters – all before a jeering, blood-thirsty crowd.

Little wonder Newton was feared and reviled in equal measure both by his prisoners as they awaited their execution and by those he sought to monitor going about their illegal trade. Nor is it surprising that he received warnings and death threats from insiders. On 16 September 1698 a warning letter arrived at the Mint from a 'chirurgeon' (surgeon) treating inmates at Newgate, relating what he had overheard:

> He [a prisoner] said that about a month ago being in Newgate for debt he there heard Frances Ball of Ashbourn complain of the Warden of the Mint for his severity against coiners and say damn my blood I had been out before now but for him, and

Whitfield who was also then in prison made answer that the Warden of the Mint was a rogue and if ever King James came again he would shoot him, and then Ball made answer: God damn my blood so will I, and though I don't know him yet I'll find him out. And the said Whitfield said further that the Warden had troubled his wife for a little bit of clippings found upon her coat and spent 5 hundred guineas to get her off.[23]

As well as exchanging his scientific explorations for his position at the Mint, there had been another radical change in Newton's life. From some time soon after he moved to the capital – possibly as early as 1696 – he shared his home with a woman, his young half-niece, Catherine Barton.

In December 1677 Newton's half-sister Hannah Smith had married a clergyman, Robert Barton. In 1693 he died suddenly, leaving her almost destitute with a young son and daughter. The boy, also called Robert, grew up to become an army officer and was killed in 1711 in Quebec; his sister was Catherine Barton.

Born in 1679, the year her grandmother Hannah Ayscough-Newton-Smith had died, it is most probable that Catherine arrived in London as she turned seventeen, towards the end of 1696, a few months after Newton had bought the house in Jermyn Street. There is, in fact, no firm evidence to pinpoint the exact date of her arrival, and some commentators have placed it as late as 1700, when she was twenty years old. However, there is one relatively minor event which points to the earlier date. In January 1697, in an effort to embarrass the man who was seen throughout the world as the greatest mathematician of the age, a jealous Gottfried von Leibniz and the Swiss mathematician Johann Bernoulli had unearthed a problem later known as the 'brachistochrone'. They had set its solution (which neither of them could deliver) as a challenge to any mathematician in Europe. They offered no financial reward but declared that 'we offer a prize suited a man of free birth, a prize composed of the honour, the praise and the applause with which we mean . . . to celebrate the perspicacity of our great Apollo-like seer'.[24]

Newton received his copy of the problem towards the end of January 1697, during his most turbulent time at the Mint, and, having already heard about Leibniz's conundrum, he was tempted to let it pass him by, realising the motive behind it. However, we know that

Catherine Barton was there when the letter arrived, because she recounted her uncle's response to the challenge when the man she later married, John Conduitt, began collecting material for his planned memoir of Newton during the 1720s. 'Sir I. N. was in the midst of the hurry of the great recoinage,' she recalled, '& did not come home until four [in the afternoon] from the Tower very much tired, but did not sleep till he had solved it which was by 4 in the morning.'[25]

This story demonstrates two things. First, the massive lead Newton had over all other mathematicians of the time. He had developed the mathematical technique that would solve this particular problem over a dozen years earlier, though prominent mathematicians, including both David Gregory and John Wallis, were unable even to begin to find an answer. Leibniz had tried a solution and embarrassed himself through his jealousy for Newton, and lesser mathematicians had thrown up their arms without making a serious attempt. The other, unrelated, matter the story illustrates is that, unless Newton told Catherine the story in later years, she must have been living with him by January 1697.

Catherine was of a quite different temperament from her uncle. She was gregarious and confident, excitable and flirtatious – though she was reputed also to be highly intelligent and intellectually curious. Rumours that Newton acted as her teacher during her first years in the capital are not supported by fact. If she was seventeen at the time of her arrival she would already have received an education suited to a seventeenth-century woman of her class, and it would have been considered unseemly for her to have furthered her education beyond that point. Her role in Newton's life was one of house-keeper for his new London home, and this would also have afforded her the opportunity to broaden her social sphere and to eventually meet a suitable husband.

Although there is nothing to suggest that Newton had ever met his niece before her arrival in London, they quickly became close, despite their very different personalities. The only surviving letter to have passed between them dates from the summer of 1700, when Newton wrote to Catherine when she was in the country, convalescing after suffering smallpox:

I had your two letters & am glad the air agrees with you, & though the fever is loath to leave you yet I hope it abates, & that the remains of the smallpox are dropping off apace. Sir Joseph [Tily] is leaving Mr Toll's house & it's probable I may succeed him. I intend to send you some wine by the next carrier which I beg the favour of Mr Gyre & his lady to accept of. My Lady Norris thinks you forget your promise of writing to her, & wants a letter from you. Pray let me know by your next how your face is and if your fever be going. Perhaps warm milk from the cow may help to abate it.

Your very loving Uncle.[26]

Newton's concerns for his niece's face were not simply those of a fond uncle. As well as being intelligent and witty, Catherine was renowned for her great physical beauty, and Newton was certainly not the only man to notice it. A visiting (and married) French bureaucrat, Pierre Rémond de Montmort, fell in love with Catherine the moment he first saw her at a dinner party hosted by Newton at his Jermyn Street house. 'Ever since I beheld her,' de Montmort oozed in a letter to his friend Brook Taylor, a fellow of the Royal Society, 'I have adored her not only for her great beauty but for her lively and refined wit. I believe there is no danger in betraying me to her. If I had the good fortune to be near her, I would henceforth become as awkward as I was the first time we met.'[27]

No portraits of Catherine Barton survive,* but descriptions of her appeared in the work of Jonathan Swift, whom she met through her uncle. At the time he made her acquaintance, around 1710, Swift was a celebrated author and one of the linchpins of London society. Although there is little to suspect they were ever lovers,† they quickly became close, prompting him to write, 'I love her better than anybody here, and see her seldomer.'[29]

Catherine's name appears frequently in Swift's *Journal to Stella* – the collected letters he addressed to Esther Johnson (Stella) and

---

* The Victorian Newton biographer David Brewster was said to possess a faded photograph of a long-lost cameo of her and wrote to a colleague of Catherine's great beauty. This photograph has since been lost.[28]
† Swift harboured a lifelong, unrequited passion for Esther Johnson, the real-life Stella to whom his *Journal to Stella* was addressed. Swift was later buried beside Esther Johnson in St Patrick's Cathedral, Dublin.

Rebecca Dingley. These references reveal that she enjoyed a rich social life and met a wide range of influential people of the time. One day she might be found dining with Lady Worsley or the great beauty Anne Long; the next would be spent with Swift and Lady Betty Germain, daughter of the Earl of Berkeley. The parson's daughter had come far.

But there was a coarser, country-girl aspect to her character that Swift was clearly delighted by. Over tea, he would prompt her to gossip, and would then filter her tales into literary history. On one occasion Catherine told him:

> An old gentle-woman died here two months ago, and left in her will, to have eight men and eight maids bearers, who should have two guineas apiece, ten guineas to the parson for a sermon, and two guineas to the clerk. But bearers, parson and clerk must all be true virgins; and not be admitted until they took their oaths of virginity so the poor woman still lies unburied, and so must do till the general resurrection.[30]

Elsewhere in *Journal to Stella*, Swift gently rebuts the criticisms of Stella towards Catherine after she had heard scandalous tales about her, joking, 'I'll break your head in good earnest, young woman, for your nasty jest about Mrs Barton.' ('Mrs' was the common form of address for a single woman at the time.) But then, with a hint that he too sometimes doubted Catherine's manners, he adds, 'Faith, I was thinking yesterday, when I was with her, whether she could break them or no, and it quite spoiled my imagination.'[31]

On the evening in 1711 when Catherine learned of her brother's death during the ill-fated attempt to take Quebec, she was at a private dinner with Swift. His report of the meeting reveals a darker side to her character. 'I made her merry enough,' Swift relates, 'we were three hours disputing on Whig and Tory. She grieved for her brother only for form, and he was a sad dog.'[32]

The relationship between Swift and Catherine was intense but short-lived. Four years after moving to London, Swift returned to his native Ireland, and in subsequent years they corresponded only rarely. Of far greater significance was her relationship with Newton's close friend Charles Montagu.

Montagu was around forty when he and Catherine first met. From portraits, he was clearly not a handsome man and was rather plump and fleshy-faced. But, as Chancellor of the Exchequer and a politician close to King William, he was one of the most powerful men in England, and in 1700 he was created Baron Halifax. Newton was so fond of him that, according to a visiting French abbé, he kept a portrait of him in his room.[33] As well as possessing both wealth and charm, he had a reputation as a wit and a womaniser who took his pleasures as seriously as his work. A decade earlier, he had married a woman twice his age, the wealthy Countess Dowager of Manchester, who had recently left him a rich widower. As others had been before him, Montagu was smitten by Catherine the first time he saw her.

How and when they met is unknown, but he visited the house in Jermyn Street frequently, and Newton, with his highly tuned social awareness, may have realised the kudos to be gained by showing off such a beautiful young relative. It is most probable the two met very soon after Catherine had settled in London – perhaps during a dinner party at Newton's new home. What is certain is that by 1703, she and Montagu were lovers.

There are no surviving letters between Catherine Barton and the Chancellor, but there is a wealth of evidence to show they shared an intense relationship which lasted until Montagu's death in 1715. The earliest sign of his feelings may be found etched into a toasting-glass at his gentleman's club, the rowdy Kit Kat Club, where it was customary for members to inscribe lines of poetry about their muse with a diamond before offering a toast to the gathering. Halifax's turn came during a Kit Kat Club dinner in 1703, when he wrote:

Beauty and wit strove each in vain
To vanquish Bacchus and his train
But Barton with successful charms
From both their quivers drew her arms
The roving god his sway resigns
And awfully submits his vines . . .
Stamps with her reigning charms, this standard glass
Shall current through the realms of Bacchus pass;
Full fraught with beauty shall new flames impart,
And mint her shining image on the heart.[34]

Once renowned for his literary wit, Montagu's talent seems to have escaped him on this occasion, but the identity of his muse is not difficult to find. 'But Barton with successful charms' leaves little room for doubt, while 'And mint her shining image on the heart' can only refer to a certain 'very loving Uncle'.

But Halifax's feelings for Newton's beautiful niece went far deeper than mere dinner-table banter. In 1706, three years after writing this poem in her honour, Halifax changed his will by adding a codicil to the effect that she was to receive all his jewels and £3,000, 'as a small token of the great love and affection I have long had for her'. Six months later he added a lifetime annuity of £200 for her, and on 1 February 1713 he changed the will again. After a bequest of £100 to Newton, 'as a mark of the great honour and esteem I have for so great a man', he increased Catherine's inheritance to £5,000 and a lifetime wardenship of his estate of Bushy Park and his house in Surrey, Apscourt Manor. 'These gifts and legacies,' the will reads, 'I leave to her as a token of the sincere love, affection, and esteem I have long had for her person, and as a small recompense for the pleasure and happiness I have had in her conversation.'[35]

Today it is accepted that Catherine lived with Halifax for at least part of the time she was supposed to be her uncle's housekeeper, and indeed this was a poorly kept secret at the time.

In the numerous references to Catherine in *Journal to Stella*, Swift never once described any social occasion where Halifax and Catherine Barton were together, but this may have been to protect his friendship not only with Catherine but with Montagu and Newton.* Indeed, in one letter, written in 1711 (the period during which Catherine and Halifax would most likely have been living under the same roof), Swift refers to her explicitly as 'my near neighbour'.[36] This was only two weeks after he had moved to St Martin's Street, Leicester Fields – Newton's address since September 1710.

The strongest evidence to support the idea that Catherine lived with Montagu comes from a simple, touching note she sent to her

---

* Although Swift was a great Tory sympathiser, he was, for a time, friendly with Montagu. Interestingly, even though he fell out with the Chancellor before returning to Ireland in 1714, Swift never did reveal anything he may have known about Montagu's relationship with Catherine.

uncle immediately after her lover's death; it read, 'I desire to know whether you would have me wait here . . . or come home . . . Your obedient niece and humble servant. C. Barton.'[37] But even this is ambiguous. Was Catherine simply at her sick lover's bedside, and does her reference to 'home' imply that this was really St Martin's Street and not Lord Halifax's London home?

Some have taken the matter further and formalised the whole affair by having Charles and Catherine secretly married. In the 1880s, long after all concerned were dead, Augustus De Morgan, the first Professor of Mathematics at the University of London, wrote a little book entitled: *Newton: His Friend: And His Niece*,[38] in which he took a single line from a letter of Newton's and constructed a conspiracy theory to show that the Chancellor and the mathematician's niece had sanctified their relationship before God.

The letter in question had been sent to Newton's distant relative Sir John Newton on 23 May 1715, four days after Montagu's death. Apologising for missing a scheduled meeting, Newton had written, 'The concern I am in for the loss of my Lord Halifax & the circumstances in which I stand related to his family will not suffer me to go abroad until his funeral is over.'[39]

However, the fact that Newton had been a friend of Montagu's for some thirty-five years, had helped him draw up his will many years earlier, and was even left a small token by the former Chancellor is certainly enough to explain his intimate turn of phrase. Furthermore, when Catherine took her marriage vows and became Mrs Catherine Conduitt only two years later, she signed a wedding certificate confirming her status as a spinster.[40]

The idea that Catherine married Lord Halifax can be safely dismissed, but she was almost certainly his common-law wife. The secrecy maintained by all concerned is not surprising. Because they had been born into very different classes, Halifax could not have married Catherine: although she was Newton's niece, she was still the daughter of a poor country clergyman, whereas he was the son of the Honourable George Montagu. Then, after Halifax's death, the truth had to remain hidden because Newton would have been unwilling to suffer embarrassment by association. Even after her uncle's death Catherine kept her past life private, through a desire to preserve the family name and to protect the honour of her husband, John Conduitt.

Newton must have been privy to the arrangement, but well-meaning Victorian hagiographers tried to sweep the whole affair under the carpet. They placed this rather difficult matter in the same category as the aberration of alchemy and ignored it because it raised too many awkward questions. Even the respected twentieth-century historian Christopher Hill has claimed that Newton was unconcerned by the illicit affair between his niece and one of his closest friends 'simply due to his not noticing'.[41]

The relationship between uncle and niece was not damaged by the liaison – Catherine returned to live with Newton for a short time after Montagu's death. We can therefore assume that, even if Newton did not condone the affair, for a number of reasons, he was quite able to tolerate it.

As a devout Puritan, Newton would presumably have set great store by the sanctity of marriage; but he had been able to relax the censorious attitudes of his faith before. During his early days at the Mint he had defended drunken workers in his ongoing battles with Lord Lucas, and he had paid expenses to agents for alcohol consumed during their investigations in public houses. At one stage during his postgraduate days in Cambridge he had himself flirted briefly with gambling and drinking, before returning to Spartan isolation. Montagu was not only one of Newton's closest and oldest friends: he was a very powerful man who had facilitated the appointment at the Mint. He was also a widower and Catherine a single woman. Given sufficient cause, Newton was quite able to grant a certain fluidity to his own moral structure.

In his Freud-influenced biography of Newton, Frank Manuel suggests that Newton could tolerate the relationship between Halifax and his niece because it satisfied his own deep-seated sexual needs.[42] Catherine was the granddaughter of Newton's mother, and by allowing his friend to have a sexual relationship with her he was vicariously making love to his own mother – something his deep subconscious had wanted since his desertion as a child.

Whether or not one subscribes to Freudian psychology, the validity of this convoluted hypothesis is, at the very least, questionable. Newton was a man driven by childhood pain, but his subconscious method of dealing with this angst was to accumulate power and to gain dominance over others. His scientific impetus was the need to know: in Newton's eyes, knowledge was power. Similarly, he was driven to succeed as a civil servant by acquiring power over those weaker than himself. He was clearly capable of adapting his internal ethical framework for what his inner self would have considered a higher cause.

As much as the protagonists tried to maintain complete secrecy, rumours about the illicit relationship between the niece of England's most esteemed and puritanical mathematician and the Chancellor of the Exchequer began to seep into society dinner-party conversation almost immediately after the couple had met. And the rumours became more exaggerated as the relationship continued, fuelled by those who, for their own reasons, cared little for discretion.

Flamsteed, the Astronomer Royal, who was in the midst of a fierce battle with Newton at the time of Halifax's death (see Chapter 12), openly mocked the former Chancellor's benevolence, declaring in a letter to a friend, 'If common fame be true, Halifax died worth £150,000; out of which he gave Mrs. Barton, Sir I. Newton's niece for her *excellent conversation*, a curious house, £5,000, with lands, jewels, plate, money, and household furniture, to the value of £20,000 or more.'[43]

Even Voltaire, who was a great proselytiser of Newton's science, was not above tittle-tattle and gossip. While popularising Newtonian mechanics throughout the Continent, he also peddled Newtonian rumour. It was Voltaire who publicised the apple story, but he was also responsible for the vicious and quite unfounded rumour that Newton had acquired his position at the Mint only because of the relationship between Montagu and Catherine:

In my youth I thought that Newton had made his fortune by his great merit. I had supposed that the Court and the City of London had named him Master of the Mint by acclamation. Not at all. Isaac Newton had a very charming niece, Madame Conduitt; she greatly pleased the Lord of the Treasury, Halifax. The infinitesimal calculus and gravity would have availed nothing without a pretty niece.[44]

This statement is false in every particular that matters. Newton had been appointed Warden, not Master of the Mint; he succeeded to the mastership only upon Neale's death in December 1699. Newton and Halifax had been friends for over a decade before the appointment, and the two had been working together to secure a public position for Newton for some time during the early 1690s. But, most significantly, when Newton was made Warden, Montagu and Catherine Barton had certainly never met.

Voltaire is thought to have been misinformed by a scurrilous book which appeared in 1710, a *roman-à-clef* called *Memoirs of Europe, Towards the Close of the Eighth Century. Written by Eginardus, Secretary and Favourite to Charlemagne; and done into English by the Translator of the New Atlantis*. The author was an extreme-Tory political writer named Mary de la Rivière Manley.

Manley, the daughter of a Cavalier, loathed the Whig leadership and Halifax in particular, and wrote a series of immensely popular books which were little more than thinly disguised polemic bundled up in tales of extreme fancy.

After the publication of her previous book, *Secret Memoirs and Manners of Several Persons of Quality, of Both Sexes. From the New Atlantis* (1709), she had been arrested and held without bail on charges of slander. Acquitted, she went on to write the even more vitriolic *Memoirs of Europe*. Most damaging for Newton, in her new book Manley had assembled a supporting cast to bolster the debauchery of her lead character, a parody of Lord Halifax. In an index, she even went so far as to list her chief characters as 'Bartica – Isaac Newton's niece' and 'Julius Sergius – Lord H*f-x'.

Early in the narrative there is a description of an orgy involving Bartica and Sergius, and later a character relates how Sergius has described his affections for Bartica:

'I think my Lord Julius Sergius,' continued I, addressing more closely to his Lordship, 'It is hard that in all this heavenly prospect of happiness, your Lordship is the only solitary lover: what is become of the charming Bartica? Can she live a day, an hour, without you? Sure, she's indisposed, dying or dead.' 'You call tears into my eyes, dear Count,' answered the hero sobbing, 'She's a traitoress, an inconsistent proud baggage, yet I love her dearly, and have lavished myriads upon her, besides getting her worthy ancient parent a good post for connivance.'[45]

At another point, Manley has more unkind words to say about Catherine through the anti-hero Sergius:

She has other things in her head, and is grown so fantastic and high, she wants me to marry her, or else I shall have no more of her, truly: 'She was ever a proud slut' . . . He told me, if he pin'd himself to death, he was resolved not to marry her whilst she was so saucy.[46]

The real-life Halifax was certainly no saint. He drank and womanised, gambled and accepted readily the many pleasures and privileges of office. Along with Lord Somers, the Lord Chancellor, he was impeached for malpractice in 1701, but was acquitted. However, the portrait Manley paints is a grotesque caricature designed for rather obvious political ends. As for Catherine's true character, little survives upon which to base an accurate analysis. If Swift's letters are anything to go by, she was certainly a very modern and liberated woman for her time, but the suggestions that she was stringing Halifax along and using him for her own ends and that her famous uncle was doing the same have no basis in fact.

Equally tenuous, but just as intriguing, is the suggestion that during the early years of the eighteenth century Newton may have enjoyed a form of romantic liaison with a woman named Lady Norris (née Elizabeth Read). The suspicion arises from a letter of proposal supposedly copied by Conduitt from an original by Newton. Across the top is written, 'A copy of a letter to Lady Norris', and another hand has added, 'A letter from Sir I. N. to —'.[47] The implication is that Conduitt found a letter of proposal written by Newton and copied it, but the addition identifying it as being from Newton was mysteriously added later.

Lady Norris was an oft-widowed acquaintance of Newton's. Her third husband, Sir William Norris, who was known to Newton at Trinity, died in 1702. She was mentioned in the sole surviving letter written by Newton to his niece, quoted earlier, in which he refers to 'My Lady Norris'.

If Newton did propose marriage, it is strange that no other evidence of the relationship has survived. It is also odd that the offer was evidently declined: by this time Newton was a wealthy and famous figure. We can only wonder what might have passed between the world-renowned mathematician who was then Master of the Mint and the mysterious Lady Norris, and how much may have been kept hidden about their possible relationship by the ever-secretive Newton.

As these romantic dramas were playing out, Newton's career was evolving along new avenues. By the time Thomas Neale died, on 23 December 1699 (two days before Newton's fifty-seventh birthday), the Warden had accrued all the responsibilities and privileges of the mastership of the Mint save the name. His succession was seen as little more than a technicality. As well as finally acquiring the £500 annual salary Montagu had originally promised him for the warden-ship, Newton received a royalty. But this was based upon the number of coins struck at the Mint, and by 1700 output had dropped dramatically. Ironically, Neale's negligible effort had brought him far more from the job than Newton would ever see.

Meanwhile, beyond the walls of the Tower, political life was again changing fast and, with his newly acquired status as Master of the Mint, Newton began to realise his social and political ambitions.

His first move was to run for office as MP for Cambridge University. In November 1701 he was re-elected to Parliament, and three weeks later he resigned his Lucasian Professorship, cutting another once-treasured link with academe.

Again he contributed nothing at Westminster, apparently seeing the Commons as little more than a means to acquire still greater social influence. The Whigs were still in government, and the power and esteem associated with a parliamentary seat brought with them many opportunities. During his first term in office, during the Convention Parliament, Newton had been a respected academic doing his duty for the university; now he was an established civil servant with powerful contacts and friendships reinforced by success in office. His greatest asset was still his enduring scientific reputation, but by combining his carefully cultivated image as a peerless intellectual with an element of political clout he hoped to create a formidable platform from which he could accrue yet more influence and power.

That, at least, was the plan; but, unfortunately for Newton, his latest political advance was quickly curtailed. His ally King William III died in March 1702, and Anne, who succeeded him to the throne, had no sympathy for the Whig cause.* Newton, who was always

---

* William actually had no sympathy for Whig politics either, but supported the Whigs throughout his reign because he knew they were the only party who would back his wars against their mutual ideological enemy, the Catholic zealot Louis XIV.

acutely aware of the political tide and the effect it could have upon him, immediately became concerned for his position at the Mint.

Although he had signed a contract with the State on the first anniversary of his becoming Master, 23 December 1700, he was concerned that his suddenly unpopular political views might leave him vulnerable. When Parliament was dissolved by Queen Anne in May 1702, Newton did not stand for re-election, opting instead for a low profile until circumstances were more favourable.

It was probably the most astute political move he ever made. Anne was a hard-line religious traditionalist, and no one was surprised when the entire Whig leadership, known as the Junto, lost office at the election and the Tories were swept to power. Overnight such figures as Lord Somers, who had once been the King's closest adviser, and the Chancellor of the Exchequer, Lord Halifax, were cast into the political wilderness (though in Halifax's case it meant only a temporary loss of political power).

As the political scene turned temporarily against Newton other, long-dormant, opportunities began to offer themselves. Throughout the time he had lived in the capital he had attended the Royal Society only rarely. His official excuse was his commitment to the Mint and the demands of recoinage, but the primary reason had been the overbearing presence there of his old enemy the Secretary, Robert Hooke.

There is no record of how Newton reacted to the news of his great rival's death in early March 1703, but one can safely assume he did not grieve. Indeed, he quickly realised it had happened with fortuitous timing. With his hopes of gaining political power temporarily dashed by the ascension of Anne and with the workload at the Royal Mint falling to little more than the level before recoinage had begun, he was perfectly placed to re-establish his links with the scientific community. Now that Hooke was out of the way, the opportunity to acquire influence and power within the Royal Society was his for the taking, and he moved swiftly.

By 1703 the Royal Society was in a sorry state. From an average of 200 during the 1670s, by the end of the century membership had almost halved, and bad management had brought the society to the edge of bankruptcy. To complete their troubles, the loyal fellows were about to lose their meeting-place and the site of their library in Gresham College. The late Secretary, Robert Hooke, had promised to leave the bulk of his estate to the society, but in typical style this came to nothing because he had died intestate.

Since 1688, the society had met in Hooke's rooms in Gresham College, having moved back from its temporary home at Arundel House, and its impressive library had been housed in a different part of the same building. In 1701 the college trustees had voted to pull down the decaying Elizabethan edifice in order to build a new college on the same site. They had then reached an agreement with the learned gentlemen of the Royal Society whereby the latter would be provided with rooms in the new building as long as the library remained at the disposal of the college. Only Hooke opposed the move and blocked the necessary enabling legislation through Parliament. But upon his death the plans of the trustees resurfaced, precipitating an accommodation crisis for the Royal Society that was forestalled only when a fellow and Gresham College Professor of Physic, Dr John Woodward, offered his rooms as a temporary replacement.

Despite the enthusiasm of such figures as Hooke, for at least the past decade the Royal Society had suffered from uninterested leadership, and much of the original ethos of the society had been slowly abandoned. This was partly thanks to an ill-conceived experiment on the part of the council, who had offered the presidency to leading political figures who knew little if any science. The scheme had been intended to help curry favour and to obtain funding, but it had backfired. Charles Montagu held the presidency between 1695 and 1698, but attended only one council meeting during the entire three years. He was succeeded by fellow Junto member Lord Somers, who chaired a total of three meetings during his five years as President (a period that ended with the abrupt collapse of his political career).

Numerous petty feuds between prominent members caused further division and enmity, and natural philosophy was being sidelined by the fellows' increasing preoccupation with pseudo-medical research. Hans Sloane, who along with Hooke was one of the two secretaries during the 1690s, was a practising doctor and an enthusiastic medical experimenter, as was the society's new host, John Woodward. Along with these influential members, court physicians such as John Arbuthnot and Richard Mead saw the society as little more than a venue for testing their more extreme ideas. If Newton had casually perused the pages of the *Transactions* during his early tenure at the Mint or drawn himself away from the presses to attend a meeting during his frantic first two years in London, he would have been forgiven for thinking the Royal Society had become an exclusively medical fraternity, obsessed with prodigies and miscellaneous medical curios.

The *Journal Book of the Royal Society* reported in May 1699 'that cow's piss drank to about a pint, will either purge or vomit with great ease'[48] – a conclusion that then led to a flurry of intense interest in the medical uses of bovine urine. The same pages were filled with learned discussions on the poisons used in murder cases, vivisections carried out on crocodiles, and the best time of the day to smell flowers.[49]

On 30 November 1703, the rot was finally stopped. At the annual elections to the council, Newton was voted a member and then made the society's latest President. It was a long-overdue move which had previously been blocked by personalities now departed and distractions now diminished, and, although many of Newton's numerous enemies came to regret the appointment, it is seen by historians as a turning-point in the fortunes of the Royal Society. If that evening the vote had gone the other way, the institution would have almost certainly disintegrated within a decade.

The Royal Society was saved not only by Newton's impressive administrative powers (the positive side of his incendiary ego and hunger for power) but also by his own example as a paradigm for how science should be conducted. And he set the tone almost immediately.

On 16 February 1704 the *Journal Book* carried a report from Dr Sloane which read, 'The President presented his book of Optics to the Society. Mr Halley was desired to peruse it and give an abstract of it to the Society. The Society gave the President their thanks for the book and for his being pleased to publish it.'[50]

Based largely upon manuscripts composed some thirty years earlier, the *Opticks* was a book dealing with matters as far removed from bovine urine or the distorted anatomies of medical freaks as one could wish for.

It was delivered without fanfare, and its production and publication were handled by Newton alone. He did not dedicate the book to the Royal Society nor to any individual. Unlike the *Principia*, it was written in English rather than Latin, and it contained very little mathematics. Locke, who read it shortly before his death in October 1704, told Newton he had studied its three volumes 'with pleasure . . . acquainting himself with everything in them'.[51]

But the first edition of the *Opticks* was actually a heavily censored version of what Newton had been working on between the 1670s and his leaving Cambridge. It had been Hooke's comments about the 'Theory of Light and Colours', delivered in 1672, that had driven Newton to silence on the subject for over three decades. Only now, with Hooke in the ground and the presidency of the society his, could Newton offer his second masterpiece to the world.

Even so, he still could not openly admit the true reasons for his thirty-year delay. The closest he came was in a passage in the 'Advertisement' – a sort of preface to the book – in which he alluded to the negative influence of his recently deceased colleague, claiming that 'To avoid being engaged in disputes about these matters, I have hitherto delayed the printing and should still have delayed it, had not the importunity of friends prevailed upon me.'[52]

To some degree he *had* been encouraged by friends to produce the *Opticks*. David Gregory was privy to the material that eventually appeared in the book, and two years before its publication he recorded that Newton had 'promised Mr Roberts, Mr Fatio, Capt. Halley & me to publish his quadratures [calculations of the area under a curve], his treatise of light & his treatise of the curves of the 2d genre'.[53] But, even so, once Newton had decided to publish the work, there were two reasons for his heavy censorship.

First was his realisation (following the pendulum experiments of 1684) that a corporeal ether was unnecessary for the operation of gravity. This suggested to Newton that the ether might also be dispensed with as a medium for the passage of light in his optical theory. However, at the time, this was merely an hypothesis and quite unprovable. Second, and crucial to the eventual shape of the book, was Newton's complete failure to create the unified theory of matter and force for which he had striven during a quarter-century at the alchemist's furnace.

The *Opticks* was originally to consist of four books. The last of these was to be a description of a grand unification of the optical phenomena demonstrated in the first three books with the mechanical theories described in the *Principia* – a theory bringing together all the known forces of Nature. Newton's failure to produce an all-embracing synthesis of the microcosmic and the macrocosmic between 1687 and 1696 meant that in the proposed fourth book he would be able only to suggest hypotheses rather than verifiable theories. He could not settle for this, as he explained within the first pages of the new treatise: 'My design in this book', he wrote, 'is not to explain the properties of light by hypotheses, but to propose and prove them by reason and experiments.'[54]

Using a blend of alchemy, mathematics and physics, Newton had already shown how universal gravitation *operated*, but he was never able to explain the *mechanism* by which it worked.* His hoped-for Book IV of the *Opticks* would have offered an explanation of how all phenomena were based upon the interaction of fundamental forces – a development of the scrapped 'Conclusio' of the *Principia*:

> How the great bodies of the Earth, Sun, Moon and planets gravitate towards one another, what are the laws and quantities of their gravitating forces at all distances from them & how all the motions of those bodies are regulated by their gravities I showed in my *Mathematical Principles of Philosophy* to the satisfaction of my readers. And if Nature be most simple & fully consonant to herself she observes the same method in regulating the motions of smaller bodies (including the corpuscles of light) which she does in regulating those of the greater.[55]

This staggering hypothesis – that the laws governing forces and matter on the larger, macrocosmic, scale ought to be reflected in the behaviour of subatomic particles (Newton's 'smaller bodies') and microcosmic forces – went the way of Newton's alchemical papers and much of his speculations upon religion and mysticism – lost to all but a protective cadre of disciples for almost 250 years. By late 1693, soon after their composition, sections of the text originally intended for Book IV of the *Opticks* were concealed from almost all contemporaries. David Gregory, who visited Newton in Cambridge in May 1694, was shown only three books of the great work,

---

* Newton's failure to accomplish such a unification should come as little surprise: scientists such as Stephen Hawking, Steven Weinberg and many others are still trying to achieve this today.

but even then he was prompted to comment that 'if it were printed it would rival the *Principia Mathematica*'.[56]

Although the *Principia* became one of the most important of all scientific works, the *Opticks*, with its myriad topics and almost visionary stance, devoid of jargon and unpalatable mathematics, provided a far broader range of influences for future generations.

It begins with explanations of refraction and reflection, the phenomena of rainbows and interference, and the behaviour of mirrors and prisms – all based upon the vast series of experiments that Newton had begun in 1664. From here it broadens rapidly to encompass such diverse issues as the phenomenon of gravitation, metabolism, sensation, the workings of the eye, and even such exotica as the Great Flood and the Creation.

Yet, despite the determination to never offer hypotheses and the knowledge that he had failed to attain his dream of unification, Newton could not resist dipping his toe into the murky waters of speculation. At odd points in the three surviving books of the *Opticks* he let slip his conviction that light operates within the same parameters as gravitation – via forces acting at a distance. In Book II (which is principally concerned with reflection and refraction), while arguing against the idea that light could be made up of invisible particles, he says:

> And this problem is scarce otherwise to be solved, than by saying, that the reflection of a ray is effected, not by a single point of the reflecting body, but by some power of the body which is evenly diffused all over its surface, and by which it acts upon the ray without immediate contact. **For that the parts of bodies do act upon light at a distance shall be shown hereafter.**[57]

Despite these revealing lapses, throughout the text Newton trod a delicate path, staying just the verifiable side of hypothesis. But because, as he put it, 'I . . . cannot now think taking these things into further consideration',[58] he decided that in place of Book IV he would compose a collection of questions he called the 'Queries'.

The purpose of the Queries was to offer his more extreme ideas without fear of ridicule or risk of criticism by presenting them in the form of questions. There were sixteen queries presented as 'Addenda' to the first edition of the *Opticks*, and in Query 1 he dived in with what would have been considered by many to be a semi-occult proposition:

> Query 1. Do not bodies act upon light at a distance, and by their action bend its rays; and is not this action . . . strongest at the least distance?[59]

Occult this may have been to the early-eighteenth-century mind, but almost exactly 200 years later the bending of light by the gravitational effect of a star lay at the heart of Einstein's general theory of relativity, according to which the distortion of space-time causes light to follow a curved course.*

A little further on, Newton asks:

> Query 5. Do not bodies and light act mutually upon one another; that is to say, bodies upon light in emitting, reflecting, refracting and inflecting it, and light upon bodies for heating them, and putting their parts into a vibrating motion wherein heat consists?[60]

Again this is not far from the foundations of quantum mechanics (a branch of physics which describes the behaviour of matter and energy on a subatomic scale) first put forward at the start of the twentieth century. Today it is a basic tenet of quantum mechanics that, on a subatomic level, light and matter interact.

In 1905, almost exactly two centuries after the first edition of the *Opticks*, Einstein described the photoelectric effect, which demonstrates how light can affect the inner structure of atoms. It has also been recognised that what we perceive as the colour of a body is determined in part by the structure of the molecules of which the body is made. All of which shows that Newton's question 'Do not

---

* General relativity predicted that the light from distant stars passing near the Sun would be bent towards it by gravity – an effect created by the distorting of space-time near the massive Sun. This was proven to be true by a series of observations made by the British astronomer Arthur Eddington in early 1919, and it was this proof that overnight catapulted Einstein and relativity to global fame.

bodies and light act mutually upon one another' demonstrated extra-ordinary foresight.

Within two years the English version of the *Opticks* was followed by a Latin edition and an opportunity to add more Queries – seven in all – with another three speculative passages tacked on to the Addenda. This was then succeeded, a few years later, by a second English edition, containing a final batch of eight new Queries.* The most striking of all these additions – indeed the most advanced specu-lation Newton ever allowed to be published – was Query 31, which appeared in the first Latin edition of 1706:

> Have not the small particles of bodies certain powers, virtues or forces, by which they act at a distance, not only upon the rays of light for reflecting, refracting and inflecting them, but also upon one another for producing a great part of the phenomena of Nature? For it's well known, that bodies act upon another by the attractions of gravity, magnetism, and electricity; and these instances show the tenor and course of Nature, and make it not improbable but that there may be more attractive powers than these. For Nature is very constant and conformable to herself.[61]

This single paragraph encapsulates perfectly Newton's thinking on the concept of unification – what we now see as a unified theory which would draw together the threads of the microcosmic world (now exemplified by quantum theory) and the macrocosmic (gov-erned by relativity). It was the closest he ever came to stating that a simple collection of forces was responsible not only for the phenom-enon of gravity but for almost all physical characteristics of the observable universe.

And the influence of the *Opticks* went beyond science. The poly-mathic Erasmus Darwin placed Newton at the head of his poetic pantheon of science (a virtual community of the great intellects of the past), writing:

---

* The seven Queries added to the first Latin edition are today numbered Queries 25–31. The eight Queries added in the second English edition are called Queries 17–24.

> NEWTON's eye sublime
> Mark'd the bright periods of revolving time;
> Explored in Nature's scenes the effect and cause,
> And, charm'd, unravell'd all her latent laws.[62]

The poet Alexander Pope was also impressed by Newton's ability to systematise the universe, but was concerned about the road along which humankind was now heading. In his *Essay on Man*, published in 1733, six years after the scientist's death, Pope expressed the view that Newton might represent the best of humanity but, in the eyes of the immortals, was still little better than ape.

> Superior beings, when of late they saw
> A mortal Man unfold all Nature's law,
> Admir'd such wisdom in an earthly shape,
> And shew'd a NEWTON as we shew an Ape.[63]

Pope's opposition to Newtonianism sprang from a rich vein of religious bigotry. He saw the mathematics of the *Principia* and the experimental descriptions and queries of the *Opticks* as rationalism diminishing the spirit and significance of human existence. Many other writers and artists of the eighteenth and early nineteenth centuries felt the same way. Berkeley and Coleridge were both anti-Newtonian, and William Blake was the most vocal of all the scientist's posthumous opponents, stating in *Jerusalem*:

> I see . . . humanity in deadly sleep . . .
> For Bacon and Newton, sheath'd in dismal steel, their terrors
>   hang
> Like iron scourges over Albion . . .
> I turn my eyes to the schools and universities of Europe
> And there behold the loom of Locke, whose woof rages dire,
> Wash'd by the water-wheels of Newton: black the cloth
> In heavy wreathes fold over every nation: cruel works
> Of many wheels I view, wheel within wheel, with cogs tyrannic
> Moving by compulsion each other, not as those in Eden, which
> Wheel within wheel, in freedom revolve in harmony and
>   peace . . .[64]

Yet Blake also harboured a certain admiration for Newton's key role in creating the *Zeitgeist* of the eighteenth century. His painting *Newton* has elicited voluminous comment since it was first seen, primarily because of its inherent contradictions. In one version of the painting Blake depicted Newton as embedded in a rock (presumably the immovable obstacle of rationalism), while in another the scientist is perched on the seabed. In each he is using a pair of compasses to measure some unknown parameter, and a page of diagrams is visible which looks like something extracted from Book I of the *Opticks*. Yet while Newton is seen as a rigid figure – symbolic of the inflexibility of pure rationalism – he is also noble, even beautiful. It has even been suggested that the face of the image bears more than a passing resemblance to that of the young Blake himself.[65]

Blake too was interested in ancient teachings, and many of his etchings illustrate this preoccupation. His *Joseph of Arimathea* is thought to have been inspired by William Stukeley's *Stonehenge: A Temple Restored to the British Druids*,* and the poet's most recent biographer draws a strikingly familiar portrait when he says, 'All his life Blake was entranced and persuaded by the idea of a deeply spiritual past, and he continually alluded to the possibility of ancient lore and arcane myths that could be employed to reveal previously hidden truths.'[66]

So, the expansion of Newton's personal world, the trajectory from monastic Cambridge to the dinner-tables of London, brought with it a flowering of influence that would continue for at least two centuries after the conversations had been stilled and the dreams of clippers and kings had faded away. In under a decade Newton had transformed himself. Having stepped back from the brink of oblivion, he had thrown body and soul into a new career and, by reaching back and plucking the *Opticks* from obscurity, he had achieved further levels of fame and respect. Striding through his seventh decade, he still did not miss a step. Ahead lay fresh mountains to scale and still more blank sheets upon which to make his mark.

---

* As well as writing the first biography of Newton, Stukeley was one of the most accomplished antiquarians of the eighteenth century and published several books on the Druids and ancient lore.

Having now garnered the power and influence he had sought throughout his life, his ascension to the status of icon appeared unstoppable.

CHAPTER 12

Old Men's Battles

*He saith among the trumpets, Ha, ha; and he smelleth the battle
afar off, the thunder of the captains, and the shouting.*

JOB 39:25

Isaac Newton was never a handsome man, and as he grew older
the lines and hollows of his face became more pronounced; the
sharp nose and brooding dark eyes that should have been soft-
ened by the fat of years continued to dominate his features, reflecting
perfectly the harshness of his personality. The accumulation of power
and veneration, it might be argued, became manifest in that face –
in the black, piercing eyes and in the stern gaze. We can trace the
changing face of Newton through the numerous portraits com-
missioned during the period from the beginning of his public life to
his final years.

From the first painting of 1689, when he was forty-six and starting
to emerge from his cocoon, to the final portrayals of the octogenarian
demigod, his image reflects the progress of his worldly manœuvres.
The first portrait, by Sir Godfrey Kneller (to whom Newton would
return), shows the man of pure intellect, wigless and wrapped in a
plain black scholar's gown; it is the eternal face of Newton, the image
from school textbooks – the creator of the *Principia*, unraveller of
Nature's innermost secrets. He looks distracted, his gaze far distant,
as if he cannot spare the time to sit (an impression probably not far
from the truth). His grey hair falls in thick curls that frame a pointed,
angular face and the eyes and nose of a bird of prey, a hawk – the
troubled face of a neurotic.

Later the image is subtly altered. The Kneller portrait of 1702
shows a far richer, better-fed specimen, still in black, but wigged and
regal – the bureaucrat; the object of reverence. Then, a year later,
the organic process of change takes another turn. Charles Jervas's

portrait of 1703, a full-length painting, offers an aspiring member of the nobility, just two years before his knighthood. Wigged and expensively dressed, Newton has gained still more weight. This is the worthy image to be displayed upon the future £1 note, a face to adorn countless issues of postage stamps around the world.

In the final set of paintings, however, we are witness to the gradual stripping away of the trappings of greatness. Sir James Thornhill's painting of 1710, when Newton was sixty-seven, returns to the honesty of Kneller's first effort. Newton clutches to himself a plain robe; his white hair is thin, and the black eyes again gaze into the middle distance, as though contemplating some far-off visage. Kneller's third and final portrayal is almost a cartoon by comparison and shows a smiling Newton – a rare thing indeed. To complete the set, John Vanderbank's painting of 1726, when the eighty-three-year-old Newton was a year from death, gives us the scientist turned supreme bureaucrat contemplating a copy of his own masterpiece, his lined and puffy face almost benign – like a gentle grandfather who might read fairy tales to children gathered at his knee.

By the time he had become Master of the Mint during the closing days of the seventeenth century, Newton had long adopted the trappings of a wealthy city gentleman. His house was well furnished and tastefully decorated, and receipts still exist for four landscapes and twelve Delft plates.[1] He kept a coach and horses – strictly the preserve of the wealthy – and could sometimes be seen in a sedan chair. At one time, he employed six servants to run the house (which presumably left Catherine Barton little to do in the way of 'housekeeping'), and the inventory of his possessions after his death tells us he owned a dinner service which included forty plates, a full set of silver flatware, about two dozen glasses, and six and a half dozen napkins.[2] All of which would confirm Conduitt's assertion that 'He always lived in a very handsome generous manner though without ostentation or vanity, always hospitable & upon proper occasions gave splendid entertainments.'[3]

Ostentation and vanity did, however, stretch to his ablutions, for the inventory also records two solid-silver chamberpots. There is evidence too of the continuing fascination with the colour crimson which had become apparent during Newton's postgraduate days at Trinity. So ubiquitous was this colour in the Newton household that it was the only one mentioned in the inventory of his possessions, leading one of his biographers, Richard de Villamil, to remark that Newton 'lived in an atmosphere of crimson'.[4]

Although Conduitt mentions occasional 'splendid entertainments', a visitor from the court of Louis XV, the Abbé Alari, who called on the scientist in 1725, found Newton's table anything but satisfactory, calling the meal he was offered 'awful' and the wines poor, suspecting they had been given to Newton as presents.[5] If Newton's interest in wines was limited, he also appeared to have little taste for fine art or contemporary literature. According to Conduitt, he never diverted himself with music or art. Yet this was certainly an exaggeration, because Newton told Stukeley he had attended the newly fashionable opera – albeit only once. 'The first act,' he recalled, 'gave me the greatest pleasure, the second quite tired me: at the third I ran away.'[6]

He could also spare little thought for poetry; from one report, he quoted and seemed to approve of Isaac Barrow's comment that 'poetry was a kind of ingenious nonsense'.[7] A little later he went on to describe his friend the Earl of Pembroke, whose collection of statues was renowned throughout Europe, as 'a lover of stone dolls'.[8] This is surprising, since in his youth Newton composed poems and was reputed to be a keen and capable artist. And if the contents of his library give an accurate indication of his reading habits, he displayed little interest in classic literature: the inventory records the complete absence of Shakespeare, Chaucer, Spenser and Milton. In fact, almost all the works of fiction in the library were contemporary and have since been attributed to Catherine Barton's reading tastes.[9]

The first decade of the eighteenth century was another turning-point for Britain's most celebrated scientist. Within three years of Queen Anne's accession to the throne in 1702, the political climate had again reverted to one which complemented Newton's own ideals. Despite Marlborough's great victory at Blenheim in 1704, by the following spring the Tories were split over the war with France and losing their political grip. Anne quickly realised that to avoid trouble she had to embrace the Whigs, and she was soon to be seen dining with once ostracised politicians such as Lord Russell, a former prominent member of the Whig Junto. She was also wise enough to realise that the most effective way to draw together both sides of the political divide was to give prominent Whigs Establishment honours. Hence Newton's knighthood in May 1705, conferred during a ceremony in Cambridge in which James Montagu was also honoured despite his brother Charles having been politically marginalised by Anne herself only three years earlier.

Yet there remained ghosts in Newton's closet, and these might have damaged his reputation if they had not been ruthlessly suppressed either by Newton himself or by his growing, but close-knit cadre of disciples.

Nicholas Fatio de Duillier had remained in England after the split with Newton and had continued to hover on the periphery of the Establishment, attending the Royal Society occasionally, where from time to time he would offer idiosyncratic ideas and creations for comment. In 1704 he displayed a set of jewelled watches in which he had replaced some of the metal parts with rubies to improve the mechanism, and he later presented Newton with two specimens which the President vouched to be 'very fine'.[10]

The two men spoke in public only rarely, although in 1702 David Gregory recorded in his diary a conversation involving himself, Newton and Fatio, and such talk always revolved around formal Royal Society business.[11] Their only surviving correspondence after the break of 1693 concerns Fatio's watchmaking endeavours and is all from him to Newton. However, their relationship was not completely extinguished, for they were known to have had private conversations on the subject of prophesy as late as 1706.[12] What effectively ended any chance of a reunion was de Duillier's involvement in 1707 with a group known as the Camisards, or the French Prophets.

The Camisards were mainly exiled Huguenots who had fled to England from the persecution of Louis XIV. They were a radical Protestant sect who believed the Day of Judgement was imminent, and that they could speak in tongues and raise the dead – convictions not far removed from those of the early Rosicrucians. They took their name from the smock of the French peasant, the *camise*.

For Fatio, who had long since assimilated Newton's Arianism and possessed a natural inclination towards both anti-Catholicism and occultism, the Prophets acted like a magnet. He joined them within weeks of their arrival in England and rose through their ranks with ease to become their secretary and bookkeeper, as well as composing a stream of pamphlets and leaflets about the group. A widely circulated anti-Camisard pamphlet even suggested that Fatio was engineering the whole enterprise as a first step towards creating his own religion. According to the anonymous author, de Duillier conceived this religion as a 'Spiritual Catholicon', or an idiosyncratic blend of those aspects of other religions that suited him.[13]

As the only established philosopher to become involved with the venture, Fatio came in for particular scorn. In a contemporary account written by one R. Kingston, Fatio's belief is treated with incredulity:

> I was very much dissatisfied with cavalier's experiment [a public display by the Prophets] that I was surprised to find any men of sense affected with it; much more to see Mr Fatio so diligent in penning everything that was said, with as much care as if it had been delivered from Mount *Sinai*.[14]

It did not take long for the scientific Establishment to regard Fatio as a pariah. The antiquary Thomas Hearne, infamous as a gossip in scholarly circles, came down on him hard, writing a character assassination in which he said:

> It has been always observed of him that he is sceptical in religion, a person of no virtue, but a mere debauchee. He was formerly a director to the Duke of Bedford, whilst he was of Magdalen College, Oxford, who, by his means, imbibed odd principles, grew a great gamester and spend-thrift . . . During the time Fatio was with him, he got by his insinuation and cunning a vast sum of money from the Duke, & made all the provisions possible for his future advantage.[15]

While little of this has been substantiated, it shows the depth of feeling within the academic community over the whole affair. Not only were the French Prophets perceived as subversives: Fatio's involvement and possible leadership threatened to undermine the credibility of the scientific community itself. And it is not hard to imagine the dazzling and disturbing effect this group had upon the public imagination. Even in an age where the only form of publicity was pamphleteering or oratory, for a short time at least the Camisards created an enormous stir. They demonstrated in the streets and members went into convulsions, claiming they were possessed by the ancient prophets of the Old Testament. They also made outrageous proclamations deliberately to incite the Establishment. On one occasion they predicted that the Lord Chief Justice, Sir John Holt, would explode while serving on the bench, his blood running from every vein. Pamphlets, both pro and anti, were quickly circulated, and a well-known playwright of the day, Thomas D'Urfey, produced a spoof called *The Modern Prophets*, which drew huge crowds for several months. But, within the sensitive religious climate of the time, such radicals could only go so far before attracting the unwelcome response of the authorities. They may have been the object of ridicule for the majority of people, but they were also alienating the moderate Protestants who had fled France to seek sanctuary in England. When, in the autumn of 1707, the Prophets announced publicly that a second Great Fire was soon to engulf London, the Queen was forced to move against them.

Some of Fatio's friends had tried to save his reputation just as the net was closing around him and his associates, but he had ignored advice that he should escape to Switzerland. His family back home in Geneva were baffled and alarmed by the news coming from England, and Fatio's brother Jean-Christopher, whom de Duillier had managed to get elected to the Royal Society only a year earlier, reminded him of the biblical imperative concerning false prophets; but it was all to no avail.

In November 1707 the organisers of the group were put on trial, accused of spreading terror among Her Majesty's subjects. Found guilty, the leaders – identified as Jean Daudé, Elias Marion and Nicholas Fatio de Duillier – were sentenced to stand on two successive days in the pillory at Charing Cross. Here they were forced to wear a notice detailing their crimes, while anyone who felt so inclined could pelt them with anything that came to hand. After that, nothing more was heard of the French Prophets, although Fatio remained a believer until the day he died.

Newton did nothing to help his erstwhile friend. Perhaps he believed that Fatio should fight his own battles, but the most likely reason was self-preservation – especially as he was said to be secretly sympathetic towards the ideas espoused by de Duillier and the Camisards. In a book published in 1820, the Reverend Joseph Spence offered two separate reports mentioning Newton in the same breath as the Camisards and the extraordinary events of 1707. The first, a letter from a Dr Lockier, the Dean of Peterborough, declared:

It is not at all improbable that Sir Isaac Newton, though so great a man, might have had a hankering after the French Prophets. There was a time when he was possessed with the old fooleries of astrology; and another when he was so far gone in those of chemistry, as to be upon the hunt after the Philosophers' Stone.

The second account came from Michael Ramsey, who was a friend of many of Newton's younger followers:

Sir Isaac himself had a strong inclination to go and hear these prophets, and was restrained from it, with difficulty, by some of his friends, who feared he might become infected by them as Fatio had been.[16]

It can only be assumed that when nineteenth-century biographers came to write their accounts of Newton's life they were either unaware of these observations or else passed over them as unsuitable for use. However, that second account does sound rather doubtful. Newton valued his own public image above most things, and it is hard to imagine him seriously considering attending a meeting of the French Prophets. Yet there may be a grain of truth beneath the hyperbole, for it was quite in his nature to be fired up by an inspired idea, to go along with it part of the way, and then, realising he was being foolish, not only to draw back but to rebound some considerable distance in the opposite direction. Blending as it did a heady mixture of prophetic zealotry and powerful anti-Catholic sentiment, Newton might at first have been impressed by the conceptual foundation of the Prophets. He may have discussed the activities of the sect with Fatio privately and even flirted with the idea of going to a public meeting – merely out of curiosity; but then, just as well-meaning friends warned him off the relationship with Fatio a decade earlier, he was made to see the danger lying ahead and instantly cut the link, banishing the misguided de Duillier from his mind, even as his once intimate friend stood for two days smeared with the juice of rotten vegetables and egg yolk.

Fatio was to live a further half-century, retaining his enthusiasm for the ideals of the then defunct Prophets, an outcast from the scientific community. He died in poverty in 1753, still fantasising that he was a member of the élite and citing intimacy with Newton, Halley and Locke as his pathetic credentials.

For Newton, fear of exposure of his unorthodox personal views did not disappear with Fatio's disgrace. No sooner had the scandal of the French Prophets faded from public scrutiny than he was confronted with further trouble from a former follower.

In 1694 the twenty-seven-year-old mathematician William Whiston had so impressed Newton with a manuscript entitled 'New Theory of the Earth' that he was soon employed as the Lucasian Professor's assistant at Trinity. Then, when he resigned from the Lucasian chair in 1701, Newton smoothed the way for Whiston's election to the position.

Soon after taking the post of Newton's assistant, Whiston had become interested in Arianism and was encouraged and guided by Newton. But, despite sharing similar religious views, the two men responded to their faith in polar opposite ways. Whereas Newton was secretive and self-protective, once guided into primitive Christianity by his master's own enthusiasm and tutelage Whiston became positively evangelical. Whereas Newton chose to keep his faith to himself and to forbid publication of his religious ideas, Whiston threw caution to the wind and quickly acquired a degree of notoriety. By 1708 – less than twelve months after Fatio's disgrace – Whiston was disturbing his friends and drawing the censure of the authorities by preaching Arianism openly.

Things came to a head for the young professor in October 1710, when a Cambridge University commission stripped him of his position. He moved to London, with his family, supplementing his meagre savings by part-time mathematics teaching while continuing to proselytise his extreme religious views. His friends – men such as the preacher James Pierce and his Cambridge colleague and fellow Arian Richard Bentley – rallied round, offering both material and moral support, but from the man who had set Whiston on this road to oblivion, his intellectual guru, Newton, came only silence.

Fearful of guilt by association, Newton said absolutely nothing in Whiston's defence. He made no comment about his successor's dismissal from the Lucasian Professorship and offered not a single word of support during his ongoing battle with the ecclesiastical authorities in Cambridge and, later, London. But that, it seems, was not enough: by 1716 Newton's passive dissociation had slipped into active betrayal. When that year Whiston applied for membership of the Royal Society, Newton threatened to resign if the nomination was accepted.[17] Whiston later commented sarcastically that 'had I known his mind, I would have done nothing that might bring the great man's grey hairs with sorrow to the grave'.[18]

Whiston was well aware that his actions would cast him into the scientific and social wilderness. He was also under no illusions about Newton's fear of publicly proclaiming his own religious views. 'But perceiving that I could not do as his other darling friends did,' he noted in his memoirs written after Newton's death, '. . . he could not, in his old age, bear such contradiction; and so he was afraid of me the last thirteen years of his life.'[19]

Whiston's view of the Royal Society's President was understandably lacking in affection, and described Newton as having 'the most fearful, cautious, and suspicious temper, that I ever knew'. He explained that he had forgone publicising what he knew of Newton's ideas about biblical chronology while the President was alive – material published in 1733 as *Observations upon the Prophesies* – 'because I knew his temper so well, that I should have expected it would have killed him'.[20]

Yet the fear that Whiston describes must have worked both ways, for even he, whose life had been transformed and his career destroyed by unalloyed evangelical Arianism, could not betray the man who had led him to that fate. None of Newton's disciples, many of whom were aware of his beliefs, said or wrote a word about the President's Arianism until after his death in 1727. The most Whiston could do as an act of rebellion was to dedicate his heretical *Astronomical Principles of Religion, Natural and Reveal'd* (1717) to Newton and the council of the Royal Society.

To a degree, one can understand Newton's reaction. He had always refrained from publicising his theological ideas, and Arianism was viewed as heresy – it was specifically excluded from the Toleration Act of 1689, which offered religious freedom to all faiths with the marked exceptions of Catholicism and Arianism. But there were other pressures. Fatio's impropriety had worried Newton, and, as Whiston headed off into the social and academic wilderness, another of Newton's disciples, the philosopher Samuel Clarke, was becoming known as an Arian (and escaped Whiston's ignominious fate only by publicly recanting his views before the Court of Convocation in 1714).

There is an undeniable whiff of self-preservation surrounding Newton's reaction to both the Fatio and the Whiston scandals – a scent not unfamiliar to those who came close to Newton and then, by their own volition, stepped too close to the edge – but these incidents also illustrate the social tight-rope that Newton himself walked for most of his life. It was not merely paranoia that motivated Newton, nor simply fear that 'he might become infected . . . as Fatio had been', as Michael Ramsey saw it. Newton was a man with real enemies, some of whom certainly suspected his interests and would have delighted in exposing the illustrious Royal Society President to the unforgiving glare of public scorn. Thomas Hearne, who had lambasted Fatio soon after the French Prophets had had their day, dropped vague hints about his knowing of Newton's secrets, but only after the scientist's death. 'Sir Isaac, though a great mathematician, was a man of little religion,' he wrote, 'in so much that he ranked with the heterodox men of the age.'[21]

And, after a dispute at the Royal Society during which Newton had claimed that Flamsteed had called him an atheist, the Astronomer Royal wrote angrily in his diary, 'I never did, but I know what other people have said of a paragraph in his *Opticks*; which probably occasioned this suggestion . . . I hope he is none.'[22] The paragraph to which Flamsteed referred was the conclusion to Query 28, in which Newton had written of infinite space as the 'sensorium of God'. This, Newton believed, was an expression of divine intervention in Nature, but others saw it as implying that God was not omnipotent but needed sense organs to perform his wonders.

Another opponent of Newton, Gottfried von Leibniz, was suspicious of Newton's entire concept of gravity, referring to it mockingly as 'the rebirth in England of a theology that is more than papist and a philosophy entirely scholastic since Mr Newton and his partisans have revived the occult qualities of the school with the idea of attraction'.[23]

Although nothing was said openly, rumours concerning Newton's unorthodox religious perspective – based upon his known links with dissenters such as Fatio, Whiston and Clarke – were circulating during his lifetime. In 1719 an anonymous Arian tract entitled *The History of the Great Athanasius* was thought by some to be

Newton's work, but no one who was genuinely close to him could seriously have contemplated the President of the Royal Society risking publication of such a document. Indeed, it might be argued that Newton deliberately and carefully exaggerated his abrasive image specifically to deter anyone who might be tempted to try to chip away at his reputation while he still breathed. And if anything his dictatorial manner became even more pronounced as he aged.

As his fame grew and deepened within the philosophical community, Newton's domination of the Royal Society had far-reaching effects, both positive and negative. The *Opticks* and the revitalisation of the Royal Society count greatly in his favour, but against these must be set his dictatorial manner as President and the string of conflicts he initiated. From the beginning of the 1700s until his death almost three decades later, policy battles and acrimonious disputes dominated his life.

He was quickly seen to favour certain friends at the Royal Society and to boost their careers while suppressing those of whom he disapproved. Within a few years of becoming President, he had come to dominate the decision-making body, the council, to such a degree that it was made up of sympathisers almost to a man. Flamsteed – one of the few men brave enough to speak his mind – said of this manipulation, 'Sir I. Newton, sees now that he is understood.'[24]

And the patronage extended beyond the walls of the Gresham College meeting-place. In 1704 the deserving Edmund Halley (who had endured two traumatic years at the ill-fated Chester Mint) was finally rewarded for his work on the first edition of the *Principia* by being appointed Savilian Professor of Geometry at Oxford at Newton's behest. Flamsteed, who despised Halley, remarked disingenuously that 'Sir I. Newton has put our Royal Society into great disorder by his partiality for E. Halley.'[25]

If Newton knew of Flamsteed's opinion, he took absolutely no notice of it. Two years later the new Plumian Professorship at Cambridge, for which Newton wrote most of the rules, was awarded to the young mathematician Roger Cotes, who later edited the second edition of the *Principia*. Other disciples were also found prestigious jobs: Edward Paget, a fellow of Trinity, was made Master of the Mathematical School of Christ's Hospital in London, and David Gregory was appointed Savilian Professor of Astronomy at Oxford, both at Newton's insistence.

This nurturing of his young followers can be seen as merely a perk of his position – a practice so common among high-ranking officials of any era that it would have seemed odd if Newton had behaved differently. Such use of power would not have raised an eyebrow. It was within the Royal Society itself that Newton's powerful personality and manipulative manner could be best witnessed, and where it came in for criticism as well as praise.

Newton's 1960s biographer Frank Manuel has made the astute point that, whenever Newton identified himself with an institution, that institution became an extension of his own personality.[26] This was true for the Mint, and it became even more apparent during Newton's presidency of the Royal Society.

One of Newton's ambitions almost from the moment he took office was to break the society's dependency on Gresham College and to purchase a site of its own. But there were stumbling-blocks. When Newton took the helm, in 1703, the Royal Society was almost bankrupt; but then the new President was a good man with money. In 1706 he and the Secretary, Hans Sloane, pushed through the council the fund-raising measure of charging members an admission fee secured by a bond. This fee was to be back-dated for up to a year. Although there were more than a few disgruntled fellows, with the help of a paid clerk instructed to collect arrears 'whensoever the President shall direct', the fees were paid.[27]

The council also invested in some property (but eventually had to take legal action to retrieve long overdue rents), and selling society stock in the East India Company and the East Africa Company topped up the coffers. By September 1710 the financial position was transformed, allowing Newton and Sloane to convene a special meeting of the council on 16 September to propose the move to a property they had found in Crane Court.

If there were any on the council who had yet to experience Newton's overbearing presence and powers of manipulation at first hand, by the conclusion of this extraordinary meeting they could have been left in no doubt about his methods. His passion for securing a permanent home for the society is clear from a scathing anonymous thirty-two page pamphlet circulated among the fellows shortly after. It was entitled *An Account of the Late Proceedings in the Council of the Royal Society. In Order to Remove From Gresham College into Crane Court, in Fleet Street. In a Letter to a Friend,* and it complained that the meeting had been called merely to rubber-stamp the President's ambitious plan: 'The President was not prepared . . . to enter upon the debate,' it declared: 'But freely (though methinks not very civilly) replied that he had good reasons for their removing which he did not think proper to be given there.'[28]

Sloane then intervened by pointing out that the Gresham professors no longer wanted the society to meet in their rooms, and that a move would give the society room for future expansion. The council responded by saying that such an important decision should not be made lightly and the meeting should reconvene in November, allowing time to digest the idea. Newton would have none of it: 'his scruples were immovable,' the pamphlet continued, 'so that some of the gentlemen with warmth enough asked him to what purpose then had he called them thither?'[29] At which point the President rose and the meeting was adjourned. Within weeks the council had purchased the property.

The authenticity of this account is supported by the minutes of the meeting, and the fact that after the pamphlet was printed and distributed, in late November 1710, there were no attempts to deny its contents lends it further weight.

The new premises needed complete refurbishment. The purchase price was £1,450, before repairs of £310 and the construction of a gallery designed by Wren costing a further £400. A loan of £900 at 6 per cent was obtained for the purchase, but this was paid off within six years by large contributions from the wealthier members – Newton donated £120, and the council took up the offer of a loan of £464 from the Clerk, Henry Hunt.★

The new building was rather plain, constructed in the modern style and set at the end of a long, dreary courtyard, but it offered more than enough space for the members and their expanding library, and Wren's gallery, although far less grandiose than the one he had originally proposed, was a sympathetic addition. Ironically, the best description we have of the building (the Royal Society removed to Somerset House under the presidency of Sir Joseph Banks in 1780) comes from a pamphlet written in opposition to the move:

★ When Hunt died in 1713, he held IOUs from the Royal Society to the value of £650, which were met by Newton and Sloane.

The approach to it I confess, is very fair and handsome, through a long court: but then they have no other property in this than the street before it; and in a heavy rain a man can hardly escape being thoroughly wet before he can pass through it. The front of the house, towards the garden, is about 42 ft long; but towards Crane Court not above 30 foot. Upon the ground-floor there is a little hall, and a direct passage from the stairs into the garden, about 4 or 5 foot wide; and on each side of it, a little room about 15 ft long, and 16 ft broad. The stairs are easy which carry you up to the next floor. Here there is a room fronting the court directly over the hall, and of the same bigness. And towards the garden is the meeting-room, which is 25½ foot long, and 16 foot broad. At the end of this room there is another (also fronting the garden) 12½ foot long, and 16 broad. The three rooms upon the next [third] floor are of the same bigness with those I have last described.[30]

The move was delayed until August 1711 because of extensive building work, but even before the fellows had vacated their old home Newton had put a great deal of thought into changing the image and the internal workings of the institution. He employed liveried doormen bearing the society's coat of arms in silver, and had the old porter pensioned off. To establish greater formality, he drew up the 'Orders of the Council' – a set of rules, first announced at the council meeting of 20 January 1711. These stated:

1. That nobody sit at the table but the President at the head and the two Secretaries towards the lower end one on the one side and the other, except some very honourable stranger, at the discretion of the President.
2. If any paper be read before the Society, it shall be minuted, and all the principal parts entered in the Journal, to be read to the Society the next meeting which will give an opportunity to the Society to debate the particulars.
3. That no person or persons talk to one another at the meetings, or so loud as to interrupt the business of the Society but address themselves to the President.[31]

At some point, possibly at the same time, Newton also introduced the practice of having the ceremonial mace placed on the council table only when the President was in the chair. (When Sloane was voted Newton's successor, in 1727, his first action was to dispense with this rule.)

Beneficial to the society though it obviously was, by pushing through his plan to buy Crane Court Newton was also fulfilling a subconscious desire to wipe the slate clean. Hooke, his oldest rival, was dead, but, for Newton, the man's spirit still haunted Gresham College. The President wanted a fresh start, to eradicate the bitter memories. Perhaps by the merest coincidence, the only existing portrait of Robert Hooke, along with many of the instruments he fashioned as Curator of Experiments, was mysteriously lost in the move, never to resurface.

Predictably, Newton was not content with giving the Royal Society a new lease of life and silencing his detractors while he was still alive and totally in control: he also made sure that the image he cultivated would seep into recollections and memoirs long after he had been replaced. The ever-loyal Stukeley, for instance, presents us with a suspiciously rosy picture:

> Whilst he presided in the Royal Society, he executed that office with singular prudence, with the grace and dignity – conscious of what was due to so noble an institution – that was expected from his character . . . Sir Isaac was very careful of giving any sort of discouragement to all attempts of improvement in natural knowledge. There was no whispering, talking, nor loud laughter. If discussions arose in any sort, he said they tended to find out truth, but ought not to arise to any personality . . . Everything was transacted with great attention and solemnity and decency . . . Indeed his presence created a natural awe in the assembly; they appeared truly as a venerable *consessus Naturae Consiliariorum* [assembly of Nature's advisers], without any levity or indecorum.[32]

Although he was the only biographer who knew Newton personally, Stukeley, it should be recalled, was also a disciple, and this account was exactly how the President would have wanted his incumbency recorded. Naturally, others, who for whatever reason got on the wrong side of Newton, viewed him with a little less 'natural awe'.

The Gresham professor John Woodward (whose rooms had been the venue for the society's regular meetings after Hooke's death) was an irascible individualist who, Newton believed, was trying to establish his own power-base within the institution. Woodward was certainly an unpredictable and argumentative man – he later fought a famous duel with Newton's friend Dr Richard Mead – but whether he truly was a threat or merely a victim of the President's thinly veiled paranoia is a matter for conjecture. He respected Newton's intellectual achievements, and even dedicated a brilliant treatise on the classification of fossils to him, but he was also known to dislike him on a personal level. And, if Woodward felt distaste for Newton, he positively loathed the society's Secretary, Hans Sloane.

The root of this animosity is not clearly known, but it came to a head at a regular meeting of the Royal Society on 8 March 1710, which happened to be one of the rare occasions when Newton was absent.★

The discussion centred around the translation of an account from the French Académie des Sciences on the subject of gallstones. According to the society minutes, as Sloane read the paper aloud, Woodward interrupted by telling him to 'Speak sense or English and we shall understand you. If you understand anatomy you would know better.' There then followed a barbed exchange during which, according to the official account, Sloane conducted himself in a gentlemanly manner while Woodward raved. Sloane, so the official record recounts, merely 'made grimaces with a laughter',[33] Woodward and his supporters countered that the fiery doctor had actually only responded to Sloane's provocation, and that anyone with medical knowledge could not accept that gallstones were responsible for colic, as Sloane had been suggesting.

Whatever the nature of Woodward's attack, for him the writing was already on the wall. Unknown to anyone at the time, save perhaps Sloane himself, Newton had drawn up a 'hit list' – a list of council members he wanted to see defeated at the next election.[34] It could

★ During his twenty-four years as President, Newton attended 161 of 175 sessions of the council, and most of his absences date from the last two years of his life.

THE LAST SORCERER

have been no coincidence that these members – seven in all – were all supporters of Woodward (who was himself a council member).

By the end of May that year, a council meeting was arranged to hear the complaints against Woodward. Newton made it known that he expected every member to be there, and as a result it was the best-attended council meeting in the society's history, with only three of the twenty-one members absent. Woodward was read the complaints, carefully worded beforehand by Newton himself.* The proud Woodward played straight into the President's hands by refusing to apologise to the council, and by the conclusion of the meeting a vote had expelled him. The minutes read: 'Dr Woodward for creating disturbances by the said reflecting words after a former admonition upon the Statute of Ejection and for restoring the peace of the Society be removed from the Council.'[35]

Not content with ousting his opponent, Newton concluded the meeting with the sanctimonious comment 'We allow you to have *natural* philosophy, but turn you out for want of *moral*.'[36]

An incensed and proud Woodward then tried to bring a legal action against the Royal Society to force his reinstatement, but the Court of Queen's Bench denied his petition.

Others who opposed Newton met a similar fate. John Harris, a friend of Woodward, had been elected in 1709 as a second Secretary, to sit beside Sloane, but he and Sloane did not get on. Harris made the mistake of supporting Woodward over his dispute with Sloane, and his was another name on Newton's hit list. He lost both his secretaryship and his seat on the council during the elections of November 1710.

Again, Flamsteed was quick to respond, remarking to a friend that 'Dr Harris has lost all reputation by actions not fit for me to tell you.'[37] On another occasion he highlighted the President's political manœuvring with the quip that Newton was 'much talked of, but not much to his advantage. Our Society is ruined by his close politic and cunning forecast, I fear past retrieving . . .'[38]

---

* Newton was no real friend of Sloane's, but saw this as an opportunity to oust Woodward and others of whom he disapproved. Three years later, in 1713, Sloane resigned his secretaryship but remained on the council throughout Newton's tenure.

By the election of November 1710 – two months after Newton and Sloane had trampled the council under foot over the purchase of Crane Court – all seven fellows on the hit list either had resigned or had failed to be re-elected to the council, including such long-standing and highly visible members as Walter Clavell, Dr William Cockburn and Sir John Percival. All had at some point crossed Newton in one way or another, from offering mild criticism to supporting members whom the President disliked or distrusted.

As Newton was rewarding friends and foes in the hallowed halls of the great universities and the council room of the Royal Society, he was fighting two other battles which together stretched through almost two decades of his life – with John Flamsteed and with Gottfried von Leibniz.

Isaac Newton and John Flamsteed came from the same social class, and the striking similarities between their early lives have been noted by a number of writers. One biographer called them 'Two troubled creatures, one the son of a tradesman, the other the son of a yeoman, now great officers of the Crown and rivals for world fame',[39] and in the opening of his autobiography, Flamsteed echoes Newton's own perception of life when he writes:

God suffers not man to be idle, although he swim in the midst of delights; for when he had placed His own image (Adam) in a paradise so replenished (of his goodness) with varieties of all things, conducing as well to his pleasure as sustenance, that the earth produced of itself things convenient for both, – He yet (to keep him out of idleness) commands him to till, prune and dress his pleasant verdant habitation; and to add (if it might be) some lustre, grace or convenience to that place which, as well as he, derived its original from his Creator.[40]

Flamsteed was the younger by four years, puritanical and methodical, serious in manner, and irritable in old age. Most significantly, his mother had died when he was three years old. This left an emotional scar as deep as that suffered by his rival, and, mirroring Newton's own troubled youth, Flamsteed never forgave his father's later remarriage. In his 'Self-Inspections of J.F.' he commented acerbically that 'my father could so well digest [the death of his wife] as to accept a second marriage'.[41]

Another radical change to Flamsteed's life came at the age of fifteen, when he developed a rheumatic fever and was bedridden for months. It left him a cripple, but the protracted illness also opened the door to an autodidactic course similar to Newton's. For the young Flamsteed, astronomy became an obsession. Even before entering Jesus College, Cambridge, in 1670, he had submitted an essay to Oldenburg, and the then Secretary of the Royal Society was so impressed that the work was passed on to Sir Jonas Moore, Master of Royal Ordinance.

Moore took a shine to the young man and followed his university career with interest, finally persuading King Charles II, a keen patron of science, to create for him the position of Astronomical Observer, which soon became known as the office of Astronomer Royal. The King had the observatory in Greenwich Park constructed at a total cost of £520, and Flamsteed took up the post in the summer of 1675.

Unfortunately, Charles's generosity was exhausted by the expense of the building, and Flamsteed had to survive on a salary of £100 per year. Furthermore, the observatory, although beautifully designed by Wren and ideally located on a hill overlooking the village of Greenwich, was little more than an empty shell. Flamsteed was forced to take a pastorship at Burstow, some twenty miles away, and to take in students to pay for equipment and the salary of one of his two assistants.

But, despite these limitations, during his first fifteen years at the observatory Flamsteed managed to conduct more than 20,000 separate observations and he was the first astronomer to create an accurate catalogue of the heavens.

Newton and Flamsteed may have encountered one another at Cambridge, but it is unlikely: the only possible point of contact would have been Newton's ill-attended lectures. Flamsteed began his undergraduate studies soon after Newton had been appointed to the Lucasian chair, but there is no record of their introduction in any form before meeting as fellows of the Royal Society sometime during the late 1670s. We know that by 1681 they were corresponding on cordial terms, and Newton was picking the brains of the Astronomer Royal for information on the recent comet appearances.

Flamsteed was critical of the first edition of the *Principia* and defensive about his generosity in contributing material that had backed Newton's theoretical offerings. '1687. His *Principia* published,' he commented in a memoir some twenty years later. 'Little notice taken of Her Majesty's observatory (with very slight acknowledgements of what he had received from the observatory).'[42] Written at the height of their dispute, in 1705, this statement is tinged with retrospective bitterness, but Flamsteed had said nothing during the entire time he had provided Newton with information.

Whether Flamsteed's sentiments concerning the *Principia* reached its author is unknown, but by September 1694 an exchange of forty-two letters had begun that lasted until January 1696. They all centred around one subject – data concerning the movement of the Moon gathered during Flamsteed's long nights at the telescope. Newton needed this information to help prove an aspect of his theory of gravity concerning the orbit of the Moon – a piece of work he saw as the centre-piece of his planned second edition of the *Principia*.

The letters began politely, with flattering exchanges and statements of mutual respect. Flamsteed was trying to ingratiate himself with a man he saw as a rising star; in turn, Newton appealed shamelessly to Flamsteed's vanity in order to get the data he needed:

And for my part I am of opinion that for your observations to come abroad thus with a *theory which you ushered into the world* & which by their means has been made exact would be much more for their advantage & your reputation than to keep them private until you die or publish them without such a theory to recommend them. For such a theory will be a demonstration of their exactness and make you readily acknowledged the most exact observer that has hitherto appeared in the world. But if you publish them without such a theory to recommend them, they will only be thrown into the heap of the observations of former astronomers until somebody shall arise that by perfecting the theory of the moon shall discover your observations to be more exact than the rest.[43]

Flamsteed had no objection to collaborating with Newton, and at this stage he had no reason to distrust the Lucasian Professor, but he did hold a grudge against the astronomer Edmund Halley, whom he knew to be an intimate of Newton's.

Animosity between Flamsteed and Halley had sprung partly from work they had conducted simultaneously during the late 1670s. Flamsteed had published a table of tides in which Halley had found a series of errors. Halley had then published a rival table in the Royal Society's *Transactions*. But Flamsteed's ill feelings towards the younger man were not based upon professional jealousy alone. The strait-laced and puritanical Flamsteed disliked Halley's flamboyance and disapproved of his reputed libertine lifestyle. The Astronomer Royal made no secret of his feelings, and wrote to Newton in 1692:

> I have no esteem of a man who has lost his reputation both for skill, candour & ingenuity by silly tricks ingratitude & foolish prate ... I value not all or any of the shams of him and his infidel companions being very well satisfied that if Christ and his apostles were to walk again upon the earth, they should not escape free from the calumnities of their venomous tongues ...[44]

Even so, despite his concern that any material he passed on to Newton would eventually find its way to Halley, Flamsteed communicated everything that was asked of him. Self-interest and a desire for reflected glory had overridden his anxieties.

It was a trying time for Flamsteed. He was already stretched to breaking-point with his official responsibilities, hamstrung by incompetent assistants, and required to maintain two other jobs to help finance the observatory. He worked hard to meet Newton's growing demands for specific data concerning the motion of the Moon, but almost inevitably mistakes crept in.

Just as inevitable was Newton's response. He never failed to place the most exacting demands upon himself, and therefore expected near-perfection from others. Pouncing on errors and lashing out at the person he believed responsible, he ignored Flamsteed's pleas that mistakes had been made by his assistant. Testing their association almost from the moment it began, Newton constantly criticised the data while at the same time demanding more and more material.

Finally Flamsteed snapped back in irritation, but, still afraid of upsetting Newton and losing his place in history, he accused Halley of the rudeness: 'I never took anything of any for communicating of my skill or pains, except of those who forced themselves upon me to devour my time ... pray therefore lay by any prejudicial thoughts of me, which may have crept into you by malicious suggestions.'[45]

Flamsteed was certainly a difficult man. He was suspicious and self-interested, sensitive to criticism and prone to self-pity, but, as one biographer has commented, 'Whatever Flamsteed's faults however, Newton was the primary cause of the unhappy episode that followed.'[46]

At this point a dispute that wasted the energies of both men could have been averted. Flamsteed was tiresome, but the Lucasian Professor's diplomatic skills were negligible and his superior manner and complete inability to read the characters of those he could have worked with made his reactions intolerable. Thinking Flamsteed was being either lazy or slapdash, he made the dreadful mistake of offering the astronomer money for the data. Flamsteed responded to the offer with predictable disgust. 'All the return I can allow or ever expected from such persons with whom I corresponded is only to have the result of their studies imparted as freely as I afford them the effect of mine or my pains,' he moaned.[47]

This reveals an important aspect of the Flamsteed–Newton dispute and, through it, the characters of the protagonists. Initially Flamsteed held Newton in the highest regard, not only as a scientist but also on a personal level, and he was keen to remain on the right side of him for the reflected glory the association would offer. He would rather have had Newton as a friend than as an enemy, and it was only after their relationship had broken down irretrievably that Flamsteed grew to despise him. As late as 1700 he was able to write to a friend, 'I believe him to be a good man at the bottom.'[48] And, even after they had slid into open aggression and public rivalry, Flamsteed never questioned his opponent's greatness as a scientist, regarding him professionally with untainted, if grudging, esteem until the day he died.

Newton, on the other hand, saw Flamsteed as little more than a technician – a mere gatherer of data for Newton's free use. He regarded himself as the genius who translated raw observed facts into all-consuming theories capable of changing the way people thought, peeling back the layers of the past to reveal the lost secrets of the ancients. It was Flamsteed's duty, so Newton believed, to offer up his findings willingly and unquestioningly.

Still bristling from Newton's insult, Flamsteed proceeded to make things worse. Needing to prove himself, he carried out some calculations and passed on the results instead of the data. Unfortunately, he or the assistant to whom he entrusted the task got the calculations completely wrong. Feeling no sense of responsibility for the deterioration in their relationship, Newton responded coldly:

> I want not your calculations but your observations only . . . If you like this proposal, then pray send me first your observations for the years 1692 & I will get them calculated & send you a copy of the calculated places. But if you like it not, then I desire you would propose some other practicable method of supplying me with observations, or else let me know plainly that I must be content to lose all the time & pains I have hitherto taken about the Moon's theory . . .[49]

Chastised, but clearly embarrassed, Flamsteed snapped back again, and the relationship foundered once more. Each responded to the other's letters with ill-disguised bitterness and resentment. Flamsteed insisted he had done everything asked of him; Newton raged back that he himself had given the Astronomer Royal material 'of more value than many observations'.[50] In his private notes, Flamsteed called Newton 'hasty, artificial, unkind, arrogant'.[51]

Then, as suddenly as the exchange had begun, it ended. By the summer of 1695 Newton appears to have lost all interest in the Moon; he failed to reply to Flamsteed's latest letter for almost two months, until September, when he informed him, 'I have not yet got any time to think of the theory of the Moon nor shall I have leisure for it this month or above: which I thought fit to give you notice of that you may not wonder at my silence.'[52]

Unknown to all but Newton's most intimate friends, the Lucasian Professor's attention had turned away from the Moon and towards London and the Mint. But, although the issue of the lunar theory was to lie fallow for almost a decade, his frustration over what he saw as Flamsteed's incompetence increased rather than diminished with time.

When they did eventually meet again, at the Royal Society, the two men said little openly but the disharmony between them bubbled just beneath the surface. David Gregory – another member of the society whom Flamsteed disliked, once calling him 'no friend of mine'[53] – was clearly echoing Newton's own feelings when he remarked quite unfairly that 'On account of Flamsteed's irascibility the theory of the Moon will not be brought to a conclusion, nor will there be any mention of Flamsteed, nevertheless Newton will complete to within four minutes what he would have completed in two, had Flamsteed supplied his observations.'[54]

Gregory was one of the exclusive coterie who could discuss personal affairs with Newton, and he was privy to the plans for a new edition of the *Principia* – a matter which resurfaced around the time Newton was appointed President of the Royal Society, in 1703. But, for a second edition to be worth the effort, Newton still needed the lunar data as well as a detailed catalogue of the stars against which the movement of the Moon could be plotted accurately. There was only one man in England privy to this information, and Newton quickly convinced himself that Flamsteed had everything he needed at his fingertips. Knowing that too many thinly veiled insults had passed between them, it soon became clear that the only way he could gain access to the data was by machiavellian scheming.

By 1704 Newton's reach had grown long indeed. Gone was the introverted professor of mathematics whose contact with the world was limited to malicious correspondence with the likes of Hooke: Newton's influence now stretched to the royal court itself, where he was even able to bend to his will the Queen's husband, Prince George.

George was not renowned for his intellect – a century and a half after his death, Queen Victoria described him as 'the very stupid and insignificant husband of Queen Anne'[55] – but, through Halifax, Newton had learned that the prince had recently developed an interest in astronomy. What, therefore, could be more natural, Newton concluded, than for the Astronomer Royal to honour the enthusiastic prince by producing a book that contained a catalogue of the heavens and details of planetary movements? A commission from George himself could not be turned down, and it would be quite proper for the Royal Society to supervise the process. As a result, Flamsteed's life's work would be the President's to use as he saw fit.

On paper the scheme was simplicity itself, but Newton had quite underestimated Flamsteed's intelligence and his suspicious character. This is surprising, given Newton's own inability to trust others, and more than any of Newton's later actions it illustrates how obsessed he had become with the issue of the lunar data.

The process was stalled further by Flamsteed's natural reluctance to place his work within the public domain until he was ready. This at least Newton should have understood, but his desire for the observations that would further his own publication was so intense that he probably never even considered his hypocrisy in pushing his colleague into print.

However, Flamsteed could do little to escape Newton's trap. Once Prince George had agreed to commission the book, there was no turning back: the best he could hope for was to produce a text as close as possible to the one he would have delivered in due course, completed in his own time and in his own way. Forced into a corner, his only revenge was to make sure that Newton paid for every scrap of information he might gain from its publication.

The project was begun in 1705 and proved to be a disaster from the very start. Newton's first move was to set up an editorial committee composed of fellows of the Royal Society: Wren, Gregory, Francis Roberts, John Arbuthnot and himself – pointedly excluding the author, Flamsteed. 'With these persons,' the downhearted Flamsteed declared, 'Isaac Newton began to act his part, and carry on his designs.'[56]

The book, to be entitled *Historia Coelestis Britannica – A British History of the Heavens* – was conceived by Flamsteed as a complete catalogue of the celestial sphere, showing the position of every star observed from Earth. At the outset, he claimed that George had supported this idea to the value of £1,200 (over twice the cost of constructing the Royal Observatory thirty years earlier).[57] But Newton was immediately suspicious of Flamsteed's expansive ideas, and believed (probably with justification) that such ambition would jeopardise the chances of ever completing the project. Needing only the bare bones of Flamsteed's proposed tome for his own purposes, he persuaded the Prince to limit his investment to £863, thereby forcing Flamsteed to lower his expectations and to complete the task far faster.

If that was not enough to inflame further Flamsteed's already raw feelings, Newton next proceeded to commission an expensive publisher who was eventually paid £1.14s. per page, while Flamsteed, who had at some point convinced himself he would be paid £2,000 by the Crown for his labours, was told he would receive nothing at all. One can sympathise with Flamsteed's lament that 'It is very hard, it is extremely unjust that all imaginable care should be taken to secure a certain profit to a bookseller and his partners, out of my pains, and none taken to secure me the reimbursement of my large expenses in carrying my work above 30 years.'[58]

Flamsteed's hands were tied, but he complained at every turn, criticising each proof page as it came off the press, and filling his private diaries with vitriol aimed at Newton, Halley and Gregory. He cunningly supplied material pertinent to the project but not that which he knew Newton wanted most – the star catalogue.*

And so it went on for the next two years. The printing process moved at glacial pace, until the presses and the verbal exchanges came to an abrupt halt in October 1708, when Flamsteed's patron and Newton's stooge, Prince George, died suddenly, aged fifty-five.

For a time at least, Newton's schemes were poleaxed, but by the beginning of 1711 he had contrived a fresh plan. Bolstered perhaps by his recent successes in manœuvring the Royal Society into purchasing Crane Court, he began a two-pronged final effort to acquire the observational data he needed.

Working again through his connections at court, Newton persuaded Anne to patronise the publication of *Historia Coelestis* in place of her husband, while he simultaneously applied pressure on Flamsteed by convincing the Queen of the need for a committee under his charge to inspect and oversee the running of the Royal Observatory at Greenwich.

At first, Newton distanced himself from this shady operation by employing a friend and member of the Royal Society council, the Royal Physician, Dr John Arbuthnot, to act as his intermediary. Arbuthnot visited Flamsteed at Greenwich to impart the news about the new committee and tried simultaneously to encourage a sense of urgency in the astronomer by emphasising that the Queen insisted his masterpiece be published with the greatest haste.

But Flamsteed was quite unimpressed by Arbuthnot's pretences. Beside himself with fury, he saw through the plot immediately,

---

* Although he never told Newton, Flamsteed knew he could not provide the complete star catalogue at this time, as he had not fully processed his observations, so his stalling was prompted by both bitterness and practicality. Newton always believed that Flamsteed knew more than he actually did and that he had at his disposal a greater body of information than he cared to admit to.

recounting in his autobiography that 'I was afresh disturbed by another piece of Sir Isaac Newton's ingenuity.'[59] Ignoring Arbuthnot's patronising attempt to rouse his sense of duty, he told the doctor he would not be hurried and immediately protested to the Secretary of State, Henry St John, about Newton's interference at the observatory. But Flamsteed was too late: the Astronomer Royal's anger was met with the terse comment that 'the Queen would be obeyed'.[60]

Newton was probably unaware of Flamsteed's protests and the subsequent rebuff from the Secretary of State. Wishing to guarantee Flamsteed's compliance, he now instructed Arbuthnot to write him a heavy-handed letter which was meant to appeal to the astronomer's patriotism and duty: 'I am the more fully persuaded you will comply with so reasonable a request,' it read, 'because of the regard you have for the memory of the Prince, as well as for your own reputation, both which are interested somewhat in this performance.'[61]

Rather than having the desired effect, this clumsy effort served only to anger Flamsteed further, and he responded with a letter guaranteed to draw the puppet-master from behind the curtain. He invited Arbuthnot to dine at Greenwich to discuss ways in which the book could proceed 'free from such hindrance and delays as have formerly retarded its progress'.[62]

If indeed it was Flamsteed's intention to draw Newton out, it worked. Newton was so outraged by the Astronomer Royal's suggestion that it was he who had slowed the book's progress that, in a hastily drafted letter, he stopped just short of accusing Flamsteed of treason:

> The observatory was founded to the intent that a complete catalogue of the fixed stars should be composed by observations to be made at Greenwich & the duty of your place is to furnish the observations . . . You are therefore desired either to send the rest of your catalogue to Dr Arbuthnot or at least to send him the observations which are wanting to complete it, that the press may proceed. And if instead thereof you propose anything else or make any excuses or unnecessary delays it will be taken for an indirect refusal to comply with Her Majesty's order. Your speedy & direct answer & compliance is expected.[63]

And so, with the presses restarted, the sorry dispute dragged on through the summer of 1711. Flamsteed complied but dug his heels in, matching Newton's attacks blow for blow and giving him what he wanted as slowly as he could. He knew he was in the inferior position: that Newton not only had the scientific Establishment at his feet, but also had the greater influence at court. His one consolation was that he possessed something for which Newton was desperate, and while that situation lasted he could not be completely trampled under foot as others had been. However, it was an uneasy stalemate, and it lasted no more than a few months.

In May 1711 Newton, via the Royal Society, had commanded the Astronomer Royal to report his observations of the forthcoming solar eclipse of 4 July. When Flamsteed deliberately ignored the order, he was called before the council to explain why. The meeting took place at Crane Court on 26 October, and at last the years of frustration and bitterness spilled over into a public scene the likes of which had never before been witnessed at the Royal Society. Flamsteed recounted in his memoirs that:

Dr Halley met me as I entered and would have had me drink a dish of coffee with him. I refused: went straight up to the house; my man helped me upstairs [Flamsteed was suffering from severe gout] where I found Sir I. Newton, Dr Sloane, and Dr Mead. These three were all the Committee that I found there: and the two last, I well knew were the assertors of the first, in all cases, right or wrong.[64]

There followed a brief, ill-tempered discussion about the observatory, towards the end of which Flamsteed snapped that he thought it was the duty of the Royal Society to encourage his labours and not to hinder them.

Newton asked Dr Sloane what I had said: who answered that I had said something about encouragement. Whereupon I told him . . . *I was robbed of the fruits of my labours*: that I had expended above £2000 in instruments and assistance. At this, the impetuous man grew outrageous; and said, 'We are, then robbers of your labours?' I answered, I was very sorry that they owned themselves to be so. After which, all he said was in a rage: he called me many hard names; puppy was the most innocent of them. I told him only that I had all imaginable deference and respect for Her Majesty's order, for the honour of the nation etc: but that it was a dishonour to the nation, Her Majesty, and that Society (nay to the President himself), to use me so. At last, he charged me, with great violence (and repeated it), not to remove any instruments out of the observatory: for I had told him before that, if I was turned out of the observatory, I would carry away the sextant with me. I only desired to keep his temper, restrain his passion, and thanked him as often as he gave me ill names: and, looking for the door, told him God had blessed all my endeavours hitherto, and that he would protect me for the future: that the wisdom of God was beyond the wisdom of men.[65]

Following this scene, the two sides never clashed openly again, but the dispute simmered away until it eventually relapsed into acrimonious stalemate.

Queen Anne died in 1714, and Newton's committee of observatory inspectors was then retired. When Newton himself had inspected the observatory in 1713, he had commented spitefully upon the sorry condition of the instruments and recommended that the instruments be removed and repaired by ordinance officers, but Flamsteed again pointed out that he owned the instruments and would permit no such thing. After Newton refused to return some books of Flamsteed's he had kept for six years, Flamsteed initiated legal proceedings against him, and when Flamsteed allowed his Royal Society dues to lapse in 1709 Newton had his name erased from the list of fellows immediately.

The unauthorised version of *Historia Coelestis* was eventually published in 1712 under Halley's experienced guidance, and the material which had by then become available to Newton was used to demonstrate successfully the validity of his lunar theory, which appeared in the second edition of the *Principia* a year later, in 1713. But Flamsteed, who referred to the 1712 version of *Historia Coelestis* as 'corrupted and spoiled' and to Halley as a 'lazy and malicious thief',[66] gained little by it: Newton had almost every reference to the astronomer's name scrubbed from the second edition of his masterpiece.

Flamsteed was eventually to find a modicum of consolation. After

Lord Halifax died in 1715, Newton lost his most influential contact at court and, with the political changes brought about by the death of the Queen in 1714, Flamsteed's stock rose. In particular he became friends with the Lord Chamberlain, Charles Paulet, the Duke of Bolton, who told Flamsteed that he could acquire all available copies of the *Historia Coelestis*. Flamsteed could hardly contain his enthusiasm and bought the entire stock – some 300 copies in all. He then piled them high in the grounds of the Royal Observatory and 'made a sacrifice of them to heavenly truth'.[67]

Flamsteed was putting the finishing touches to his own version when he died on New Year's Eve 1719, but the work was completed by two friends, Joseph Crosthwait and Abraham Sharp, and was published in 1725 as Volume I of a trilogy.

Before his death, Flamsteed had prepared a preface to the book which contained a scathing attack on the President of the Royal Society:

> His design was . . . *to make me come under him* . . . force me to comply with his humours, and flatter him, and cry him up as Dr Gregory and Dr Halley did . . . He thought to work me to his ends by putting me to extraordinary charges . . . *Those that have begun to do ill things, never blush to do worse to secure themselves.* Sly Newton had still more to do, and was ready at coining new excuses and pretences to cover his disingenuous and malicious practices. I had none but very honest and honourable designs in my mind: I met his cunning forecasts with sincere and honest answers, and thereby frustrated not a few of his malicious designs . . . I would not court him . . . For, honest Sir Isaac Newton (to use his own words) would *have all things in his own power*, to spoil or sink them; that he might force me to second his designs and applaud him, which no honest man would do nor could do; and, God be thanked, I lay under no necessity of doing.[68]

Although ailing fast, Newton was very much alive in 1725 and presided still over a Royal Society where such honesty could not be tolerated. The preface was suppressed, and even when Francis Baily's biography of Flamsteed appeared in 1835, more than a century after the President's death, there were serious attempts to force the author to expurgate the many derogatory references to Newton.

Throughout this battle, Newton had maintained the dominant position. Flamsteed was a thorn in his side and hindered the development of his scientific work, but he never presented a threat, nor did he at any point succeed in undermining Newton's reputation. But, as this row developed from a misunderstanding to open hostility, Newton had to fight another battle simultaneously – a dispute that grew out of deeper prejudices and anxieties and became a clash that threatened his status as the world's leading thinker, challenging the authority he had spent a lifetime acquiring.

CHAPTER 13

A Question of Priority

With the dead there is no rivalry.
THOMAS BABINGTON MACAULAY[1]

The German mathematician and natural philosopher Gottfried Wilhelm von Leibniz was born in Leipzig in 1646. The son of a professor of moral philosophy, he grew up in an environment of strict Lutheran piety and maintained a traditional religious outlook throughout his life. His father died young and, like Flamsteed, Newton and many intellectuals of the period, Leibniz followed an autodidactic course before taking up his official studies. He attended the University of Leipzig to take a law degree, and was so successful that he qualified for his doctorate by the age of twenty. According to the rules of the university, however, he could not be awarded the degree, because he was too young. Embittered, he left Leipzig for the free city of Nuremberg, at whose university in 1666 he completed a dissertation entitled 'De Casibus Perplexis' – 'On Perplexing Cases' – which was so brilliant that it not only gained him his doctorate but also brought an offer of a professorship. This he declined, choosing instead to pursue his scientific interests.

Even before attending the University of Leipzig, Leibniz had become fascinated with mathematics and science. Well-read in the classic works of Galileo, Kepler and Descartes, like the young Newton he had continued his private researches throughout his orthodox degree course. His keenest interest had been in the field of logic, and as a teenager he had composed a paper called 'De Arte Combinatoria' – 'On the Art of Combination' – which is now seen by some scholars as providing an early theoretical model for the modern computer.

Leibniz is considered by many to have been Newton's equal. One Newton biographer has gone as far as to say that they were 'Two of the greatest geniuses of the European world, not only of their own time but of its whole long history.'[2] Another writer has described Leibniz as 'One of the greatest polymaths in history.'[3]

Like Newton, Leibniz was also multi-talented. He was adept at administration, and was employed for a short time as a diplomat and lawyer for the Elector of Mainz, Johann Philipp von Schönborn. He was manually dextrous, and constructed a functioning calculating-machine which earned him membership of the Royal Society when he first visited London in 1673. But above all he was a master of logic and pure mathematics.

The conflict that developed between Newton and Leibniz was no struggle between two mismatched enemies contesting over the rights to use scientific material, as in the Newton–Flamsteed dispute; nor was it based upon jealousy and petty rivalry. Leibniz was a world-class intellectual who had blossomed from an early age, and, as Christiaan Huygens's analytical powers began to fade, he rose to become the leading natural philosopher outside England – 'the Continental Newton'. It may be argued that there was insufficient room for two such geniuses living simultaneously and that a conflict was inevitable. If, as happened, it had not arisen over the calculus it would perhaps have stemmed from some other source.

After his visit to England, Leibniz returned to Paris, where he had been working for the Elector. Arriving there, however, he found himself out of a job – in his absence, von Schönborn had died. Instead of returning immediately to Mainz or Leipzig to seek fresh employment, Leibniz now decided he would give himself time to follow through his intellectual pursuits, preferring to live on the verge of poverty in France than to take immediately a menial post serving a local governor or regional dignitary.

Thus began his own *anni mirabiles*. In a two-year period between 1673 and 1675, working in almost complete isolation, Leibniz mirrored Newton's own development of a decade earlier and produced a revolutionary canon of higher mathematics, including a technique called the infinite series and, most crucially, a version of the calculus.

During his visit of 1673, Leibniz had been introduced to Oldenburg at the Royal Society, and through him he had begun to correspond with the collector and publisher John Collins, who in turn had maintained sporadic contact with Newton and still hoped one day to publish the Lucasian Professor's work. Leibniz and Collins had not met during the visit, but, as he sat in a garret in Paris reconstructing modern mathematics, Leibniz needed a line of communication with the broader scientific community. He was in regular correspondence with Huygens, but through Collins and Oldenburg he could access the rich network of European mathematicians and philosophers who corresponded with the Royal Society.

As his ideas flowed, Leibniz sent them to Collins, who, although no mathematician, could understand enough (with Oldenburg's help) to see that the young German was making significant advances. Collins encouraged him, sending him the latest ideas circulating within the Royal Society – an act Newton later saw as a betrayal. This correspondence was crucial in later public battles between the two mathematicians, and Newton used it as evidence that Leibniz had *stolen* from him. Yet, for the most part, it consisted of nothing more helpful than non-technical résumés of current mathematical debates. Although in April 1675 Collins sent Leibniz a catalogue of recent developments of the infinite series by Newton, Gregory and others, this did not contain a single demonstration and arrived some time after Leibniz had developed his own method. Because he was untrained in mathematics, Collins's letter consisted merely of gossip and occasional vague references to the work of many different mathematicians.[4]

Newton and Leibniz had not met in London, and the Lucasian Professor probably did not learn about his rival until 1675. But within twelve months of Leibniz's visit both Collins and Oldenburg were beginning to see potential conflict brewing. They were both aware of Newton's mathematical work, and both knew he had developed the method of infinite series and the calculus in the 1660s. Leibniz's notation was different, but they realised that his and Newton's techniques were fundamentally the same. They also knew from the conflict with Hooke just how touchy the Lucasian Professor could be. Remembering Newton's threat to resign from the Royal Society in 1673, Oldenburg at least was aware of Newton's volatile nature. Convinced that Leibniz was about to publish his discoveries, the two of them tried to press Newton into publishing first.

Collins had boxed himself into a corner. He could not tell Newton he had passed on information to Leibniz, even if it was of minimal use: to admit that he was communicating the professor's thoughts to another mathematician without permission would certainly have produced outrage and the breaking of all ties. Yet he also saw that this recent turn of events could be a Heaven-sent opportunity: if Newton could be persuaded to publish without knowing about the correspondence with Leibniz, the work would almost certainly go to him.

Unfortunately for Collins, and later for the scientific world, this intrigue was unfolding just when Newton was in conflict with Hooke over the 'Theory of Light and Colours'. Consequently, the Lucasian Professor would not have entertained the idea of having his mathematical work published by Collins or anyone else – requests for permission were met with either stony silence or rejection.

Some five years earlier, after failing to persuade Newton to publish what was then his newly conceived calculus, an exasperated Collins had written to James Gregory (David Gregory's uncle) declaring, 'I desist and do not trouble him [Newton] any more.'[5] And in September 1675 he repeated himself, telling Gregory, 'I have not written to or seen Newton this 11 or 12 months, not troubling him as being intent upon chemical studies and practices, and both he and Dr Barrow are beginning to think mathematical speculations grow at least nice and dry, if not somewhat barren.'[6]

Eventually it was left to Oldenburg to persuade Newton to write directly to Leibniz, on the pretext that the German had a number of mathematical queries that only he was qualified to answer fully. With reluctance, Newton finally complied.

His responses were far more than simple letters. Of the two most important, that later known as the 'Epistola Prior' – 'The First Letter' – written in June 1676, was eleven pages long, and the 'Epistola Posterior' – 'The Later Letter' – composed a few months later, in October, stretched to nineteen pages. Together they summarised Newton's mathematical discoveries and were designed to show Leibniz that he had arrived at a version of the infinite series and other breakthroughs many years earlier. But even then, nervous that others would steal his ideas, Newton pointedly left out any mention of the calculus. Instead, he added an encrypted version of the material in the form of a code – what might today be considered a form of patent. 'I cannot proceed with the explanation of the fluxions [the calculus] now,' he wrote, 'I have preferred to conceal it thus: 6accdae13eff7i3l9n404qrr4s8t12vx.' Translated, this has been interpreted as: 'given any equation involving any number of fluent [varying] quantities, to find the fluxions, and vice versa'.[7] In other words, this coded message defined the meaning of the calculus – what mathematicians call the techniques of differentiation and integration: methods used to find maxima and minima of curves, gradients, areas under curves and other quantities. (See Chapters 4 and 5.)

In a covering letter to Oldenburg, Newton then tried irritably to draw a veil over the correspondence with the comment 'I hope this will so far satisfy M. Leibniz that it will not be necessary for me to write any more . . . For having other things in my head, it proves an unwelcome interruption to me to be at this time put upon considering these things.'[8] Even then, still unaware of the depth of Leibniz's own achievement and with a typically wary glance over his shoulder, in a letter to Oldenburg written only two days after the 'Epistola Posterior' was sent Newton insisted, 'Pray let none of my mathematical papers be printed without my special licence.'[9]

Because of bad communications with the Continent, the 'Epistola Posterior' did not reach Leibniz until June 1677, eight months after it had been sent, but in the interim he had acquired an official position at the court of Johann Friedrich, Duke of Brunswick Lüneburg, in Hanover. He had also visited London again, where he met Collins for the first time.

Remarkably, during the visit the publisher allowed his guest free access to his collection of papers and correspondence. Again, when Newton discovered this years later, he accused Leibniz of theft and Collins of complicity. Today it is generally agreed that Leibniz did not find in the collection much that he did not already know, and it is significant that the papers on the calculus were practically ignored – adding weight to the theory that he had already devised his own method sometime earlier.[10]

Oldenburg died later that year, and for a time there was no further communication between Newton and Leibniz. As the Lucasian Professor concentrated on alchemy and began to work on the *Principia* following Halley's visits to Cambridge, the question of the calculus and the obscure German mathematician whom Oldenburg had pushed into the spotlight slipped to the back of his mind. But then in October 1684, two months after Halley had sparked off Newton's interest in celestial mechanics, Leibniz delivered his first paper on the calculus, published in the *Acta Eruditorum*, a learned journal produced by the University of Leipzig.

Newton's immediate response is not recorded, but the degree of duplication with his own work (what he certainly assumed straight-away to be plagiarism) must have come as an incredible shock. Working feverishly to complete the *Principia* may have cushioned the blow, but from now on there could never be civilised communication between the two men. Newton's first move to claim priority was to add a passage to the manuscript of the *Principia*, the 'Scholium' to Book II, Section II, Proposition VII, which reads:

> In letters which went between me and that most excellent geom-eter, G. W. Leibniz, ten years ago, when I signified that I was in the knowledge of a method of determining maxima and min-ima, of drawing tangents, and the like, and when I concealed it in transposed letters involving this sentence [the encryption from the 'Epistola Posterior'] . . . that most distinguished man wrote back that he had also fallen upon a method of the same kind, and communicated his method, which hardly differed from mine, except in his forms of words and symbols.[11]

Simultaneously, Newton began to feed pieces of his methods into the mathematical community, in order to encourage awareness of his claim.

Oblivious to the fury his paper was to produce in Cambridge, Leibniz had no reason to be other than magnanimous about the question of priority. In July 1684, a few months before his calculus paper was published, Leibniz received a letter from a friend, Otto Mencke, a professor of philosophy at Leipzig, who warned him that the invention of the calculus would be attributed to Professor New-ton.[12] This did not seem to trouble Leibniz, who felt sure it would be possible to announce the work as a parallel invention.

> As far as Mr Newton is concerned, I have letters from him and the late Mr Oldenburg in which they do not dispute my quadrature [calculation of the area under a curve] with me, but grant it. Nor do I believe that Mr Newton will ascribe it to himself, but only some inventions about infinite series which he in part also applies to the circle. Mr Mercator, a German, first came upon this and Mr Newton developed it further, but I arrived at it by another way. Meanwhile, I acknowledge that Mr Newton already had the principles from which he could well

have derived the quadrature, but one does not come upon all
the results at one time: one man makes one contribution, another
man another.[13]

Sadly, the concept 'one man makes one contribution, another man
another' was completely alien to Newton, which accounts for the
passion with which he defended his claim and attacked his adversary.
Superficially, the conflict appeared to stem from Leibniz's desire to
publicise his work clashing with Newton's fear of publicity; but the
roots of the problem extended far deeper.

Sixteen years earlier, in 1668, Newton had been infuriated by the
publication of the mathematician Nicholas Mercator's *Logarithmotechnia*, because it contained material that Newton had
already derived. And this had not been an isolated reaction: many
years later he referred grudgingly to Christiaan Huygens's work on
centrifugal forces with the comment 'What Mr. Huygens has published . . . about centrifugal forces I suppose he had before me.'[14]

Newton maintained an obsessive belief in his own uniqueness: he
was convinced there could be only one Christ-like interpreter of
divine knowledge in the world at any one time, and he never doubted
that he was the chosen one. The idea that others could independently
acquire the same insights and accomplish the same breakthroughs
as he had was simply unacceptable to him. So Leibniz was a thief
who had stolen the knowledge Newton had unveiled, and had then
profligately displayed this material to the world.

When Newton learned of Leibniz's work, he automatically linked
it with the material he had shown Oldenburg and Collins many years
earlier and concluded that his ideas had been transmitted secretly to
his rival. His immediate reaction was to go on the attack, to undermine his enemy.

Without ever understanding the deeper reasons for their master's
reaction to Leibniz, Newton's young supporters were as keen to
fight this battle as they were in attempting to bring the unfortunate
Flamsteed to heel. During a dispute that dragged on through four
decades, until Leibniz's death in 1716, the rows between Newton's
defenders in London and their rivals in Europe grew into a partisan
battle of ideologies. What had begun as an argument between two
mathematicians over an issue of priority led eventually to a schism
in philosophical thought and mathematical practice that lasted for

generations. Leibniz's superior notation was quickly adopted
throughout Europe but deliberately ignored by British mathematicians and scientists.* British mathematicians led the world during
Newton's lifetime, but their refusal to accept the Leibniz notation
meant that they lost this advantage during the subsequent fifty
years.

Soon after the *Principia* was published and Newton's reputation
blossomed, the issue of the calculus quickly became a *cause célèbre*.
British mathematicians such as John Wallis fuelled the controversy
by claiming Newton's clear priority. In 1693 Wallis published the
first instalment of his three-volume *Mathematical Works*, in which
he made only scant reference to the issue of the calculus. But then,
after hearing fresh rumours from the Continent concerning the dispute, he regretted not having said more and wrote to Newton asking
permission to publish both the 'Epistola posterior' and the 'Epistola
prior' in the preface to his Volume II, adding acerbically, 'I had
intimation from Holland . . . that . . . your notions of fluxions pass
there with great applause, by the name of *Leibniz's Calculus Differ-
entalis*.'[15]

Newton, of course, agreed to the use of his work, but it did little
good. Rather than persuading his European rivals of his priority, this
reference to his version of the calculus merely agitated them further.
Leibniz's friend and supporter Johann Bernoulli was even moved to
write to Leibniz suggesting that it was Newton who was the plagiarist[16] – a notion even Leibniz began to yield to gradually. 'I could easily
believe that Newton possessed some very remarkable knowledge at
that time [1676] which, in his usual way, he had greatly polished up
in the subsequent period,' Leibniz replied to Bernoulli's overenthusiastic claim.[17]

Vocal support for Newton came from Fatio de Duillier. In 1699,
while struggling to maintain a reputation as a philosopher and exist-

---

* The difference between Newton's notation and Leibniz's was ease of use for
the mathematician. Because Newton worked in isolation he did not design a
notation that would be easy to communicate but instead created a system that
he personally found efficient. Leibniz, who communicated far more with other
mathematicians and natural philosophers, wanted to express his mathematical
ideas in a universally understandable format. His notation has become part of
standard mathematical language and is used universally today.

ing on the fringes of the Establishment, he attacked Leibniz's claim publicly.

Over a dozen years earlier the German mathematician had dismissed Fatio's own version of the calculus, and de Duillier decided the time was now right to take his revenge. In a mathematical treatise bearing the grandiose title *Lineae Brevissimi Descensus Investigatio Geometrica Duplex – A Two-fold Geometrical Investigation of the Line of Briefest Descent* – he announced:

> But I now recognise, based upon the factual evidence that Newton is the first inventor of this calculus, and the earliest by many years; whether Leibniz, the second inventor, may have borrowed anything from him, I should rather leave to the judgement of those who have seen the letters of Newton and his other manuscripts. Neither the silence of the more modest Newton, nor the remitting exertions of Leibniz to claim on every occasion the invention of the calculus for himself, will deceive anyone who examines these records as I have.[18]

Some questions have been raised concerning the extent to which Newton may have instigated this attack, but this was vitriol delivered by Fatio alone. By now, Newton and de Duillier were no longer intimate and probably met only during official Royal Society business. No correspondence between them from this time survives, but the indications are that, rather than being encouraged or pleased by this passage, Newton was actually rather embarrassed by it. Although Fatio was merely voicing the private thoughts of many – not least Newton himself – at this stage of the dispute his attack was simply too forthright.

Throughout the many conflicts between scientists of the era, one common thread is the hypocritical politeness of their personal letters. Until their battles became public at least, Newton and his rivals referred to each other as 'esteemed friend' or 'honourable colleague'; littering their letters are compliments such as 'I value your friendship very highly . . .'[19] Yet, just beneath the surface, emotions seethed and rivals fought tooth and nail for both their contemporary reputations and the images they would leave behind them. Like most of his colleagues, when Newton was in direct communication with an enemy he invariably passed over the sharpened rapier, preferring instead the concealed dagger, the verbal garrotte. Fatio's attack was unsubtle. It was inserted into a published work and so did not require traditional niceties, but it went too far, too soon.

If Fatio was Newton's most flamboyant champion, the title of most vociferous and effective supporter must go to a mathematician named John Keill. As a young lecturer at Oxford under David Gregory, Keill was a devout Newtonian and became the first lecturer anywhere in Britain to teach experimental philosophy, delivering talks on hydrostatics, dynamics and optical phenomena based upon Newton's two great works. Soon after Gregory's death in 1708, Keill came to Newton's attention via a paper containing a direct refutal of Leibniz which he had offered for publication in the *Transactions*. 'All of these [laws] follow from the now highly celebrated arithmetic of fluxions which Mr Newton, without any doubt, first invented, as anyone who reads his letters published by Wallis can readily determine,' he wrote; 'yet the same arithmetic, under a different name and method of notation, was afterwards, published by Mr Leibniz in the *Acta Eruditorum*.'[20]

Although this was an overtly aggressive restatement that Newton was the first to the calculus, it avoided sliding into Fatio's crassness because Keill refrained from implying that Leibniz was a thief. Also, by this time, almost a decade after Fatio's libel, such thinly veiled attacks were seen in some quarters as fair game – Leibniz had already upped the stakes with an 'anonymous' review of Newton's *Opticks* for the January 1705 *Acta Eruditorum* in which he had had the temerity to accuse Newton of plagiarism. Leibniz had fooled no one concerning the identity of the reviewer with such statements as:

Instead of the Leibnizian differences [Leibniz's version of the calculus] then, Mr Newton employs, and has always employed *fluxions*. He has made elegant use of these both in his *Mathematical Principles of Nature* and in other publications since, just as Honoré Fabri in his *Synopsis Geometrica* substituted the progress of motions for the method of Cavalieri.[21]

What really angered Newton and his supporters about this review was what they claimed to be a comment clearly equating Newton with one Honoré Fabri, a notorious plagiarist of the time – a charge

Leibniz later dismissed as 'the malicious interpretation of a man who was looking for a quarrel'.[22]

Although it had been published in the October 1708 issue of the *Transactions*, Leibniz did not see a copy of Keill's paper until late in 1710, but when he did he was no less angry about it than Newton had been about the review of his *Opticks*. He swiftly drafted a stiff letter to the Royal Society, demanding an apology. Leibniz's grievances were then aired at a meeting of the society on 5 April 1711, when his letter was read to the gathering and Keill was asked to write an apology.

At this point it is likely that Newton began to see that Keill's enthusiasm could be employed to his own advantage. Keill, like many other young followers of Newton, was certainly keen for preferment. He had been passed over as Gregory's successor to the Savilian Professorship at Oxford, and was more than willing to be manipulated by the great Isaac Newton. The apology sent to Leibniz six weeks after the meeting therefore bears the master's stamp:

> I suggest only this, that Mr Newton was the first discoverer of the arithmetic of fluxions or differential calculus; however as he had in two letters written to Oldenburg (which the latter transmitted to Leibniz) given pretty plain indications to that man of most perceptive intelligence, whence Leibniz derived the principles of that calculus or at least could have derived them; but as that illustrious man did not need for his reasoning the form of speaking and notation which Newton used, he imposed his own.[23]

For all Keill's fine words and sugary phrases, this was of course no apology at all. Here we see Newton's vindictiveness – a clear accusation that Leibniz could not have formulated the calculus independently, but was simply fed the basics by Oldenburg (who was long dead and could not contradict the charge). It was little more than a declaration that the German had merely rearranged the material and 'imposed his own' notation.

Not surprisingly, Leibniz would have none of it. His response arrived at the Royal Society early in 1712 and was read before the meeting of 31 January. 'What Mr John Keill wrote,' he declared,

'. . . attacks my sincerity more openly than before; no fair-minded or sensible person will think it right that I, at my age and with such a full testimony of life, should state an apologetic case for it, appearing like a suitor before a court of law.'[24]

At last the gloves were off, and Newton had the perfect opportunity to bring things to a head. Stretching his privilege as President to the absolute limit, and inspired perhaps by Leibniz's reference to his 'appearing like a suitor before a court of law', Newton decided to create a committee to investigate the dispute.

The odds were stacked against Leibniz from the start. Beneath the veneer of officialdom and impartiality, Newton guided his marionettes with the long experience of the master manipulator. Described as 'a numerous committee of gentlemen of several nations',[25] Newton's inquisitors were a panel of eleven, at least half of whom were Newton devotees – in particular Edmund Halley, Dr Arbuthnot and Abraham Demoivre.[26] Newton supervised every stage of the investigation and wrote the committee's report – the *Commercium Epistolicum*, published only six weeks after the formation of the committee (and one week after the final three members had been appointed). Indeed, the investigation and the report were such a sham that the committee members did not even sign it and their identities were unknown until they were unearthed from the society's archive during the nineteenth century.[27]

After detailing the reasons for Newton's claim and the shaky premiss of Leibniz's contribution, the report concluded, 'For which reasons we reckon Mr Newton the first inventor and are of opinion that Mr Keill in asserting the same has been in no way injurious to Mr Leibniz . . .'[28]

Some years later, Newton specifically denied any involvement with the committee. In a letter written to the Abbé Varignon in 1719, he said, 'I was so far from printing the *Commercium Epistolicum* myself that I did not so much as produce the letters in my custody . . . lest I should seem to make myself a witness in my own cause.'[29] Perhaps he had deluded himself into believing that this really was the case, yet it is a rare blatant lie from Newton, the most pious of men. For evidence we need look no further than his own 'Account of the *Commercium Epistolicum*', published anonymously a few years after these events and taking up all but three pages of the *Philosophical Transactions of the Royal Society* for January and February

1715.* There is also a draft of the committee's report written in Newton's hand, thus proving his culpability.[31]

More than any other document written by Newton, his 'Account of the *Commercium Epistolicum*' reveals the depth of his determination and the lengths he would go to in order to destroy anyone who crossed his path. Leibniz was his most despised opponent because his claims cut the deepest, endangering Newton's self-image and personal esteem. And, if the **Commercium Epistolicum** itself is an unalloyed attack upon Leibniz's intellect, the 'Account' is a character assassination in which Newton piles on the accusations and brings forth evidence with every resource utilised to the point of exhaustion. He seems to have become so caught up in hatred and bitterness for Leibniz that he slides into naked hypocrisy – a state reaching its apogee when he writes of his enemy, 'But no man is a witness to his own cause.'[32] At no point did Newton seem to become aware of how far he was going, how blindly he was passing judgement upon another with statements that could be far better applied to himself.

Even after Leibniz was in his grave, Newton could not leave the poor man alone – adding comments to the drafts of the 'Account' such as 'Second inventors count for nothing.'[33] Some of these documents run to a dozen drafts and fill more than 500 folios in collections held by the university library in Cambridge, the Mint archive and a private collection kept at Shirburn Castle. The **Commercium Epistolicum** and the 'Account of the *Commercium Epistolicum*' were bludgeons with which Newton attempted to hammer the relatively powerless Leibniz into the ground. Although he never succeeded in breaking the man's will and singularly failed to rewrite history, Newton convinced himself that he had. In the final months of his life, he crowed to his doctor, Samuel Clarke, that 'He had broke Leibniz's heart with his reply to him.'[34]

He had done no such thing. Contrary to Newton's expectations, news of the **Commercium Epistolicum** was received stoically by Leibniz. The document was published in January 1713 and

---

* We know Newton was the author of this treatise because it contains numerous technical references and details of his mathematical insights that only he could have written. According to one source, 'His identity [as the author] is unmistakable.'[30]

distributed to academic centres throughout Britain and Europe. Leibniz first heard about it two months later, when Johann Bernoulli, whose nephew Nikolaus had picked up a copy in Paris, wrote to him detailing its contents. '. . . you are at once accused before a tribunal consisting, as it seems, of the participants and witnesses themselves,' he reported, '. . . then documents against you are produced, sentence is passed: you lose the case, you are condemned.'[35]

Rather than being silenced or even humiliated, Leibniz was offended – wounded by what he saw as the treachery of the Royal Society, a body that had once honoured him and received his work and ideas favourably. But gradually hurt turned to anger. 'I have not yet seen the little English book directed against me . . . those idiotic arguments which (as I gather from your letter) they have brought forward deserve to be lashed by satirical wit,' he replied to Bernoulli. Then, holding nothing back, he tore into Newton's own claim to the calculus: 'He knew fluxions, but not the calculus of fluxions which he put together at a later stage after our own was already published. Thus I have myself done him more than justice, and this is the price I pay for my kindness.'[36]

Within three months Leibniz and Bernoulli had struck back. (Bernoulli was happy to help as long as he remained incognito.) They composed a pamphlet, the *Charta Volans* (published on 29 July 1713), that followed the *Commercium Epistolicum* to the academic centres of Europe. Although it was anonymous and contained no indication of where it had been published, its source was unmistakable. Leibniz had gained something of a reputation for 'anonymous' compositions, and the artifice of referring to himself in the third person throughout the pamphlet fooled no one – least of all the object of his attack.

'Newton took to himself the honour due to another,' the author of the *Charta Volans* declared. '. . . he was too much influenced by flatterers ignorant of the earlier course of events and by a desire for renown . . . Of this Hooke too has complained, in relation to the hypothesis of the planets, and Flamsteed because of the use of his observations.'[37]

One can imagine Newton reading his copy and growing steadily more furious with each attack. But in truth Leibniz was floundering, sinking fast into fatal illness and neglect – a man equal to Newton intellectually but outclassed and outgunned in all other respects.

Newton presided at the Royal Society as 'Perpetual Dictator'[38] upheld by a large group of supporters and young disciples who hung on his every word. In stark contrast, Leibniz was neglected in his position as archivist for the Elector of Hanover (the future King George I of England). He too had once been president of an academic society – though it had been in the cultural backwater of Berlin – and he was not without supporters of his own; but even his most determined supporter, Johann Bernoulli, preferred his attacks to remain anonymous and went in for snide remarks behind Newton's back rather than head-on assaults.

Leibniz mouldered at the Hanoverian court. He was forced to earn a meagre salary by writing an interminable history of the house of Brunswick-Hanover, and was denied superior postings and academic positions despite his many talents. Even when his patron, the great-grandson of James I, acceded to the English throne in 1714, Leibniz's bid to move to England to become Royal Historian was rejected and he was forced to spend the rest of his days in what had become a deserted court. 'We dwell here in a kind of solitude since our court has gone to England,' he wrote to a friend soon after.[39]

To add insult to injury, after making regular visits to the Royal Society and engaging in lengthy talks with Newton, Leibniz's former pupil Princess Caroline, King George's daughter-in-law, turned away from her former tutor's philosophy and became a Newtonian.*

There were some efforts to reach an amicable solution. The chief mediator was a young cleric and keen philosopher, the noble-born Abbé Antonio-Schinella Conti, who had visited England as part of a delegation of European philosophers to witness the solar eclipse of 1714. He decided to stay on in London, and gradually ingratiated himself with Newton and others at the Royal Society.

Although he was a powerful man (he was a friend of both Leibniz and Princess Caroline), Conti's claims to have Newton's ear were overblown through self-interest. He succeeded in gaining a degree of Newton's trust, but his efforts at bringing together the two sides

---

* Although Leibniz and Newton clashed over their almost identical mathematical techniques and shared many scientific and mathematical convictions, Leibniz lived and died a Cartesian and never fully accepted Newtonian physics, viewing many of Newton's ideas, including action at a distance, as 'occult'.

and acquiring kudos as a peace-broker failed utterly. The dispute was never truly settled: the most that can be said for the final years of the conflict is that the public show of animosity faded and diminished as the rivals grew older and Leibniz's health began to fail.

One of the most unusual aspects of Newton's character was that his anger increased rather than diminished as the years passed. For him, time did not heal: his bitterness and resentment merely festered. He had almost no capacity for forgiveness – particularly over a matter so integral to his own persona and self-image. But, given that Leibniz's patron was now King of England, the dispute could not continue under public scrutiny and a semblance of resolution came in 1716 when King George instructed Newton to write a letter of reconciliation to Leibniz via Conti.

The King could command his servants to communicate, but he could not make them say what they did not believe to be true, and the letter sent by Newton in February 1716 was little more than a slightly milder reiteration of the arguments presented in the *Commercium Epistolicum*. 'But as Leibniz has lately attacked me with an accusation which amounts to plagiarism: if he goes on to accuse me, it lies upon him by the laws of all nations to prove his accusation on pain of being accounted guilty of calumny,' Newton declared. And, still unrelenting in his insistence that the Royal Society had acted fairly, he attacked Leibniz's claim to the contrary: 'the *Commercium Epistolicum* contains the ancient letters and papers . . . collected & published by a numerous committee of gentlemen of several nations appointed by the Royal Society'.[40]

Ill and now almost alone at the Hanoverian court, Leibniz had little energy to comment further and dubbed the letter a '*cartel de défi*' – an act of defiance.[41]

Although the personal arguments gradually drained away, the debate continued to rage on beyond the confines of letters and slanderous verbal attacks. Leibniz died in November 1716, his funeral being attended by the single servant who had remained with him at the Hanoverian court. But despite official neglect, his later supporters – no less patriotic and determined than Newton's – ensured that their master's work survived him. For generations, European scholars refused to accept that Newton had invented the calculus, and a form of settlement was achieved only during the nineteenth century. Since then it has been accepted that the two great rivals who fought a priority battle spanning half their lives had each achieved independently the mathematical breakthroughs that the other had claimed as his own.

Newton never acknowledged Leibniz's contribution and never forgave him for what he saw as a criminal attempt to intrude upon his sacred domain, his divine mission. Going the way of Hooke and Flamsteed, Leibniz's name was ritualistically deleted from the third edition of the *Principia*, which appeared in 1726.

Any assessment of the results of the major conflicts of Newton's long life would have to conclude that he had gained ground with each of them. By the end of the series of disputes with Hooke during the 1670s and '80s Newton's work had been fully vindicated, and in some respects the Curator of Experiments had been humbled by the *Principia*. Throughout the twenty-year battle with Flamsteed, the Astronomer Royal had stood his ground against the rapacious onslaught of his Royal Society superior, but he had been bludgeoned into submission over publication of his *Historia Coelestis Britannica*, only regaining some dignity by completing his own edition which was published posthumously. Newton had acquired the information he needed to verify his lunar mechanics, and used it in the second edition of the *Principia*. It was this work that lay at the heart of computer programs employed by NASA scientists guiding the first spaceships to the Moon almost 300 years later. Flamsteed's contribution to this achievement is now largely forgotten.

From our perspective, the forty-year feud between Newton and Leibniz ended in a draw. Newton is rightly seen as the first man to have devised the calculus; Leibniz has been cleared of plagiarism, and his notation is used throughout the world. But Newton never stopped believing he had been wronged: to have been first was not enough. He, more than anyone, realised that science and mathematics were universals, beyond the passing whim of kings and queens, beyond the fabrications of men and petty self-interest. His discoveries, he believed, were those of a demigod, they were received universal wisdom – a gift more precious than the divine right of kings. Newton's God knew who had first devised and given the world the calculus, but it was not enough to take comfort from that belief: Newton never stopped trying to force others to know it too. Perhaps this is why, even during his own final days, he insisted he had broken Leibniz's heart – it was the only way he could live with his neuroses, and the fact that he was not totally unique after all.

CHAPTER 14

Joining the Ancients

*I do not know what I may appear to the world; but to myself I
seem to have been only like a boy, playing on the sea shore, and
diverting myself, in now and then finding a smoother pebble or a
prettier shell than ordinary, whilst the great ocean of truth lay all
undiscovered before me.*

ISAAC NEWTON[1]

E ven as the great battles of Newton's life reached their most
acrimonious stage, he did not neglect the other aspects of his
life. The Royal Society and the Mint may have been the twin
pillars of his daily existence, but he also continued to pursue his
theological researches and to maintain his network of society acquaintances and friends. Also, as he grew older and wealthier, his family
increasingly turned to him for guidance and assistance, and he began
to enjoy the status of patriarch. The boy once deserted by his mother
and forced to live with his grandparents was now regarded with awe
by his relatives; at last he was depended upon.

Some of them were gold-diggers – distant members of the family
were attracted by his fame and wealth. Surprisingly, he tolerated the
spongers with infinitely more patience than he had shown clippers
or scientific rivals. Soon after his nephew Robert (Catherine Barton's
brother) had been killed in Canada in 1711, Newton spent £4,000
purchasing an estate for his widow and three children (only to discover later that it was actually worth less than half of this). He gave
£500 to a descendant of his mother's family, Ralph Ayscough, loaned
a Thomas Ayscough £100 which he remitted, and, according to one
relative, gave £800 to another Ayscough.[2]

In 1714, Katherine Rastall, the daughter of Newton's uncle, William Ayscough, wrote to him in desperation, declaring, 'Sir, I humbly
desire you that you will be pleased to give the bearer [of this letter]

something for me ... Sir, humbly begging the favour that you will be pleased to answer this I remain sir your humble servant.'[3] That he responded favourably to the request is borne out by a later letter in which Katherine gushes thanks for Newton's generosity.[4]

It appears from surviving correspondence that he was also generous to many outside his own family. There are several letters from people with whom Newton once had a brief acquaintance or from those who, although unable to understand a word of what he had discovered, felt drawn to the man who by this time had become a cornerstone of British culture.

A Mary (or Ann) Davies, who seems to have known Newton from Cambridge, wrote in 1723 thanking him for his assistance:

> Honoured Sir,
> I have made bold to trouble your honour with these few lines to return your honour thanks for the two guineas that your honour was pleased to send us by the gentleman that waited upon your honour with the letter honoured Sir we hope your honour will pardon our rudeness in not writing before but my mother and I have been very bad and that was the cause of our not writing to return your honour thanks before now.[5]

Others kept coming back for small sums. A William Newton, who claimed his father's name was Isaac, wrote during 1716 to thank him for his help, and then eight months later sent another letter – this time from prison – asking for an extra pound on top of the £3.4s.6d. he had already borrowed from Newton. Two years later he wrote another begging letter, this time while in gainful employment in Whitby, Yorkshire, but still in debt.

For a man who headed the Royal Mint and administered the Royal Society, Newton often displayed surprising naïvety in his domestic affairs. He lost a reputed £20,000 in the financial disaster of the South Sea Bubble when his shares in the South Sea Company fell out of view.* And when William Whiston's nephew lived in Newton's

---

* The South Sea Company was established in 1711 as a government attempt to solve the problem of a growing national debt created by a succession of wars in which England had become involved. The shares were sold illegally to politicians and courtiers at an unrealistically low price, but soon their value

house, some time between 1710 and 1715, 100 guineas went missing from Newton's desk drawer. Although suspicion clearly fell on the young man, and Newton later mentioned privately that he believed the youth had taken the key to his desk from a pair of breeches, he decided against taking any legal action and kept the incident quiet, presumably through fear of ridicule if the story got out.[6]

A more welcome guest at the Newton home from the summer of 1717 was John Conduitt, who, after marrying Catherine Barton, went on to play a significant role in providing posterity with details of Newton's life.

The son of a wealthy farmer, Conduitt had served as commissary to the British forces in Gibraltar soon after the War of the Spanish Succession. While in the army he had discovered the site of the lost Roman city of Carteia and had sent news of the find to the Royal Society in London. Returning to England in 1717, he was invited to deliver a talk before a meeting of the society, where he first met the President. At this time Newton was still deeply involved in refining his ancient chronology, and invited Conduitt to his home. There he met Catherine Barton, who had recently completed her period of mourning for Charles Montagu.

Theirs must have been a whirlwind romance: Conduitt arrived in England in June, and he and Catherine were married on 26 August. It was also unusual in that Conduitt was only twenty-nine, whereas Catherine – still beautiful by all accounts – was almost a decade his senior.

But if John Conduitt loved his new bride, he also adored her uncle. His own scholastic life had begun and ended with his discovery of Carteia, but almost from the moment he had made Newton's acquaintance he decided to keep a record of their conversations and to collect every scrap of available information about him. From an early stage in their relationship Conduitt also decided to write a

soared – only to crash in spectacular fashion a short time later. Newton was not involved in this illegal activity and bought his shares honestly, but many Members of Parliament, both Whig and Tory, were implicated, along with several public figures and senior civil servants. Interestingly, just about the only senior politician not involved in this very public corruption was Robert Walpole, who was appointed the First Lord of the Treasury almost immediately after the scandal and became the first Prime Minister in 1721.

biography of his illustrious relative, but this ended up as little more than an outline (kept now in the library of King's College, Cambridge). He also wrote a short memoir of Newton which he sent to the French scientist Bernard le Bovier de Fontenelle as an official eulogy for presentation before the French Académie des Sciences in 1728.

Tradition has it that the couple remained at Newton's house for a short time after their wedding, but they later removed to Conduitt's estate of Cranbury Park, near Winchester. Catherine was given a large but unspecified sum by her uncle when she married, but the fortune Halifax had promised her did not survive the legal battles precipitated by the former Chancellor's irate and powerful relatives. Newton had spent the better part of 1715 and 1716 defending his niece's claims, but had eventually lost. The country estate and most of the £5,000 inheritance bequeathed to Catherine remained in the Montagu family.

John Conduitt was a wealthy man in his own right. He became an MP for Whitchurch, and took on many of Newton's responsibilities at the Mint towards the end of the scientist's life, eventually succeeding him to the position of Master in 1727. Catherine led a very comfortable life divided between the family estate in Hampshire and a town house they purchased in London. Two years after the marriage they had a daughter they named Kitty, who in 1740 married Viscount Lymington, the eldest son of the first Earl of Portsmouth. (It was through this route that Newton's papers eventually became known as the Portsmouth Collection, much of which was purchased in 1936 by Maynard Keynes for Cambridge University.) Through these connections, within three generations of Isaac Newton's own relatively humble birth into the Woolsthorpe yeomanry, the family had beaten a steeply ascending path to nobility.

Although Catherine's role in Newton's life diminished after her marriage, little else changed immediately at Newton's house in Leicester Fields. Newton became more and more preoccupied with his analysis of the chronology of ancient kingdoms, constantly reworking his scheme and adding details to the overall picture he had begun a half-century earlier. This remained his private domain. It is clear that Newton had no intention of publishing his thoughts on the subject during his own lifetime, and Conduitt played it down even after Newton's death, calling the obsession a 'divertissement'.[7]

Of the huge body of Newton material to be found in various collections around the world, much of that dealing with chronology originated from the final decade of his life: a large portion of these papers have never been deciphered or analysed in any way. Lord Keynes preserved much of this material for the library of King's College, Cambridge, and the Ekins family (who had come by them via the Reverend Jeffrey Ekins during the nineteenth century) donated four volumes containing over 1,000 folios to New College, Oxford; these are now kept in the Bodleian Library.

Some scholars see Newton's continued application to such intense study of ancient chronology and the history of the Church as being a 'daily meditation' – a fulfilment of a deep spiritual need.[8] Indeed, apart from his first decade at the Mint, Newton never stopped researching comparative theology and ancient history, and if anything he became more fixated on them as he grew older. His papers record seemingly endless refinements to his manuscripts, recalculations of dating systems and redefinitions of his notions of prophecy and biblical meaning. This activity was an expression of his inner drives, a manifestation of his desire to analyse, to draw together disparate themes as he had done with the work that had led him to the *Principia* and the *Opticks*. It also grew out of a desire to find incontrovertible evidence to support his religious instincts. During his years of scientific inactivity the spiritual importance of the chronology and the unravelling of ancient mysteries became increasingly important to him, keeping his mind alert and acting as a counterweight to the relative mundanity of the Royal Society and the Mint years after the excitement of intellectual warfare had passed.

Although Newton did not like to make known his deeper religious concerns and his interest in unprovable ideas – whether they sprang from scientific, alchemical or theological roots – many of his interests were widely realised, even if their origins were known only by a very select coterie of initiates, including, at different times, Fatio, Whiston and Clarke.

Newton's fascination with numerology, time-scales and even biblical prophecy was not so unusual in the early eighteenth century as to draw criticism or even serious adverse comment from the academic community, and few would have suspected his deeper motives. Indeed, Conduitt wrote after Newton's death that, towards the end of his life, his uncle-in-law had become quite friendly with the future

King George II and his wife and had spent many hours in private conversations with them. George's wife, Caroline, was particularly interested in Newton's chronology and apparently treasured a hand-written account she had asked him to compose for her.[9] For Newton, who was accustomed to secrecy and almost instinctively self-protective, this lowering of his guard must have been difficult – something he would not have contemplated for anyone less important.

Most significantly, after the *Principia* was completed, Newton's Arian faith gained a new meaning and greater poignancy as his deeply felt religious convictions began to merge with his scientific ideas.

Newton's scientific achievements were without doubt both revolutionary and practical, but there were still unanswered questions associated with his discoveries, and as the years passed these failures in his attempt to create an all-embracing model of the universe continued to trouble him. Principal was his inability to achieve a unification of microcosmic and macrocosmic phenomena in a grand theory that would explain all the forces of Nature. Although with each new edition of the *Opticks* he continued to pick at this problem and wrote Query after Query in an attempt to resolve the matter, he must have long since realised that a unified theory was for him an impossible dream. But a problem that became increasingly important during the final decade of his life – and, he believed, an attainable goal – was the description of a mechanism for universal gravitation: an explanation of *how* gravity worked.

By the time he composed the *Principia*, Newton had concluded that gravity operated by the inverse square law, that it was a force that seemed to act at a distance by some unknown mechanism, and that this force was not limited to planetary motion or the behaviour of comets, but that universal gravitation resulted in all matter being attracted to other matter. If an ether existed to facilitate gravitation, then it was almost a vacuum by nature and incorporeal in form. And it was this last point that most troubled him. What was the nature of this incorporeal ether, and how could it facilitate gravitation?

Newton's ideas on the subject evolved steadily from the period soon after the publication of the *Principia* until the day he died, becoming increasingly esoteric and resonating more intimately with his Arian beliefs.

By the 1690s he had concluded that the Greek philosopher and mathematician Pythagoras had been acquainted with the inverse square relationship and had described how it governed planetary motion. He reached this conclusion from his reading of Pythagorean concepts of harmony and number. The Greek philosopher had supposed that the universe operated via strict numerical relationships. For him, number was all.

Newton believed that the model Pythagoras had developed for musical scale and the harmony of certain notes was in fact a metaphor or model for the universe, and that Pythagoras had arrived at the inverse square law for planetary motion and encoded it in this model of musical structure. By comparing 'the lengths of the strings with the distances of the planets,' Newton claimed, 'he [Pythagoras] understood by means of the harmony of the heavens that the weights of the planets towards the Sun were reciprocal as the squares of their distances from the Sun.'[10]*

This may have confirmed Newton's conviction that the ancients held knowledge that he had spent his life rediscovering, but it did little to answer his main concern of how gravity actually operated. For Pythagoras and all those who followed in his footsteps, planetary motion was simply governed by divine action. According to the Roman historian Justin, Pythagoras believed that:

> God is one. And he is not, as some think, outside the world, but in it, for he is entirely in the whole circle looking over all generations. He is the blending agent of all ages, the executor of his own powers and deeds; the first of all things; the light in heaven; the Father of all; the mind and animating force of the universe; the motivating factor of all the heavenly bodies.[11]

But Newton could not settle for this answer. In the matter of gravity, where piety and curiosity met, he was compelled to find a solution that both satisfied his intellect and also fitted with his religious beliefs.

---

* Newton was by no means the first to expand upon this relationship and to study this theory linking celestial mechanics with numerical hierarchies and Nature's number patterns as exemplified by the musical scale. The Neoplatonists espoused Pythagorean concepts, and Johannes Kepler had been greatly influenced by the teachings of Pythagoras.

By the early years of the eighteenth century he was still grappling for an answer and expressing his frustration in drafts of Queries planned as additions to the *Opticks* (but invariably cut before publication). In an early version of Query 23 for the *Opticks* of 1706, he asked:

> By what means do bodies act on one another at a distance? The ancient philosophers who held atoms and vacuum attributed gravity to atoms without telling us the means unless in figures: as by calling God harmony representing him and matter by the god Pan and his pipes ... Whence it seems to have been an ancient opinion that matter depends upon a deity for its laws of motion as well as for its existence.[12]

It was almost certainly sometime after this that Newton began to formulate a link between aspects of Arianism and the means by which gravity operates. During the early 1700s he had begun a fragmentary passage in which he explored the nature of Christ's body and form before and after his earthly incarnation:

> he had after his resurrection such a body as he had before his incarnation. And therefore as his [natural] mortal body by the resurrection became an immortal body, so his immortal body by the incarnation became a mortal one. And it is easy to believe the one as the other.[13]

According to Arian doctrine, Christ stood somewhere between God and man in the universal hierarchy. Jesus was immortal and 'the first created', but theological notions were hazy when it came to Christ's form. Was he a spiritual being who could take on the mantle of physical existence, or was he material? As Newton entered the final years of his life, he began to accept the idea that Christ possessed a 'spiritual body'. Writing sometime during the late 1710s to early 1720s, he declared:

> And he who by his resurrection has changed his mortal flesh into immortal spiritual body might by his incarnation change his immortal spiritual body into a body of flesh. For whereas the Father is the invisible God whom no eye hath seen nor can see

and therefore is totally incorporeal, the Son before his incarnation and the Holy Spirit have appeared in visible shapes upon several occasions and therefore have spiritual bodies.[14]

Elsewhere he stated repeatedly that 'God does nothing by himself which he can do by another.'[15] So, he had concluded, God does not himself control directly the gravitational forces that keep the planets in motion, nor does he provide directly the medium via which universal gravitation operates. Instead, the incorporeal ether which facilitates the phenomenon of gravitation (and perhaps other forces) is actually the body or spiritual form of Jesus Christ.

Of course Newton had no means of proving this hypothesis – it was principally a faith-based concept: a notion derived from his Arian convictions – but it described neatly the way in which God could preside over his creation without dirtying his hands by direct contact with the physical world. Christ was a mediator for all action in the universe, the intermediary via which the system of the universe was maintained, God's 'commander-in-chief', his viceroy.

To clarify his thoughts on the subject, sometime around 1720 Newton wrote what he perceived as a personal credo, a form of amalgamation of science and religion – a guide, perhaps, for future explorers. This included a clear picture of the role he saw for Christ in the universal scheme of things – not least the function of the spiritual body of Jesus as the medium by which celestial mechanics was maintained. 'Jesus was beloved of God before the foundation of the world,' he wrote, 'and had glory with the father before the world began and was the principle of the creation . . . *the agent by whom God created all things in this world.*'[16]

To summarise, the spiritual body of Jesus, the first created, was the facilitator for the creation of the physical universe, provided the means via which the cosmos continued to function mechanically, and acted as a medium via which forces acted at a distance without any visible, tangible, measurable mechanism.

This was a concept that Newton refined and distilled during his final years. Although no means of proof could be elucidated, and even he had doubts at times about the details of this system, it was the fullest explanation of gravitation he could arrive at. If it did nothing else, it acted as a prop – it was a comforting model that could be neither proved or disproved but could serve to fill one of the gaps in a model of the universe that had been so successful in every practical and empirical sense.

During his final years Newton moulded himself into a grand patriarch of the Royal Society, a guru of science; on the Continent he was known as 'Le Grand Newton' (in spite of the ongoing arguments over the calculus). But, as the battles with Leibniz and Flamsteed came to an end, his razor-sharp analytical powers were growing blunt. He only very rarely tackled mathematical problems and most of his creative energies were expended on his religious interests, but he remained hale into his eighties and attended almost every meeting of the Royal Society up to the final weeks of his life.

There were further skirmishes over questions of philosophy with the likes of Leibniz's great supporter Johann Bernoulli, but these were never on the same scale as his epic fights with Hooke, Flamsteed and Leibniz. However, Newton had lost none of his ferocity, and even friends and supporters felt a flick of the lion's tail from time to time. When Stukeley applied for the position of Secretary to the Royal Society without first consulting the President, 'Sir Isaac showed a coolness toward me for 2 or 3 years,' he later reported.[17]

Even his old friend Halley treated Newton with exaggerated reverence. In 1725, a few years after the astronomer had been appointed as Flamsteed's successor at Greenwich, Newton asked him for some calculations based upon the motions of the comet of 1680–81 for his proposed third edition of the *Principia*. The sixty-eight-year-old Astronomer Royal complied willingly, but soon after sending the results he was horrified to discover an error in his working. He immediately confessed to Newton in a state of near panic:

> I was astonished to find myself capable of an intolerable blunder, for which I hope it will be easier for you to pardon me, than for me to pardon myself, who hereby runs the risk of disobliging the person in the universe I most esteem. I entreat therefore that you would not think of any other hand for this computation, and that you please allow me the rest of the week to do it.[18]

Although Newton did not go to 'any other hand', he decided against including Halley's amended calculation in the new edition.

The work of re-editing the *Principia* and the *Opticks* continued seamlessly during the first quarter of the eighteenth century, and both books have remained in print ever since. But, after the first edition of each, the bulk of the work was placed in the hands of trusted young disciples anxious to earn a footnote in the history of science. The Master of Trinity College, Richard Bentley, became the publisher of the second edition of the *Principia*, which appeared in 1713. Earlier, to edit the new edition, Newton had passed over major figures including David Gregory, who had been desperate for the honour, and had instead chosen the relatively unknown Roger Cotes. Similarly, the third edition, published in 1726, was placed in the hands of the young mathematician Henry Pemberton.

Pemberton had returned to England in 1722 after a protracted stay in Europe. He had been warned by Newton's enemies there that the President of the Royal Society had lost his wits and was unbearable to work with. Pemberton, however, found the eighty-year-old Newton quite capable of working through the mathematics he had devised over half a century earlier: 'Though his memory was much decayed, I found he perfectly understood his own writings, contrary to what I had frequently heard in discourse from many persons.' He also reported that 'Neither his extreme great age, nor his universal reputation had rendered him stiff in opinion, or in any degree elated.'[19]

Conduitt describes Newton as displaying 'An innate modesty and simplicity' in old age and tells us that 'he was blessed with a very happy and vigorous constitution'.[20] But even intellectual giants have bodies of flesh and blood, and as Newton passed into his ninth decade he began to suffer a series of debilitating illnesses. These became serious sometime towards the end of 1722, and during the first three months of 1723 he was so ill that he attended only two meetings of the Royal Society.

Employing the help of famous society doctors, Richard Mead and William Cheselden, Newton was quickly diagnosed as suffering a slackness of the sphincter which made him incontinent especially after exercise. This was aggravated by kidney stones, eventually forcing him into a Bath chair. But, according to Conduitt, his stubbornness meant that he constantly contradicted the advice of his doctors and insisted upon walking whenever he could – declaring to his nephew-in-law, 'Use legs and have legs.'[21]

By 1724 Newton's usefulness at the Mint had passed, and most of the work had fallen to Conduitt. With illness, the President's attendance at the Royal Society had also become erratic, and so his deputy, Martin Folkes, took over most of his day-to-day responsibilities. Newton loathed losing his grip on power, and after he had been forced to move to the more rarefied atmosphere of the village of Kensington, to the west of London, he insisted upon journeying into town under any pretext he could devise. He would often arrive at the Royal Society or the Mint unannounced, then be either helped to walk slowly around his old domain or wheeled into the council meeting-room or the press room at the Tower in a Bath chair.

It was during one of these impromptu trips into town early in 1727, a few weeks before his death, that, with Conduitt's assistance, Newton burned a collection of papers at his house in Leicester Fields. Conduitt reported the incident in his collected papers, describing burning 'boxfuls of informations'.[22] Orthodox historians claim that the boxes of papers were merely duplicates of Mint administration papers and copies of minutes, bills and receipts, but one obvious question remains: Was there any greater significance to this burning?

When surveying the Newton papers, it is not uncommon to encounter twenty or thirty drafts of a single document, which raises doubts about the criteria upon which Newton might have deemed material unnecessary and meriting the fire. The burned papers could have been personal and family correspondence, and indeed there is precious little surviving correspondence between Newton and any of his family from any period of his life. There is only one surviving letter from Hannah Newton to her son, but others may have been lost quite innocently during the course of three and a half centuries. The dearth of letters between Newton and Catherine Barton is a little more surprising, but considering she lived under his roof for two decades this is not really so strange. Furthermore, it seems odd that Newton should have destroyed family correspondence but saved letters to and from Flamsteed that showed him in a far from favourable light.

We will probably never know the contents of the burned papers, and it is possible that the fire may have acquired an unnecessary glaze of mystery. But there was another witness there that day, an acquaintance of Newton's named Samuel Crell – a Socinian who had fled Europe and to whom Newton had recently given financial assistance.* Although Conduitt is vague about the contents of the boxes, a few months after the incident, in July 1727, Crell wrote to a friend, Mathurin Veyssière de Lacroze, and recalled specifically that the material contained manuscripts.[23] However, the material passed down to us from Newton's alchemical pursuits, his chronology and his prophecy studies is revealing in itself. It would seem odd that Newton should destroy certain manuscripts while allowing other, often iconoclastic, material to survive. If he was going to destroy anything, it would surely have included some of the more extreme of his alchemical ideas.

But there remains one question associated with the apotheosis of Newton's work, and the burning incident may have some bearing on the conclusion we reach about this. Did Newton venture along paths leading far from his study of alchemy – paths we would now consider those of pure magic, pure heresy? Did he dabble in what might be called the 'black arts'?

Although short-lived, the nervous breakdown that Newton suffered in 1693 was, as we have seen, severe. It debilitated him for several weeks, and within three years he had given up both scientific study and alchemical experiments. Traditional scholarship is divided over the cause of this breakdown, with explanations ranging from chemical poisoning to the emotional trauma of failing to elucidate a unified theory. The breakdown might also have been precipitated by Newton's loss of emotional equilibrium after the collapse of his relationship with Fatio de Duillier.

Fatio de Duillier and Newton were certainly experimenting with alchemy together in Cambridge. Their correspondence during Fatio's illness of 1692 indicates that they had also accumulated some material or substantive set of results they were anxious to protect, and there are many other references to shared experiments as well as alchemical investigations and discussions concerning biblical prophecy. For example, in a letter written in May 1693, a few weeks

---

* Socinianism, created by the sixteenth-century Italian humanist Faustus Socinus, was a heretical Christian doctrine similar to Arianism. It differed in that believers claimed that Christ was not divine but merely a human instrument of God. Arians believed Christ to be the 'first creature', neither of the same substance as God the Father, as orthodox teaching stated, nor human as Socinians believed.

before their final separation, Fatio describes a process to produce what he calls 'mineral trees' – probably an unusual amalgam of several metals and inorganic compounds which under the correct conditions takes on the shape of a twisted branch or tree:

These matters being put in a sealed egg in a sand heat do presently swell, and puff up, and grow black and in a matter of seven days go through the colours of the philosophers. After which time there grows a heap of trees out of the matter . . . there is plainly a life and a ferment in that composition.[24]

Many years after the dissolution of the Newton–Fatio relationship, Fatio de Duillier became involved in the extreme occult group the French Prophets and ended his days in poverty and semi-madness. Would it be unreasonable to suggest that, towards the end of their relationship, he had tried to encourage Newton to use his knowledge to explore the black arts?

For all his faults, Fatio was an adventurous spirit and took readily to a number of extreme ideologies during the course of his long life. Some might say with justification that he was gullible; but he was also a highly intelligent man with a powerful intellect and possessed of considerable charisma. He, like Newton, would have believed the acquisition of knowledge to be his spiritual duty. Fatio may have been more willing to cast off the shackles of orthodoxy, but, when the situation suited him, Newton was also quite able to break the bonds of tradition and religious convention.

First, as a young man driven by a deep-rooted psychological need, Newton had dispensed with Trinitarianism. He, like any good scientist, could also discard the dogma of received wisdom about the nature of the world, and this drove him to his great discoveries. Later, when it suited his ambitions, he was able to ignore the impropriety of a close friend and his favourite relative becoming lovers. When he and his assistants at the Mint needed to go under cover to catch clippers and counterfeiters, he could relinquish his distaste for keeping company with drinkers and gamblers. And he was capable of lying, as he had done on at least one important occasion during the dispute with Leibniz.

Yet there was a dangerous disharmony in Newton's psychological make-up. He possessed a powerful impulse to discover (or, as he would have seen it, to rediscover) and to travel along any path that led him to greater knowledge. But, although encouraged by Fatio, who shared the same ideals, taken to an extreme this would have clashed with his Puritanism. It was one thing to consider the Pope the Devil incarnate and the true Church to be Arian, but quite another to step beyond the limits of Christianity and natural philosophy altogether in order to experiment with heathen practices and black magic.

Newton did not believe in evil spirits or demons, but he may have realised the power of humankind's own dark recesses. He may have even understood the potential of ritual – not because it could conjure up devils or demons, but because it could focus energies in a way not dissimilar to the ritualistic element of alchemy. The concept of ritualistic concentration of psychic energy was certainly known if not entirely understood by his alchemist predecessors. It is quite possible that Fatio tried to persuade Newton to experiment simply to learn what would happen, to explore another avenue of rediscovery at a time when Newton was desperate to elucidate a unified theory. If this was the case, the tension created within Newton's mind would have been even greater than the pressure created by the suppression of any homosexual feelings and by his known fear of exposure as a heretic. Could this tension have pushed him into temporary insanity?

Which leads us back to the papers burned at the beginning of 1727. They could have been nothing more than ream upon ream of repetitious dry accounts and office trivia, but they could also have documented some of the more extreme experiments that Newton and Fatio had devised.

The orthodox explanation for the burning begs two difficult questions: Why burn the Mint documents at this particular time? and Why supervise the burning personally?

A consideration of the alternatives offers some answers. First, the timing is highly significant. Clearly Newton would have been sensitive about this material and, realising he was approaching death, he would have wanted to protect his self-image for posterity. He would have insisted upon supervising the burning personally, and who other than Conduitt and Crell could he have trusted with the task? Anyone with a scientific training might have become curious if he had caught a glimpse of a stray note before it was ignited, but, unlike devotees such as Halley or Stukeley, Conduitt knew no science. And Crell was as much a heretic as Newton, but was not, as far as can be ascertained, a practising alchemist. He had much to hide, and his assistance would have posed little threat.

Unlike the central theme of this biography – that Newton arrived at his theory of gravity partly through his exploration of alchemy and early biblical theology – the notion that he crossed the line into black magic is not supported by any hard evidence, but the circumstantial evidence available offers an intriguing possibility.

John Conduitt spent a good deal of time with his uncle-in-law during the final years of his life, and his observations and notes provide useful insights into Newton's last days. As well as the story of the burning, they also offer the best surviving description of Newton's steady physical decline, recording that:

He ate little flesh, and lived chiefly upon broth, vegetables, and fruit, of which he always ate very heartily. In August 1724 he voided, without any pain, a stone about the bigness of a pea, which came away in two pieces; one at some distance from the other. In January 1724/5, he had a violent cough and inflammation of the lungs, upon which he was, with much ado, persuaded to take a house at Kensington, where he had in his eighty-fourth year a fit of the gout, for the second time, having had a slight attack of it a few years before: after which he was visibly better than he had been some years. The benefit he found from the air at Kensington induced him to keep the house until he died.[25]

Despite the growing list of afflictions, Newton, even in his final months, still harboured ideas of returning to scientific work. During a conversation with Halley in which he raised again the dispute with Flamsteed, he commented that he was once more considering 'another shake at the Moon'.[26] On a different occasion he vouched to his nephew Benjamin Smith, who lived with him in London from time to time, that he might yet have 'another touch at metals'.[27] But these were little more than the dreams of an old man.

He also fantasised about moving back to Woolsthorpe: when Stukeley told him in April 1726 that he was planning to move there, Newton grew interested and even asked his friend to investigate the availability of the house to the east of the church where Mrs Vincent (his teenage friend Catherine Storer) had once lived.

According to some accounts, Newton also grew sentimental during these twilight years. A young relative remembered him as being 'remarkably fond of the company of children'.[28] And Conduitt relates that 'A melancholy story would often draw tears from him, and he was exceedingly shocked by any act of cruelty to man or beast; mercy to both being the topic he loved to dwell upon.'[29]

Yet, even as late as August 1724, he was still sending counterfeiters to the gallows without any discernible sign of mercy. When an official named Lord Townshend asked whether Newton wanted a counterfeiter named Edmund Metcalf hanged as scheduled, he wrote back:

I know nothing of Edmund Metcalf convicted at Derby assizes of counterfeiting the coin; but since he is very evidently convicted, I am humbly of opinion that it's better to let him suffer, than to venture his going on to counterfeit the coin & teach others to do so until he can be convicted again, for these people very seldom leave off. And it's difficult to detect them.[30]

Such a dichotomy – sentimentality blended with ruthlessness – is not uncommon in powerful and driven figures.

The end came in the spring of 1727. Newton had visited London to attend the 2 March meeting at Crane Court. Conduitt met him there and commented that he thought Newton looked well. Smiling, the President told him he had 'slept the Sunday before, from eleven at night to eight in the morning without waking'.[31] But the strain of travelling to London had proved too much. Upon returning to Kensington he became ill and was forced to take to his bed. Dr Mead was called, and John Conduitt rushed to Newton's bedside.

During the following two weeks, Newton alternated between a semi-comatose state and relative lucidity, during which he could converse with his colleagues and members of his family who had come to pay their last respects. 'Though the drops of sweat ran down his face,' his nephew-in-law recounts, 'he never complained, or cried out, or showed the least signs of peevishness or impatience, and during the short intervals from that violent torture, would smile, and talk with his usual cheerfulness.'[32]

During one of these lucid periods Newton told his nephew-in-law that he had no intention of accepting the final rites – a fact that so disturbed Conduitt that he felt obliged to fabricate an apologia for his hero. 'It may be said his whole life was a preparation for another state,' he later wrote.[33] In other words, unlike most men, Newton, he believed, had nothing to confess – he had been at peace with his maker all his life. Although he may not have realised it, by making such a pronouncement Conduitt was setting the tone for the deliberate warping of history that followed Newton's passing. This was the first premeditated attempt to obfuscate Newton's inner conflict taken by an intimate of the great man. From this point on Newton's followers began to generate what soon became a self-perpetuating Newton myth – that of the puritanical godlike genius who had achieved the greatest scientific breakthrough in history through the application of pure intellect alone.

Newton died on 20 March 1727, and three days later the simple message 'The Chair being vacant by the death of Sir Isaac Newton there was no meeting this day' was written in the Journal Book of the Royal Society.[34]

John Conduitt acted as executor of the will. Newton's liquid assets were some £32,000, which was divided equally between his eight nephews and nieces. The manor at Woolsthorpe was to pass to his 'closest relative'. Thomas Mason, the vicar of the nearby village of Colsterworth, traced this relative as being a descendant from Isaac Newton senior's side of the family – a man named John Newton. Mason wrote to Conduitt with the details, describing the man as 'God knows a poor representative of so great a man, but this is a case that often happens.'[35] And so it proved to be this time. John Newton was a gambler and a drinker who squandered his inheritance and met an undignified end. Returning home drunk one night, he stumbled and the pipe he was smoking lodged in his throat, choking him.

Isaac Newton was buried in Westminster Abbey on 4 April 1727, his pall borne by a group of distinguished figures including the Lord High Chancellor, the Dukes of Montrose and Roxborough, and the Earls of Pembroke, Sussex and Macclesfield. Four years later, in 1731, Newton's heirs had erected the baroque monument which stands today on one side of the abbey. The monument dominates an area which has since become known as Scientists' Corner – a final resting-place that Newton shares with Charles Darwin, James Clerk

Maxwell, Michael Faraday and other illustrious British scientists.

Designed by William Kent, the monument illustrates the many facets of Newton's life. Cherubs play with a prism, a telescope and newly minted coins; a celestial globe shows the path of the 1681 comet. Dominating the monument is a relief of Newton himself, with still more cherubs in attendance. Regally, he leans on a pile of four books, labelled 'Divinity', 'Chronology', 'Optica' and 'Phil Princ. Math'. A volume entitled *Alchemy* is conspicuous by its absence.

References

Note: In the main text, the spelling and capitalisation of extracts from Newton's writings and from works by his contemporaries have been modernised for the convenience of the reader.

Abbreviations

Correspondence *The Correspondence of Isaac Newton* (Cambridge: Cambridge University Press): Vols. 1–3 ed. H. Turnbull (1959, 1960, 1961); Vol. 4 ed. J. F. Scott (1967); Vols. 5–7 ed. Rupert Hall and Laura Tilling (1975, 1976, 1977)

JNUL   Jewish National and University Library, Jerusalem

KCL   King's College Library, Cambridge

ULC   University Library, Cambridge

Introduction: The Truth Revealed

1. Maynard Keynes, 'Newton the Man', in Royal Society, *Newton Tercentenary Celebrations* (Cambridge: Cambridge University Press, 1947), pp. 27–34.
2. Michael Hart, *The 100* (London: Simon & Schuster, 1993).
3. Stukeley's book was not published in its entirety until 1936 (ed. A. Hastings White (London: Taylor & Francis)). The original manuscript is preserved in the Royal Society.
4. David Brewster, *Memoirs of The Life, Writings, and Discoveries of Sir Isaac Newton*, Vol. 2 (Edinburgh, 1855), pp. 371–6.
5. Keynes, 'Newton the Man', pp. 27–34.
6. Christopher Wren, *Parentalia* (London, 1700), p. 201.
7. See in particular David Castillejo, *The Expanding Force in Newton's Cosmos: 'As Shown in his Unpublished Papers'* (Madrid: Ediciones de Arte y Bibliofilia, 1981) and T. G. Cowling, *Isaac Newton and Astrology: The Eighteenth Selig Brodetsky Memorial Lecture* (Leeds: Leeds University Press, 1977).
8. Isaac Newton, *Sir Isaac Newton's Theological Manuscripts*, ed. H. McLachlan (Liverpool: Liverpool University Press, 1950), p. 17.

Chapter 1: Desertion

1. Alexander Pope, 'Epitaph Intended for Sir Isaac Newton, In West-minster-Abbey' (1730).

2. William Stukeley, *Memoirs of Sir Isaac Newton's Life*, ed. A. Hastings White (London: Taylor & Francis, 1936), pp. 11–16.

3. Thomas Fuller, as quoted in G. M. Trevelyan, *A Shortened History of England* (New York: Longman, 1942), p. 316.

4. Canon C. W. Foster, 'Sir Isaac Newton's Family', *Reports and Papers of the Architectural Societies of the County of Lincoln, County of York, Archdeaconries of Northampton and Oakham, and County of Leicester*, Vol. 39, pt 1 (1928), pp. 13–15.

5. Ibid.

6. Thomas Maude, *Viator, a Poem: or, A Journey from London to Scarborough by the Way of York* (London, 1782), pp. iv-v.

7. Autographed draft of Newton's pedigree for College of Heralds kept in the Babson College Library, Massachusetts, MS 439.

8. KCL, Keynes MSS 130 (10), p. 1.

9. David Brewster, *Memoirs of The Life, Writings, and Discoveries of Sir Isaac Newton*, Vol. 1 (Edinburgh, 1855), p. 4.

10. KCL, Keynes MS 130, pp. 9–10.

11. Hannah Ayscough Smith's will, dated 1672, proved 11 June 1679, at Lincoln, produced in Foster, 'Sir Isaac Newton's Family', pp. 50–53.

12. *Correspondence*, Vol. 1, p. 2.

13. R. S. Westfall, 'Short-Writing and the State of Newton's Conscience, 1662 (1)', *Notes and Records of the Royal Society of London*, Vol. 18 (1963), pp. 10–11.

14. The Pierpont Morgan Library, New York, Notebook of Sir Isaac Newton.

15. From Isaac Newton's school Latin exercise book kept in a private collection in Los Angeles.

16. Stukeley, *Memoirs*, p. 44.

17. Ibid., p. 46.

18. Edmund Turnor (ed.), *Collections for the History of the Town and Soke of Grantham* (London, 1806), pp. 178, 180.

19. KCL, Keynes MS 136, p. 5. This poem was in fact copied by Newton from a book widely read at the time called *Eikon Basilike: The Portraiture of His Sacred Majesty in His Solitudes and Sufferings*.

20. KCL, Keynes MS 130 (2), p. 18.

21. ULC Add. MS 3975, p. 178.

22. Stukeley, *Memoirs*, pp. 19, 22.
23. KCL, Keynes MS 136, p. 6.
24. Ibid., p. 7.
25. Turnor family archive, Lincoln, Lincolnshire, as quoted in R. S. Westfall, *Never At Rest: A Biography of Isaac Newton* (Cambridge: Cambridge University Press, 1980), p. 63. (There is a chance that the Isaac Newton on whom these fines were levied was not Hannah's son but someone of the same name living in one of the villages of the region.)
26. KCL, Keynes MS 136, pp. 6–7.

Chapter 2: The Changing View of Matter and Energy

1. Charles Singer (ed.), *Studies in the History and Methods of Science* (Oxford: Oxford University Press, 1917), p. 240.
2. W. C. Dampier, *A History of Science* (Cambridge: Cambridge University Press, 1984 (first published 1929)), p. 35.
3. Ibid., p. 28.
4. Ambrosian Library, Milan, 'Codex Atlanticus', fol. 109v-a
5. E. A. Burtt, *Metaphysical Foundations of Modern Science* (London and New York, 1925), p. 75.
6. Quoted in A. N. Whitehead, *Science and the Modern World* (Cambridge: Cambridge University Press, 1927), p. 218.
7. Burtt, *Metaphysical Foundations of Modern Science*, p. 154.

Chapter 3: Academia

1. KCL, Keynes MS 130 (7).
2. Rowland Parker, *Town and Gown: The 700 Years' War in Cambridge* (Cambridge: Patrick Stephens, 1983), p. 134.
3. C. H. Cooper, *Annals of Cambridge: Vol. 3, 1603–1688* (Cambridge, 1845), p. 506.
4. Parker, *Town and Gown*, p. 121.
5. Ibid., p. 132.
6. William Stukeley, *Memoirs of Sir Isaac Newton's Life*, ed. A. Hastings White (London: Taylor & Francis, 1936), pp. 50–51.
7. James Hutton, 'New Anecdotes of Sir Isaac Newton', *Annual Register*, Vol. 19 (1776), p. 24.
8. R. S. Westfall, 'Short-Writing and the State of Newton's Conscience, 1662 (1)', *Notes of the Royal Society of London*, Vol. 18, (1963), pp. 10–16.

9. *Correspondence*, Vol. 1, p. 12. The letter was probably never sent and the existing copy is to be found in the Morgan Notebook. But, even if it were merely an exercise, it illustrates Newton's attitude towards earthly pleasures as a nineteen-year-old student

10. Westfall, 'Short-Writing and the State of Newton's Conscience, 1662 (1)', p. 11.

11. Trinity College Library, Cambridge, Trinity Notebook, ref. 53 (c).

12. KCL, Keynes MS 137.

13. Ibid.

14. *Correspondence*, Vol. 7, p. 368.

15. ULC Add. MS 3996, fols. 88–135.

16. Ibid., fols. 111–112.

17. Ibid., fol. 121.

18. Ibid., fols. 88–89.

19. Ibid., fol. 89r.

20. For example, R. S. Westfall, *Never At Rest: A Biography of Isaac Newton* (Cambridge: Cambridge University Press, 1980), p. 97. A popular view is that so grand a figure as More would not have contemplated talking to a schoolboy and would not have known Newton at all as an undergraduate. There is no evidence to support this view, whereas it is likely that Clark and Babington may have provided a point of contact between More and Newton either before or soon after the former entered Cambridge University.

21. ULC Add. MS 3996, fol. 101v.

22. KCL, Keynes MS 130 (4), p. 9.

23. Cooper, *Annals of Cambridge*, p. 517.

24. L. T. More, *Isaac Newton: A Biography* (New York: Scribner's, 1934), p. 40.

25. *Correspondence*, Vol. 1, p. 92 (Newton to Oldenburg, 6 February 1672).

26. ULC Add. MS 3996, fol. 122.

27. Trinity Notebook, ref. 53 (c), fol. 103.

28. Ibid., fol. 133.

29. *Correspondence*, Vol. 3, p. 153 (Newton to John Locke).

30. Add. MS 3975, p. 15.

31. ULC Add. MS 4004.

32. Isaac Newton, *The Mathematical Papers of Isaac Newton*, Vol. 1, ed. D. T. Whiteside (Cambridge: Cambridge University Press, 1967), p. 5.

33. ULC Add. MS 4007, fols. 706–7.

34. KCL, Keynes MS 130 (10), fol. 2v.
35. KCL, Keynes MS 136, p. 8.

Chapter 4: Astronomy and Mathematics Before Newton

1. Quoted by Michael Church, *The Independent*, 24 August 1996.
2. John North, *Stonehenge: Neolithic Man and the Cosmos* (London: HarperCollins, 1996).
3. W. C. Dampier, *A History of Science* (Cambridge: Cambridge University Press, 1984 (first published 1929)), p. 15.
4. Arthur Koestler, *The Sleepwalkers: A History of Man's Changing Vision of the Universe* (Harmondsworth: Penguin, 1964).
5. Ibid., p. 72.
6. Ibid.
7. Ptolemy, *Almagest*, Book III, Ch. 2.
8. Nicholas Copernicus, *On the Revolutions of the Heavenly Spheres*, Book I, Ch. 10.
9. Quoted in Dampier, *A History of Science*, p. 130.

Chapter 5: A Toe in the Water

1. Werner Heisenberg, *Physics and Philosophy* (London: Allen & Unwin, 1963).
2. *Correspondence*, Vol. 1, pp. 9–10.
3. Frank E. Manuel, *A Portrait of Isaac Newton* (New York: Da Capo, 1968), p. 99.
4. Daniel Defoe, *A Visitation of the Plague* (London: Penguin, 1995), pp. 28–9.
5. Samuel Pepys, *The Shorter Pepys* (London: Penguin, 1993), p. 486.
6. Ibid., p. 497.
7. C. H. Cooper, *Annals of Cambridge: Vol. 3, 1603–1688* (Cambridge, 1845), p. 517.
8. Fitzwilliam Museum, Cambridge, Fitzwilliam Notebook.
9. 'Isaac Newton' entry in *Biographica Britannica*, Vol. 5 (London, 1760), pp. 32–41.
10. William Stukeley, *Memoirs of Sir Isaac Newton's Life*, ed. A. Hastings White (London: Taylor & Francis, 1936), pp. 19–20.
11. Voltaire, *Elements of Newton's Philosophy* (London, 1738), p. 289.
12. Henry Pemberton, *A View of Sir Isaac Newton's Philosophy* (London, 1728), Preface.
13. ULC Add. MS 3968.41, fol. 85r.

14. ULC Add. MS 4000, fol. 14v.

15. *Correspondence*, Vol. 1, p. 300.

16. ULC Add. MSS 3958.2, fol. 45r.

17. ULC Add. MS 3968.41, fol. 85r.

18. Sir Edward Bysshe, *The Visitation of the County of Lincoln made by Sir Edward Bysshe, Knight, Clarenceux King of Arms in the Year of Our Lord 1666*, ed. Everard Green (Horncastle: printed for the Lincoln Record Society by W. K. Morton & Sons, 1917), p. 44.

19. Isaac Newton, *Correspondence of Sir Isaac Newton and Professor Cotes*, ed. J. Edleston (London, 1850), pp. xlii-xliii.

20. Fitzwilliam Notebook.

21. 'A true and perfect inventary [sic] of all and singular the Goods, Chattels and Credits of Sir Isaac Newton', reproduced in Richard de Villamil, *Newton: The Man* (London: G. D. Knox, 1931), pp. 50–61.

22. The Pierpont Morgan Library, New York, Notebook of Sir Isaac Newton.

23. KCL, Keynes MS 135.

24. Sir Isaac Newton, *Principia Mathematica* (London, 2nd edn, 1713), 'Scholium Generale'.

25. *Correspondence*, Vol. 1, p. 13.

26. Ibid., p. 14.

27. Ibid., pp. 14–15.

28. Isaac Barrow, *Lectiones XVIII Cantabrigiae in Scholis Publicis Habitae: In Quibus Opticorum Phaenomen Genuinae Rationes Investigantur, ac Exponuntur* (London, 1669).

Chapter 6: The Search for the Philosophers' Stone

1. Isaac Newton, *Opticks or A Treatise of the Reflections, Refractions, Inflections & Colours of Light*, Book III, Pt 1, Query 30 (London, 1706; New York: Dover Publications, 1952), p. 373.

2. Specifically, Kenelm Digby's *Powder of Sympathy, and Of Vegetation of Plants* (London, 1669), published on the Continent during the late 1650s.

3. JNUL, Yahuda MS 41, fols. 6–7.

4. David Hume, 'Natural History of Religion', in *Four Dissertations* (London, 1757), p. 172.

5. Bodleian Library, Oxford, MS Don. b. 15, fol. 3.

6. From *Corpus Hermeticum*, attributed to Hermes Trismegistus, quoted in

Jack Lindsay, *The Origins of Alchemy in Graeco-Roman Egypt* (London: Frederick Muller, 1970), p. 28.

7. *Hermetica. The Ancient Greek and Latin Writings Which Contain Religious or Philosophical Teachings Ascribed to Hermes Trismegistus*, Vol. 1, trans. Walter Scott (Boston: Shambhala, 1985), p. 96.

8. Charles Mackay, *The Alchymists: From Memoirs of Extraordinary Popular Delusions* (London, 1841), p. 106.

9. Elias Ashmole, *Theatrum Chemicum Britannicum* (London, 1652; reprinted in facsimile, London and New York, 1967), p. 443.

10. Frances A. Yates, *The Rosicrucian Enlightenment* (London and New York: Routledge, 1972), pp. 200–202.

11. Ibid., pp. 30–40.

12. Quoted in Mackay, *The Alchymists*, p. 189.

13. Ibid., pp. 192–3.

14. Yates, *The Rosicrucian Enlightenment*, p. 222.

15. John Harrison, *The Library of Isaac Newton* (Cambridge: Cambridge University Press, 1978), p. 59.

16. For a full text of Newton's notes see Ian Macphail, *Alchemy and the Occult: Catalogue of Books from the Collection of Paul and Mary Mellon Given to Yale University Library*, Vol. 3 (New Haven: Yale University Press, 1968), p. 102.

17. Paracelsus, *Alchemy, The Third Column of Medicine*, ed. A. E. Waite (London, 1897), p. 44.

18. Sir David Brewster, *Memoirs of The Life, Writings, and Discoveries of Sir Isaac Newton* (Edinburgh, 1855).

19. KCL, Keynes MS 130, 'A Conduitt Notebook'.

20. Maynard Keynes, 'Newton the Man', in Royal Society, *Newton Tercentenary Celebrations* (Cambridge: Cambridge University Press, 1947), p. 29.

21. Originally entitled *The Chemical Wedding of Christian Rosencreutz* and first published in London in 1616. Christian Rosencreutz – a man whose identity has never been satisfactorily confirmed – was thought to be the semi-mythical founder of the Rosicrucian movement.

22. Quoted in Lindsay, *The Origins of Alchemy*, p. 216.

23. Cornelius Agrippa, *Three Books of Occult Philosophy*, trans. 'J. F.' (London, 1651).

24. Carl Jung, *Memories, Dreams and Reflections* (London: Collins and Routledge & Kegan Paul, 1963), p. 147.

25. C. G. Jung, *Man and His Symbols* (London: Aldus Books, 1964), p. 210.

26. KCL, Keynes MS 27, fol. 4.
27. KCL, Keynes MS 39, fol. 2r.

Chapter 7: The Sorcerer's Apprentice

1. A remark made by Albert Einstein at Princeton, 9 May 1921, and later carved over the fireplace in the common room of Fine Hall, Princeton Institute of Advanced Study.
2. KCL, Keynes MS 137.
3. Stanford University MS 538.
4. ULC Add. MS 3975, pp. 181–2.
5. Thomas Norton, *The Ordinall of Alchimy* (1477); first published in English in 1652 in London, p. 143.
6. Bodleian Library, Oxford, MS Don. b. 15.
7. *Correspondence*, Vol. 2, p. 2.
8. Thanks to Carrie Branigan for unearthing this useful information after visiting Petty's tomb in Romsey Abbey. The description of Foxcroft is quoted in B. J. Dobbs, *The Foundations of Newton's Alchemy, or The Hunting of the Greene Lyon* (Cambridge: Cambridge University Press, 1975), p. 112.
9. KCL, Keynes MS 33, fol. 5r.
10. KCL, Keynes MS 33, fol. 5v.
11. ULC Add. MS 3975.
12. Charles Nicholl, *The Chemical Theatre* (New York: Routledge & Kegan Paul, 1980), p. 10.
13. Elias Ashmole (ed.), *Theatrum Chemicum Britannicum* (London, 1652), pp. 446–7.
14. Lao tzu, *Tao Te Ching*, XXV, trans. D. C Lau (Harmondsworth: Penguin, 1963), p. 82.
15. C. G. Jung, *Psychology and Alchemy* (London: Routledge & Kegan Paul, 1979 (Vol. 12 of Jung's Collected Works)), p. 304.
16. Thomas Vaughan, 'Aula Lucis', in *The Works of Thomas Vaughan*, ed. Alan Rudrum and Jennifer Drake-Brockman (Oxford: Oxford University Press, 1984), p. 17.
17. Bodleian MS Don. b. 15.
18. Basilius Valentinus, *The Triumphal Chariot of Antimony* (London: Vincent Stuart, 1962), p. 175.
19. ULC Add. MS 3975, p. 82.
20. KCL, Keynes MS 18.
21. ULC Add. MS 3975, p. 143.

22. William Stukeley, *Memoirs of Sir Isaac Newton's Life*, ed. A. Hastings White (London: Taylor & Francis, 1936), p. 59.

23. Isaac Newton, *Correspondence of Sir Isaac Newton and Professor Cotes*, ed. J. Edleston (London, 1850), p. lxi.

24. Countway Medical Library, Harvard University, Countway MS, item 4.

25. Ibid., item 3, fol. 7.

26. JNUL, Yahuda MS 259, no. 9, p. 41.

27. Trinity College Library, Cambridge, Trinity Notebook, fol. 126.

28. Ibid., fol. 128.

29. JNUL, Yahuda MS 14, fol. 25.

30. *Correspondence*, Vol. 7, p. 387.

31. ULC Res. a. 1893, packet E.

32. KCL, Keynes MS 2, fol. XIIIv.

33. Ibid.

34. JNUL, Yahuda MS 21, fol. 2.

35. JNUL, Yahuda MS Var. I, Newton MS 41, fol. 7.

36. David Castillejo, *The Expanding Force in Newton's Cosmos: 'As Shown in his Unpublished Papers'* (Madrid: Ediciones de Arte y Bibliofilia, 1981), p. 91.

37. JNUL, Yahuda MS 7.2I, fol. 4.

38. JNUL, Yahuda MS 1.1. f 12r.

39. Charles Webster, 'Prophesy' in *From Paracelsus to Newton: Magic and the Making of Modern Science* (Cambridge: Cambridge University Press, 1982), pp. 15–47.

40. Richard H. Popkin, *The Third Force in Seventeenth Century Philosophy: Scepticism, Science and Biblical Prophecy* (Madrid, Nouvelle République des Lettres, 1983), pp. 35–63.

41. Robert Bauval and Adrian Gilbert, *The Orion Mystery* (London: Heinemann, 1994).

42. KCL, Keynes MS 33, fol. 5v.

43. Otto von Simson, *The Gothic Cathedral. Origins of Gothic Architecture and the Medieval Concept of Order* (New York, Pantheon, 2nd edn, 1962), pp. 8–38.

44. JNUL, Yahuda MS Var. I, Newton MS 17.3, fols. 8–10.

45. Ibid., Newton MS 41 (5, n. 84), fol. 6.

46. Quoted in B. J. Dobbs, *The Janus Faces of Genius: The Role of Alchemy in Newton's Thought* (Cambridge: Cambridge University Press, 1991), p. 151.

Chapter 8: Feuds

1. Leon Edel, *Henry D. Thoreau* (Minneapolis, 1970), p. 8.
2. ULC Add. MS 4002, p. 1.
3. KCL, Keynes MS 135, p. 6.
4. Quoted in D. A. Winstanley, *Unreformed Cambridge* (Cambridge: Cambridge University Press, 1921), pp. 132–3.
5. ULC Add. MS 3975, p. 12.
6. ULC Add. MS 4002, p. 22.
7. *Correspondence*, Vol. 1, pp. 3–4.
8. Conduitt's memorandum of 31 August 1726; KCL, Keynes MS 130. 10, fol. 3–3v.
9. *Correspondence*, Vol. 1, p. 73.
10. Ibid., p. 79.
11. Ibid., pp. 82–3.
12. E. N. Da C. Andrade, *A Brief History of the Royal Society* (London: Royal Society, 1960), p. 201.
13. Richard Nichols, *The Diaries of Robert Hooke: The Leonardo of London* (London: Book Guild, 1994).
14. *Correspondence*, Vol. 1, p. 110–11.
15. Ibid., p. 116.
16. Ibid., p. 100.
17. Ibid., p. 4 (an account drafted by J. Collins and appended to a copy he made of Newton's first written description of the reflecting telescope).
18. Ibid., p. 135.
19. Ibid., pp. 130–33.
20. Isaac Newton, *Isaac Newton's Papers and Letters on Natural Philosophy*, ed. I. Bernard Cohen (Cambridge, Mass.: Harvard University Press, 2nd edn, 1978), p. 106.
21. *Correspondence*, Vol. 1, pp. 159–160.
22. Ibid, pp. 171–3.
23. Ibid.
24. Ibid., p. 262.
25. Ibid., p. 282.
26. Ibid., p. 408.
27. Ibid., pp. 412, 413.
28. Ibid., p. 416.
29. Richard Waller, 'The Life of Robert Hooke', in *Posthumous Works of Hooke* (London, 1765), pp. xxvi-xxvii; John Aubrey, *Brief Lives*, ed. O. L. Dick (Ann Arbor: University of Michigan Press, 1957), p. 165.

Chapter 9: To the *Principia*

1. Louis Pasteur – address given in the inauguration of the faculty of science, University of Lille, 7 December 1854.
2. *Correspondence*, Vol. 2, pp. 433–5.
3. Ibid.
4. Ibid., p. 442.
5. Ibid., p. 442.
6. KCL, Keynes MS 137, p. 16.
7. ULC Add. MS 4007, fols. 707r-707v.
8. Isaac Newton, *Correspondence of Sir Isaac Newton and Professor Cotes*, ed. J. Edleston (London, 1850), p. lxxxv.
9. KCL, Keynes MS 130 (8).
10. Lincoln Cathedral Archives, Bishop's Transcripts of Colsterworth.
11. *Correspondence*, Vol. 2, pp. 502–4.
12. Ibid., p. 297.
13. Ibid., pp. 300–301.
14. Ibid., p. 436.
15. Ibid., pp. 300–301.
16. Ibid., pp. 304–5.
17. From the *Lectiones Cutleriane*, quoted in John Aubrey, *Brief Lives*, ed. O. L. Dick (Ann Arbor: University of Michigan Press, 1957), p. 166.
18. *Correspondence*, Vol. 2, p. 307.
19. Ibid., pp. 312–13.
20. ULC Add. MS 3965.1, fols. 1–3.
21. Abraham Demoivre, 'Memorandum on Newton's Life', Joseph Halle Schaffner Collection, University of Chicago Library, MS 1075–7.
22. *Correspondence*, Vol. 2, p. 447.
23. Robert Hooke, *The Diary of Robert Hooke, 1672–1680*, ed. Henry W. Robinson and Walter Adams (London: Taylor & Francis, 1935), pp. 459–60.
24. *Correspondence*, Vol. 2, p. 315.
25. Ibid., p. 368.
26. Ibid., p. 361.
27. Ibid., pp. 336–9 (Flamsteed to Halley, 17 February 1681).
28. Ibid., p. 360.
29. Robert Hooke, *Cometa*, reprinted in R. T. Gunther (ed.), *Early Science in Oxford*, Vol. 8 (Oxford: Oxford University Press, 1922), pp. 223–4, 247.
30. Richard Westfall, 'Newton and Alchemy', in Brian Vickers (ed.), *Occult*

and Scientific Mentalities in the Renaissance (Cambridge: Cambridge University Press, 1984), p. 330.

31. Henry More, *The Immortality of the Soul, So farre forth as it is demonstrabl from the Knowledge of Nature and the Light of Reason* (London, 1659), pp. 467–8.

32. Sotheby lot no. 113; Smithsonian Institute Library, Washington DC, Dibner Library of the History of Science and Technology, MSS 1031 B.

33. KCL, Keynes MS 19, fols. 1, 3.

34. Add. MS 3973, fols. 13, 21.

35. Add. MS 3975, pp. 108–9.

36. *Correspondence*, Vol. 1, pp. 368–70.

37. Ibid., pp. 365–6.

38. Westfall, 'Newton and Alchemy', p. 323.

39. *Correspondence*, Vol. 2, pp. 288, 291.

40. Isaac Newton, *Unpublished Scientific Papers of Sir Isaac Newton. A Selection from the Portsmouth Collection in the University Library, Cambridge*, Vol. 2, ed. and trans. A. Rupert Hall and Marie Boas Hall (Cambridge: Cambridge University Press, 1962), p. 220.

41. Ibid., p. 113.

42. Thomas Birch, *The History of the Royal Society of London*, Vol. 4 (London, 1757), p. 347.

43. KCL, Keynes MS 135.

44. *Correspondence*, Vol. 2, p. 407.

45. Ibid., p. 409.

46. Ibid., p. 413.

47. KCL, Keynes MS 135.

48. Ibid.

49. Isaac Newton, *Principia Mathematica*; trans. and ed. Andrew Motte as *Mathematical Principles of Natural Philosophy* (1729), rev. Florian Cajori (Berkeley and Los Angeles: University of California at Berkeley Press, 1934), p. 397.

50. KCL, Keynes MS 133, p. 10.

51. *Correspondence*, Vol. 3, pp. 152, 155–6.

52. Ibid., Vol. 2, pp. 431–2.

53. Ibid., pp. 435–7.

54. Ibid.

55. Ibid.

56. Ibid., pp. 441–3.

57. Ibid.
58. Ibid., p. 474.

Chapter 10: Breakdown
 1. Samuel Johnson, 'Inch Kenneth', in *A Journey to the Western Islands of Scotland* (London, 1775).
 2. KCL, Keynes MS 130. 6, Book 2.
 3. KCL, Keynes MS 130. 5, sheets 2–3.
 4. *Correspondence*, Vol. 2, p. 484.
 5. Richard Nichols, *Diaries of Robert Hooke: The Leonardo of London* (London: Book Guild, 1994), p. 197 (3 February 1689 and 3 July 1689).
 6. Isaac Newton, *Unpublished Scientific Papers of Sir Isaac Newton. A Selection from the Portsmouth Collection in the University Library, Cambridge*, Vol. 2, ed. and trans. A. Rupert Hall and Marie Boas Hall (Cambridge: Cambridge University Press, 1962), p. 333.
 7. Charles Henry Cooper, *Annals of Cambridge*, 5 vols. (Cambridge, 1842–1908), Vol. 3, p. 615.
 8. Ibid., p. 634.
 9. *Correspondence*, Vol. 2, pp. 467–8.
10. Ibid., pp. 502–4.
11. KCL, Keynes MS 116.
12. Cooper, *Annals of Cambridge*, Vol. 3, p. 631.
13. John Locke, *An Essay Concerning Human Understanding*, ed. Peter H. Nidditch (Oxford: Oxford University Press, 1975), pp. 9–10.
14. J. T. Desaguliers, *Experimental Philosophy*, Vol. 1 (London, 1734), Preface.
15. Locke, *Human Understanding*, p. 319.
16. William Seaward, *Anecdotes of Distinguished Persons*, Vol. 2 (London, 5th edn, 1804), p. 178.
17. Gilbert Burnet, *Travels* (London, 1750), p. 14.
18. Seaward, *Anecdotes*, p. 190.
19. *Correspondence*, Vol. 3, p. 45.
20. Ibid.
21. Manuscript letter in Geneva quoted in Charles Andrew Domson, 'Nicholas Fatio de Duillier and the Prophets of London: An Essay In the Historical Interaction of Natural Philosophy and Millennial Belief in the Age of Newton' (doctoral dissertation, Yale University, 1972), pp. 32–4.

22. *Correspondence*, Vol. 3, pp. 390–91.

23. Ibid.

24. Ibid., p. 170.

25. Christiaan Huygens, *Œuvres complètes* (The Hague, Haarlem, 1888–1950), Vol. 10, pp. 271–2.

26. *Correspondence*, Vol. 3, p. 230.

27. Letter from Fatio de Duillier to Huygens 18 December 1691, in Huygens, *Œuvres*, Vol. 10, p. 213.

28. *Correspondence*, Vol. 3, p. 231.

29. Ibid., pp. 231–3.

30. Ibid., pp. 241–2.

31. Ibid., p. 243.

32. Bibliothèque Publique et Universitaire de Genève, MS français 602, fol. 82v (Fatio de Duillier to his brother, 3 February 1693).

33. *Correspondence*, Vol. 3, p. 245.

34. Ibid., p. 263.

35. Ibid., p. 391.

36. The diary of Robert Hooke, reproduced in R. T. Gunther (ed.), *Early Science in Oxford*, Vol. 10 (Oxford: Oxford University Press, 1920–5), pp. 190–91.

37. *Correspondence*, Vol. 3, p. 191.

38. Ibid., p. 279.

39. Ibid., p. 281.

40. Ibid., p. 283.

41. Ibid., p. 280.

42. Ibid., p. 283.

43. Ibid., p. 284.

44. Babson College Library, Massachusetts, Babson MS 420, p. 19.

45. Ibid., p. 18a.

46. P. E. Spargo and C. A. Pounds, 'Newton's "Derangement of the Intellect": New Light on an Old Problem', *Notes and Records of the Royal Society of London*, Vol. 34, no. 1 (July 1979), pp. 116–21. See also R. Seitz and J. Y. Lettvin, 'Mercury and Melancholoy: The Decline of Isaac Newton', *Bulletin of the American Physical Society*, ser. II, 16 (1971), p. 1400.

47. *Correspondence*, Vol. 4, p. 131.

48. Ibid., p. 84.

49. *Correspondence*, Vol. 3, p. 282.

50. Ibid., Vol. 4, p. 188.

51. Ibid., pp. 193–4.
52. *Correspondence*, Vol. 3, p. 185.
53. *Correspondence*, Vol. 4, p. 195 (from the original in the Bank of England).

Chapter 11: Metamorphosis

1. Alfred, Lord Tennyson, *Locksley Hall* (1842), l. 181.
2. John Evelyn, *The Diary of John Evelyn*, Vol. 3, ed. E. S. De Beer (Oxford: Oxford University Press, 1955), p. 234.
3. KCL, Keynes MS 130 (6), Book 2.
4. *Correspondence*, Vol. 4, p. 201.
5. Public Records Office, London, Mint MS 19.1, fols. 9–10.
6. Evelyn, *Diary*, Vol. 5, p. 228.
7. Narcissus Lutrell, *A Brief Historical Relation of State Affairs*, Vol. 5 (Oxford, 1857), p. 86.
8. *Correspondence*, Vol. 4, p. 258.
9. Hopton Haynes, 'Brief Memoirs Relating to the Silver & Gold Coins of England: With an Account of the Corruption of the Hammerd Monys, And of the Reform by the Late Grand Coynage, At the Tower, & the Five Country Mints, In the Years 1696, 1697, 1698, & 1699', British Library, Lansdowne MS DCCCI, fol. 78v.
10. John Harrison, *The Library of Isaac Newton* (Cambridge: Cambridge University Press, 1978), p. 59.
11. *Correspondence*, Vol. 4, p. 208.
12. Ibid., p. 243.
13. Public Records Office, London, Mint MS 1.6, fol. 618.
14. *Correspondence*, Vol. 4, pp. 209–210.
15. Mint MS 19. 1, fol. 467.
16. Public Records Office, London, E351, Roll 2073.
17. *Correspondence*, Vol. 4, p. 217.
18. John Craig, 'Isaac Newton and the Counterfeiters', *Notes and Records of the Royal Society of London*, Vol. 18 (1963), p. 139.
19. Royal Mint Library, Depositions Book, no. 233.
20. Ibid., no. 215.
21. Ibid., no. 205 (Letter of 6 March 1699).
22. Haynes, 'Brief Memoirs', fol. 68.
23. Royal Mint Library, Depositions Book, no. 27.
24. *Correspondence*, Vol. 4, p. 225.
25. KCL, Keynes MS 130 (5).

26. *Correspondence*, Vol. 4, p. 349.

27. Brook Taylor, *Contemplatio Philosophica* (London, 1793), pp. 93–4.

28. Bodleian Library, Oxford, New College MS 361, IV, fol. 76 (David Brewster to Revd Jeffrey Ekins, April 1855).

29. Jonathan Swift, *Journal to Stella*, Vol. 1, ed. Harold Wiliams (Oxford: Oxford University Press, 1948), pp. 229–30.

30. Ibid., pp. 229–30 (3 April 1711).

31. Ibid., p. 109 (30 November 1710).

32. Ibid., Vol. 2, p. 383 (14 October 1711).

33. Isaac Newton, *Correspondence of Sir Isaac Newton and Professor Cotes*, ed. J. Edleston (London, 1850), p. lxxviii.

34. John Dryden, *Miscellany Poems*, Vol. 5 (London, 5th edn, 1727), p. 61.

35. *The Works and Life of the Right Honourable Charles, Late Earl of Halifax* (London, 1715), Appendix, pp. iv, v–vi.

36. Swift, *Journal to Stella*, Vol. 2, p. 395 (25 October 1711).

37. Catalogue of Newton Papers Sold by Order of Viscount Lymington, lot 176, p. 41.

38. Augustus De Morgan, *Newton: His Friend: And His Niece* (London, 1885).

39. *Correspondence*, Vol. 6, p. 225.

40. Joseph Lemuel Chester (ed.), *The Marriage, Baptismal, and Burial Registers of the Collegiate Church or Abbey of St Peter Westminster* (London, 1876), p. 354.

41. Christopher Hill, 'Newton and His Society', in Robert Palter (ed.), *The Annus Mirabilis of Sir Isaac Newton, 1666–1966* (Cambridge, Mass.: Harvard University Press, 1970), p. 40.

42. Frank E. Manuel, *A Portrait of Isaac Newton* (New York: Da Capo, 1968), pp. 261–2.

43. Francis Baily, *An Account of the Revd. John Flamsteed, the First Astronomer Royal* (London, 1835), p. 314.

44. Voltaire, 'Dictionnaire philosophique', in *Œuvres complètes de Voltaire*, Vol. 42 (Paris, 1787), p. 165.

45. Mary de la Rivière Manley, *Memoirs of Europe, Towards the Close of the Eighth Century. Written by Eginardus, Secretary and Favourite to Charlemagne; and done into English by the Translator of the New Atlantis*, Vol. 1 (London, 2nd edn, 1711), pp. 261–7.

46. Ibid.

47. Bodleian Library, Oxford, New College MS 361, IV, fols. 159–160 (David Brewster to Revd Jeffrey Ekins).

48. Journal Book of the Royal Society, Vol. 10, p. 128 (31 May 1699).

49. Ibid., p. 120 (25 April 1699).

50. Journal Book (Copy) of the Royal Society, Vol. 10, pp. 63–4 (16 February 1704).

51. Jean Théophile Desqueliers, *Course of Experimental Philosophy*, Vol. 1 (London, 3rd edn, 1763), p. viii.

52. Isaac Newton, *Opticks or A Treatise of the Reflections, Refractions, Inflections & Colours of Light* (London, 1706; New York, Dover Publications, 1952), p. cxxi.

53. W. G. Hiscock (ed.), *David Gregory, Isaac Newton and Their Circle* (Oxford: Oxford University Press, 1937), p. 14.

54. Newton, *Opticks*, p. 1.

55. ULC Add. MS 3970. 3, fol. 336.

56. *Correspondence*, Vol. 3, p. 338.

57. Newton, *Opticks*, p. 266 (my bold type).

58. Ibid., p. 338.

59. Ibid., p. 339.

60. Ibid., p. 340.

61. Ibid., pp. 375–6.

62. Erasmus Darwin, *The Temple of Nature; or The Origin of Society* (London, 1803), Canto I, p. 1, ll. 1–8.

63. Alexander Pope, *An Essay on Man* (London, 1733), Epistle II, ll. 31–4.

64. William Blake, *Jerusalem* (1804), in *Complete Poetry and Prose*, ed. Geoffrey Keynes (London: Reinhardt Books, 1989), pp. 574–5.

65. Peter Ackroyd, *Blake* (London: Sinclair Stevenson, 1995), p. 194.

66. Ibid., p. 58.

Chapter 12: Old Men's Battles

1. JNUL, Yahuda MS 7.3p.

2. From 'The Inventory of Newton's Goods, Chattels and Credits, Taken 21–27th April 1727', quoted in Richard de Villamil, *Newton: The Man* (London: G. D. Knox, 1931), pp. 50–61.

3. KCL, Keynes MS 129A, p. 18.

4. Villamil, *Newton: The Man*, pp. 14–15.

5. Henry St John, Viscount Bolingbroke, *Lettres historiques, politiques, philosophiques, et particulières . . . précédées d'un essai historique sur sa vie*, Vol. 1, ed. Count H. P. Grimoard (Paris, 1808), p. 156.

6. William Stukeley, *Memoirs of Sir Isaac Newton's Life*, ed. A. Hastings White (London: Taylor & Francis, 1936), p. 14.

7. Joseph Spence, *Anecdotes, Observations, and Characters of Books and Men Collected from the Conversations of Mr Pope, and Other Eminent Persons of His Time*, ed. Samuel S. Singer (London, 1820), p. 368.

8. KCL, Keynes MSS 136, p. 12.

9. John Harrison, *The Library of Isaac Newton* (Cambridge: Cambridge University Press, 1978), p. 70.

10. W. G. Hiscock (ed.), *David Gregory, Isaac Newton and Their Circle* (Oxford: Oxford University Press, 1937), p. 17.

11. Ibid., p. 14.

12. Margaret C. Jacob, 'Newton and the French Prophets: New Evidence', *History of Science*, Vol. 16, (1978), pp. 134–42.

13. Quoted in *Clavis Prophetica; or, a Key to the Prophesies of Mons. Marion, and the other Camisars, with some Reflections on the Character of these New Envoys, and of Mons. F their Chief Secretary* (London, 1707).

14. R. Kingston, *Enthusiastick Imposters, no Divinely Inspir'd Prophets, Being a Historical Relation of the Rise, Progress and Present Practices of the French and English Prophets* (London, 1707), p. 22 (Kingston's italics).

15. Thomas Hearne, *Remarks and Collections*, Vol. 2, ed. C. E. Doble et al. (Oxford, 1892), p. 244 (letter from 28 August 1709).

16. Spence, *Anecdotes*, pp. 56–7, 72.

17. William Whiston, *Memoirs of the Life and Writings of Mr William Whiston* (London, 1749), pp. 292–3.

18. Ibid., p. 293.

19. Ibid., pp. 293–4.

20. Ibid., p. 294.

21. Hearne, *Remarks and Collections*, Vol. 2, pp. 100–101.

22. Francis Baily, *An Account of the Revd. John Flamsteed, the First Astronomer Royal* (London, 1835), p. 229.

23. Gottfried von Leibniz, *Die Philosophischen Schriften*, ed. C. J. Gerhardt (Berlin, 1875–90; Hildesheim, 1960–61), Vol. 3, pp. 328–9.

24. Baily, *Flamsteed*, p. 306.

25. Ibid., p. 276 (Flamsteed to John Sharp, 14 July 1710).

26. Frank E. Manuel, *A Portrait of Isaac Newton* (New York: Da Capo, 1968), p. 283.

27. Council Minutes of the Royal Society, London, Vol. 2, pp. 185–191.

28. *Correspondence*, Vol. 5, pp. 76–8.

29. Ibid.

30. Charles R. Weld, *A History of the Royal Society*, Vol. 1 (London, 1848), pp. 393–4.

31. Council Minutes of the Royal Society, London, Vol. 2, pp. 233–4.

32. Stukeley, *Memoirs*, pp. 78–81.

33. Council Minutes of the Royal Society, London, Vol. 2, p. 214.

34. JNUL, Yahuda MS 15.7, fol. 180v.

35. *Correspondence*, Vol. 2, p. 217.

36. Stukeley, *Memoirs*, p. 67.

37. Baily, *Flamsteed*, p. 276.

38. Ibid., p. 272.

39. Manuel, *A Portrait of Isaac Newton*, p. 305.

40. Baily, *Flamsteed*, p. 7.

41. John Flamsteed, 'Self-Inspections of J. F.' (Vol. 32A, fol. 2), Archives of the Royal Greenwich Observatory at Herstmonceux.

42. Baily, *Flamsteed*, p. 60.

43. *Correspondence*, Vol. 4, pp. 87–88. (The italics indicate Flamsteed's underlining.)

44. *Correspondence*, Vol. 3, p. 203 (Flamsteed to Newton, 24 February 1692).

45. Ibid., Vol. 4, pp. 54–5.

46. R. S. Westfall, *Never At Rest: A Biography of Isaac Newton* (Cambridge: Cambridge University Press, 1980), p. 658.

47. *Correspondence*, Vol. 4, p. 58.

48. Ibid., p. 332.

49. Ibid., p. 134.

50. Ibid., p. 143.

51. Ibid., p. 134 (Flamsteed's comments added to Newton's letter of 9 July 1695).

52. Ibid., p. 169.

53. Ibid., p. 290.

54. Ibid., p. 277.

55. David Williamson, *Kings and Queens of Britain* (London: Webb & Bower, 1992), p. 152.

56. Baily, *Flamsteed*, p. 77.

57. Ibid., p. 66.

58. Ibid., p. 246.

59. Ibid., p. 270.

60. Ibid., p. 92.

61. *Correspondence*, Vol. 5, p. 99.

62. Ibid., p. 101.

63. Ibid., p. 102.

64. Flamsteed, 'Self-Inspections of J. F.', Vol. 33, fols. 104–106.

65. Ibid. (Flamsteed's emphasis).

66. Baily, *Flamsteed*, p. 323.

67. Ibid., p. 101.

68. Flamsteed, 'Self-Inspections of J. F.', Vol. 32c, fols. 78–82 (Flamsteed's emphasis).

Chapter 13: A Question of Priority

1. Thomas Babington Macaulay, 'Lord Bacon', *Edinburgh Review*, October 1833.

2. Frank E. Manuel, *A Portrait of Isaac Newton* (New York: Da Capo, 1968), p. 323.

3. David Millar et al., *Chambers Concise Dictionary of Scientists* (Cambridge: Chambers and Cambridge University Press, 1989), p. 242.

4. See Joseph E. Hoffman, *Leibniz in Paris, 1672–1676. His Growth to Mathematical Maturity* (Cambridge: Cambridge University Press, 1974), p. 291.

5. *Correspondence*, Vol. 1, p. 15.

6. Ibid., p. 356.

7. *Correspondence*, Vol. 2, p. 134.

8. Ibid., p. 110.

9. Ibid., p. 163.

10. Hoffman, *Leibniz in Paris*, Ch. 20.

11. Isaac Newton, *Principia Mathematica*; trans. and ed. Andrew Motte as *Mathematical Principles of Natural Philosophy* (1729), rev. Florian Cajori (Berkeley and Los Angeles: University of California at Berkeley Press, 1934), pp. 655–6.

12. Gottfried von Leibniz, *Sämtliche Schriften und Briefe*, Ser. I, Vol. IV (1950), pp. 475–6.

13. Ibid., p. 477.

14. ULC Add. MS 3968.41, fol. 85r.

15. *Correspondence*, Vol. 4, p. 100.

16. Bernoulli to Leibniz, quoted in Isaac Newton, *The Mathematical Papers of Isaac Newton*, Vol. 7, ed. D. T. Whiteside (Cambridge: Cambridge University Press, 1977), p. 181.

17. A. Rupert Hall, *Philosophers at War: The Quarrel Between Newton and Leibniz* (Cambridge: Cambridge University Press, 1980), p. 118.

18. Nicholas Fatio de Duillier, *Lineae Brevissimi Descensus Investigatio Geo-metrica Duplex* (*A Two-fold Geometrical Investigation of the Line of Briefest Descent*) (London, 1699), p. 18.

19. *Correspondence*, Vol. 3, p. 286.

20. John Keill, 'Epistolia . . . de Legibus Virum Centripetarum', *Philosophical Transactions of the Royal Society*, Vol. 26 (1708), p. 185.

21. 'Isacci Newtoni Tractatus Duo, de Speciebus & Magnitudine Figurarum Curvilinearum', *Acta Eruditorum* (January 1705), p. 35 (Leibniz's emphasis).

22. Pierre Des Maizeaux, *Recueil des pièces sur la philosophie, la religion naturelle, l'historie, les mathèmatiques, etc. par Mrs. Leibniz, Clarke, Newton, & autres autheurs célèbres*, Vol. 2 (Amsterdam, 1720), p. 49.

23. *Correspondence*, Vol. 5, p. 142.

24. Ibid., p. 207.

25. Joseph Raphson, *The History of Fluxions* (London, 1715), p. 100.

26. Journal Book of the Royal Society, 11 March 1712.

27. 'On a Point Connected with the Dispute between Keil and Leibniz about the Invention of Fluxions', *Philosophical Transactions of the Royal Society*, Vol. 136 (1846), p. 107.

28. Journal Book (Copy) of the Royal Society, London, Vol. 10, p. 391.

29. ULC Add. MS 4007B, fol. 617.

30. Journal Book (Copy) of the Royal Society, London, Vol. 10, p. 337.

31. ULC Add. MS 3968.37, fol. 539r.

32. Isaac Newton, 'Account of *Commercium Epistolicum*', *Philosophical Transactions of the Royal Society*, Vol. 29, (1714–16), p. 211.

33. ULC Add. MS 3968.37 fol. 539r.

34. William Whiston, *Historical Memoirs of the Life of Dr Samuel Clarke* (London, 1730), p. 132.

35. *Correspondence*, Vol. 6, p. 3.

36. Ibid., pp. 8–9.

37. Ibid., pp. 18–19.

38. First quoted in a letter from John Chamberlayne to the Royal Society; Royal Mint Library, Newton MS II, fol. 334 (Chamberlayne to Newton, 25 November 1713).

39. Gottfried von Leibniz, *Die Philosophischen Schriften*, Vol. 3, ed. C. J. Gerhardt (Berlin, 1875–90; Hildesheim, 1960–61), p. 589.

40. *Correspondence*, Vol. 6, p. 288.

41. Quoted in Gale E. Christianson, *In the Presence of the Creator* (New York: Free Press, 1984), p. 548.

Chapter 14: Joining the Ancients

1. Newton told this to his nephew, Benjamin Smith, and it was recounted in Joseph Spence, *Anecdotes, Observations, and Characters of Books and Men Collected from the Conversations of Mr Pope, and Other Eminent Persons of His Time*, ed. Samuel S. Singer (London, 1820), p. 54.

2. Sotheby Catalogue, p. 50. Babson College Library, Massachusetts, Babson MS 426.

3. *Correspondence*, Vol. 6, p. 183.

4. Ibid., Vol. 7, pp. 166–7.

5. Ibid., p. 243.

6. Recorded by Conduitt in KCL, Keynes MSS 130.6, Book 2, and 130.7, sheet 1.

7. Conduitt's memoir to Bernard le Bovier de Fontenelle, KCL, MS 129A; printed in Edmund Turnor (ed.), *Collections for the History of the Town and Soke of Grantham* (London, 1806), pp. 165–6.

8. See, for example, Frank E. Manuel, *A Portrait of Isaac Newton* (New York: Da Capo, 1968), p. 363.

9. KCL, Keynes MS 129A, pp. 20–21.

10. J. E. McGuire and P. M Rattansi, 'Newton and the Pipes of Pan', *Notes and Records of the Royal Society of London*, Vol. 21 (1966), pp. 16–17.

11. Justin Martyr, 'Exhortation to the Greeks', as quoted in S. K. Heninger Jr, *Touches of Sweet Harmony: Pythagorean Cosmology and Renaissance Poetics* (San Marino, Cal.: The Huntingdon Library, 1974), p. 202.

12. ULC Portsmouth Collection, MS Add. 3970, fol. 619r, Draft of Query 23.

13. JNUL, Yahuda MS Var. I, Newton MS 15.3 (2, n.50), fol. 66v.

14. JNUL, Yahuda MS Var. I, Newton MS 15.5 (2, n.50), fol. 96r.

15. JNUL, Yahuda MS Var. I, Newton MS 15.5 (2, n.50), fol. 67r.

16. Ibid. (my italics).

17. William Stukeley, *Memoirs of Sir Isaac Newton's Life*, ed. A. Hastings White (London: Taylor & Francis, 1936), p. 17.

18. *Correspondence*, Vol. 7, p. 302.

19. Henry Pemberton, *A View of Sir Isaac Newton's Philosophy* (London, 1728), Preface.

20. KCL, Keynes MS 129A, pp. 23–4.

21. KCL, Keynes MS 130 (6), Book 2.

22. KCL, Keynes MS 130. 7, sheet 2.

23. Mathurin Veyssière de Lacroze, *Thesaurus Epistolicus Lacrozianus*, Vol. 1 (Leipzig, 1742), p. 105 (Samuel Crell to Lacroze, 17 July 1727).

24. *Correspondence*, Vol. 3, pp. 265–7.

25. KCL, Keynes MS 129A, pp. 24–26.

26. KCL, Keynes MS 130 (6).

27. Ibid.

28. Bodleian Library, Oxford, New College MS 351, IV, fol. 194.

29. KCL, Keynes MS 129A, p. 23.

30. Isaac Newton, *Correspondence of Sir Isaac Newton and Professor Cotes*, ed. J. Edleston (London, 1850), p. 316.

31. Turnor, *Collections for the History of Grantham*, p. 166.

32. Ibid.

33. KCL, Keynes MS 130.

34. Journal Book (Copy) of the Royal Society, London, Vol. 13, p. 62.

35. *Correspondence*, Vol. 7, p. 355.

Index

Picture Credits

All Fourth Estate books are available at your local bookshop or newsagent, or can be ordered direct from the publisher.

Indicate the number of copies required and quote the author and title.

Send cheque/eurocheque/postal order (Sterling only), made payable to Book Service by Post, to:

Fourth Estate Books
Book Service By Post
PO Box 29, Douglas
I-O-M, IM99 1BQ.

Or phone: 01624 675137

Or fax: 01624 670923

Or e-mail: bookshop@enterprise.net

Alternatively pay by Access, Visa or Mastercard

Card number: ☐☐☐☐☐☐☐☐☐☐☐☐☐☐☐☐

Expiry date ....

Signature ....

Post and packing is free in the UK. Overseas customers please allow £1.00 per book for post and packing.

Name ....

Address ....

....

....

Please allow 28 days for delivery. Please tick the box if you do not wish to receive any additional information. ☐

Prices and availability subject to change without notice.